Composites Science and Technology

Series Editor

Mohammad Jawaid, Lab of Biocomposite Technology, Universiti Putra Malaysia, INTROP, Serdang, Malaysia

Composites Science and Technology (CST) book series publishes the latest developments in the field of composite science and technology. It aims to publish cutting edge research monographs (both edited and authored volumes) comprehensively covering topics shown below:

- Composites from agricultural biomass/natural fibres include conventional composites-Plywood/MDF/Fiberboard
- Fabrication of Composites/conventional composites from biomass and natural fibers
- Utilization of biomass in polymer composites
- Wood, and Wood based materials
- Chemistry and biology of Composites and Biocomposites
- Modelling of damage of Composites and Biocomposites
- Failure Analysis of Composites and Biocomposites
- Structural Health Monitoring of Composites and Biocomposites
- Durability of Composites and Biocomposites
- Biodegradability of Composites and Biocomposites
- Thermal properties of Composites and Biocomposites
- Flammability of Composites and Biocomposites
- Tribology of Composites and Biocomposites
- Bionanocomposites and Nanocomposites
- Applications of Composites, and Biocomposites

To submit a proposal for a research monograph or have further inquries, please contact springer editor, Ramesh Premnath (ramesh.premnath@springer.com).

More information about this series at http://www.springer.com/series/16333

Mohamed Thariq Hameed Sultan ·
Mohd Ridzuan Mohd Jamir ·
Mohd Shukry Abdul Majid ·
Azwan Iskandar Azmi ·
Naheed Saba
Editors

Tribological Applications of Composite Materials

Springer

Editors
Mohamed Thariq Hameed Sultan
Department of Aerospace Engineering
Universiti Putra Malaysia
Serdang, Selangor, Malaysia

Mohd Shukry Abdul Majid
Faculty of Mechanical Engineering Technology
Universiti Malaysia Perlis
Arau, Perlis, Malaysia

Naheed Saba
Laboratory of Biocomposite Technology
Institute of Tropical Forestry and Forest Products (INTROP)
Universiti Putra Malaysia
Serdang, Selangor, Malaysia

Mohd Ridzuan Mohd Jamir
Faculty of Mechanical Engineering Technology
Universiti Malaysia Perlis
Arau, Perlis, Malaysia

Azwan Iskandar Azmi
Faculty of Mechanical Engineering Technology
Universiti Malaysia Perlis
Padang Besar, Perlis, Malaysia

ISSN 2662-1819 ISSN 2662-1827 (electronic)
Composites Science and Technology
ISBN 978-981-15-9634-6 ISBN 978-981-15-9635-3 (eBook)
https://doi.org/10.1007/978-981-15-9635-3

This Springer imprint is published by the registered company Springer Nature Singapore Pte Ltd.
The registered company address is: 152 Beach Road, #21-01/04 Gateway East, Singapore 189721, Singapore

Preface

Tribology is an interdisciplinary field that has attracted increased attention from various research fields comprising of engineering to other life sciences fields. It is an important interdisciplinary field that utilizes skills from mechanical engineering, materials science and engineering, chemistry and chemical engineering and other engineering fields. The science of tribology focuses on contact physics and the mechanics of moving interfaces that generally involve energy dissipation. The contact between two materials, and the friction that one exercises on the other, causes an inevitable process of wear. Generally, tribology includes three key topics: friction, wear and lubrication. Friction is the resistance to relative motion, wear is the loss of material due to that motion, and lubrication is the use of a fluid to minimize friction and wear. Tribology is particularly important in today's world because of energy loss in the engineering components. Significant energy is lost due to friction in sliding interfaces; therefore, finding many alternatives ways to minimize friction and wear through new technologies in tribology is critical to a greener and more sustainable world.

This book provides a comprehensive overview of tribological properties of composites materials such as natural fiber-reinforced composites, non-oxide-based ceramic matrix composites, nano-HAp-polyoxymethylene composites, elastomer composites, titanium-based composite materials, etc. This book offers an understanding of the processes, materials, techniques and mechanisms related to the tribological concepts and includes information on the most recent developments in the field. With contribution from international experts, the book discusses non-oxide-based ceramic matrix composites in tribological applications; tribological performance of polymer, ceramic, metal-matrix composite, and the coating for automotive applications; solid–liquid lubricants in composite material; polymeric systems for enhanced tribological properties; friction and mechanical properties of inner banana trunk polymer matrix composites and eggshell polymer matrix composites using injection molding (IM) and vacuum-assisted resin transfer molding (VARTM); tribological properties of nano-hydroxy apatite (nHAp) added polyoxymethylene (POM) composites using a pin on disc tribometer; surface composition, structure and morphology and the related changes to hardness, microroughness and the surface energy of elastomer composites; titanium-based biocomposites in hip joint replacement; impact of fillers on processing, functions, mechanical and tribological, environmental properties; tribological analysis of developed composite coating with MoS_2-TiO_2; DLC composite hard coating as the potential tool coating; feasibility applications of polymer composite spur gears in tribological applications; influence of green solid particle (hexagonal boron nitride, HBn) enriched in the modified jatropha oil (MJO) through tribology testing using four-ball tribotester machine; friction and wear properties of natural fibre-reinforced polymer composites (NFRPC) as tribo-materials for different engineering system as well as their applications in automotive, biotechnology, biomedical and metal forming field.

We are thankful to all authors around the world who contributed their valuable research in our edited book and make our imaginary thought into reality. Editors are also thankful to Springer team for supporting and cooperation during whole project without their help we cannot complete this project.

Serdang, Malaysia — Mohamed Thariq Hameed Sultan
Perlis, Malaysia — Mohd Ridzuan Mohd Jamir
Perlis, Malaysia — Mohd Shukry Abdul Majid
Perlis, Malaysia — Azwan Iskandar Azmi
Kuala Lumpur, Malaysia — Naheed Saba

Contents

About the Editors

Prof. Ir. Ts. Dr. Mohamed Thariq Hameed Sultan is a Professional Engineer (PEng) registered under the Board of Engineers Malaysia (BEM), a Professional Technologist (PTech) registered under the Malaysian Board of Technologists, and also a Charted Engineer (CEng) registered with the Institution of Mechanical Engineers United Kingdom, currently attached to the Universiti Putra Malaysia as the Head of the Biocomposite Technology Laboratory, Institute of Tropical Forestry and Forest Products (INTROP), UPM Serdang, Selangor, Malaysia. Being the Head of the Biocomposite Technology Laboratory, he is also appointed as an Independent Scientific Advisor to Aerospace Malaysia Innovation Centre (AMIC) based in Cyberjaya, Selangor, Malaysia. He received his Ph.D. from the University of Sheffield, UK. He has about 10 years of experience in teaching as well as in research. His area of research interests includes hybrid composites, advance materials, structural health monitoring, and impact studies. So far he has published more than 100 international journal papers and received many awards locally and internationally. In December 2017, he was awarded a Leaders in Innovation Fellowship (LIF) by the Royal Academy of Engineering (Raeng), UK. He is also the Honourable Secretary of the Malaysian Society of Structural Health Monitoring (MSSHM) based in UPM Serdang, Selangor, Malaysia. Currently, he is also attached to the Institution of Engineers Malaysia (IEM) as the Deputy Chairman in the Engineering Education Technical Division (E2TD).

Dr. Mohd Ridzuan Mohd Jamir is a Professional Engineer (PEng) registered under the Board of Engineers Malaysia (BEM), a Professional Technologist (PTech) registered under the Malaysian Board of Technologists, and also a Chartered Engineer (CEng) registered with the Institution of Mechanical Engineers UK, currently attached to the Universiti Malaysia Perlis as Associate Professor at Faculty of Mechanical Engineering Technology. He obtained Diploma in Mechanical Engineering from Universiti Teknologi Malaysia (UTM) in 2006. He graduated in Bachelor of Engineering (Mechanical) and Master of Engineering (Mechanical) from Universiti Teknologi Malaysia (UTM) in the year of 2009 and 2010, respectively. He also worked as Quality Assurance Engineer in Venture

Pintarmas Sdn Bhd in Johor Bahru in 2010 before joined Universiti Malaysia Perlis (UniMAP) in the same year. He received his Ph.D. in Mechanical Engineering from Universiti Malaysia Perlis (UniMAP) in 2016. He has about 10 years of experience in teaching as well as in research. An advocate of interdisciplinary research, his research interests includes the strength of material's area with emphasis on the natural fibre composite, looking at the performance of composite structures and tribological properties, hybrid-reinforced/filled polymer composites, lignocellulosic-reinforced/filled polymer and biodegradable composites. He has published more than 100 international journal research paper with Scopus index and ISI ranked journal publications. In addition, he is also a regular reviewer for high-impacts ISI ranked journals.

Dr. Mohd Shukry Abdul Majid received his Bachelor of Engineering in Mechanical Engineering from University Manchester Institute of Science and Technology (UMIST) in 2001. Upon his return to Malaysia, he worked as a research and development (R&D) engineer at a semiconductor industry before joining Universiti Malaysia Perlis (UniMAP) as a lecturer in 2004. He completed his M.Sc. in Mechanical Systems Engineering from the University of Liverpool in 2005 and his Ph.D. in Composite Engineering from Newcastle University, UK, in 2011. Currently, he is serving Universiti Malaysia Perlis as an Associate Professor at Faculty of Mechanical Engineering Technology. An advocate of interdisciplinary research, his research interests lie in the strength of material's area with emphasis on the composite piping, looking at the performance of composite structures, NDEs of composites and natural fibre/green composites, hybrid-reinforced/filled polymer composites, lignocellulosic-reinforced/filled polymer and biodegradable composites. His aptitude for high-quality research of international standing has been supported by his 188 Scopus indexed publications with *47 (32 Q1) publications in ISI-ranked Journals having a cumulative impact factor (CIF) of 148.948*. He received his professional engineer qualification (Ir.) from Board of Engineer Malaysia (BEM) in Mac 2016 and has been a Chartered Engineer (CEng) from the Engineering Council, UK, since 2014. Dr. Mohd Shukry has been honoured with numerous local and international awards for his achievements. He is the first recipient from Technical University Network (MTUN) to have been awarded Malaysia's Research Star Award 2017, as one of Malaysia's most promising and influential researchers by the Ministry of Higher Education Malaysia (MOHE).

Dr. Azwan Iskandar Azmi received his Bachelor's Degree in Mechanical Engineering from Purdue University, USA, in 1999 and Master's in Advanced Manufacturing Technology from Universiti Teknologi Malaysia in 2003. In 2013, he completed his doctoral study at the University of Auckland, New Zealand, in Mechanical Engineering with specialization in the area of fibre-reinforced composite machining. He is currently serving the Faculty of Mechanical Engineering Technology, Universiti Malaysia Perlis (UniMAP) as Associate Professor. Dr. Azwan is a Professional Engineer (PEng) registered under the Board of Engineers Malaysia (BEM), a Professional Technologist (PTech) registered under the Malaysian Board of Technologists (MBOT), and also a Chartered Engineer (CEng) registered with the Institution of Mechanical Engineers (IMechE) in the UK. He has more than 10 years of experience in teaching, research, and industries. His area of research interests includes machining and machinability of carbon, glass and their hybrid composites. Currently, the research also covers the drilling and milling of lignocellulosic-reinforced polymer composites. His interest extends on the machinability study of difficult-to-cut metal alloys such as titanium, Inconel and nickel–titanium (NiTi) alloys. So far, he has published more than 50 research articles in renowned international journals, proceedings and review papers. He has been awarded with research funds from Ministry of Higher Education Malaysia (MOHE) and Ministry of Science Technology and Innovation (MOSTI). Due to the research outputs, Dr. Azwan has served as technical reviewer for a number of reputable high-impact ISI ranked journals and international conferences.

Dr. Naheed Saba completed her Ph.D. (Nanocomposites) with Distinction from Laboratory of Biocomposites Technology, Institute of Tropical Forestry and Forest Products (INTROP), Universiti Putra Malaysia, Serdang, Selangor, Malaysia, in 2017. She completed her Masters in Chemistry and also completed her Postgraduate Diploma in Environment and Sustainable Development from India. She has published over 50 scientific peer review articles in international journal and 4–5 articles come under top-cited articles in Construction and Building Materials, and Polymers Journal during 2016–2019. She edited six books from Elsevier and also published more than 25 chapters in Springer, Elsevier and Wiley publication. She attended few international conferences and presented research papers. Her research interest areas are nanocellulosic materials, fire-retardant materials, natural fibre-reinforced polymer composites, biocomposites, hybrid composites and nanocomposites. She is also recipient of International Graduate Research Fellowship and Graduate on Time (GOT) Award from Universiti Putra Malaysia, Malaysia. Presently, she is Editor-in-Chief of the Journal of Composites and Advance Materials (ISSN 2716-8018) is a peer-reviewed bi-annual journal. She is also reviewer of several international journals such as Cellulose, Constructions and Building Materials, Composite Part A, Composite Part B, Journal of Polymers and the Environment, Journal of Energy Storage, Journal of Elastomers and Plastics, Journal of Materials Research and Technology, BioResources and Carbohydrate Polymers. Her H-index=23 (Scopus); H-Index=26 (Google Scholar).

Tribomechanical Behaviour of Non-oxide Ceramic Matrix Composites in Dry Sliding

Subhrojyoti Mazumder, Hendrik Simon Cornelis Metselaar, Nazatul Liana Sukiman, and Nurin Wahidah Mohd Zulkifli

Abstract Non-oxide ceramics have been widely used in many tribological applications under a wide range of operating conditions over many decades. Nevertheless, a comprehensive review of tribomechanical characteristics of such potential ceramics relating to modern trends is not articulated anywhere before. Therefore, this chapter reviews studies regarding the tribological and mechanical performance of non-oxide based ceramic matrix composites (CMCs) carried out by various researchers. This work concerns only the tribological investigation of the ceramic composite in dry sliding/erosive condition and hence the characteristics under liquid lubricant or any other liquid medium is beyond the scope of this writing. Widely used boride, carbide and nitride based ceramics are considered as the matrix phase for these ceramic composites. The tribological and mechanical effects of reinforcements and dopants, such as hard or soft nano-micro particles (SiC, ZrC, B_4C, WC, hBN, graphene etc.), nanowire/nanotube/fibre (SiC/CNTs/C_f etc.) and some rare earth compounds (Sm_2O_3, Y_2O_3, La_2O_3, Nd_2O_3, Yb_2O_3, Lu_2O_3 etc.) into the non-oxide based ceramic composites are described under sliding conditions from room temperature to elevated temperature. In addition to this, the key applications of such composites in a wide range of operating conditions are discussed. This chapter also highlights the advantages and disadvantages of this class of ceramic matrix composites.

S. Mazumder · H. S. C. Metselaar · N. L. Sukiman
Centre of Advanced Materials, Department of Mechanical Engineering, University of Malaya, 50603 Kuala Lumpur, Malaysia
e-mail: subhrojyoti@iitdalumni.com

H. S. C. Metselaar
e-mail: h.metselaar@um.edu.my

N. L. Sukiman
e-mail: nazatul@um.edu.my

N. W. M. Zulkifli (✉)
Centre for Energy Sciences, Department of Mechanical Engineering, University of Malaya, 50603 Kuala Lumpur, Malaysia
e-mail: nurinmz@um.edu.my

M. T. Hameed Sultan et al. (eds.), *Tribological Applications of Composite Materials*, Composites Science and Technology, https://doi.org/10.1007/978-981-15-9635-3_1

Keywords Non-oxide · Ceramic matrix composites · Fabrication process · Mechanical property · Tribological characteristics

1 Introduction

Currently, the trend of using ceramic matrix composites (CMCs) in crucial engineering components is growing tremendously due to their promising characteristics such as high hardness and fracture toughness, chemical inertness, thermal stability, oxidation resistance, superior wear resistance in a wide range of operating environments. For instance, aviation industries require some mechanical components viz., roller/ball bearings and power transmitting gears which are to be sustained for long term operating conditions at elevated temperature (above 500 °C). Apart from that, manufacturing industries are continuously looking for potential materials for metal removing tools and dies for forming operations having low wear rate and high thermal stability at high temperatures. Automotive industries use cylinder liners/piston rings for the engines which require significant thermal shock resistance and stability at elevated temperature. Using ceramic composites/coatings can be an effective solution to these applications as they provide superior wear resistance and desirable thermal properties at such critical operating conditions (Torres et al. 2018; Hasan et al. 2019; Zhu et al. 2019). To make these ceramic composites more suitable for wear resistance applications, different classes of solid lubricants are introduced into the composites to act as lubricating agents at the contacting interfaces. Among these solid lubricants, BaF_2, CaF_2, CuO, MoO_x, MoS_2, etc. are widely used as they can provide lubricity at a wide range of temperatures (Kong et al. 2012; Mazumder et al. 2019a, b, c; Zhang et al. 2010; Moazami and Nemati 2018). However, most of these solid lubricants are added to the oxide based ceramic matrices such as Al_2O_3, ZrO_2, etc. The inclusion of solid lubricants into non-oxide based ceramics is still an area of active study. On the other hand, a number of works of tribo-mechanical property evaluation of non-oxide ceramics are performed on composite coatings. For example, a coating of cesium silicate can provide low friction and wear rate at 650 °C on Si_3N_4 ceramic matrix along with enhances the endurance life of the matrix (Rosado et al. 2000). In other experiments, some self-lubricating nitride based coatings are fabricated by Mo_2N-MoS_2-Ag, TiN-Ag, ZrN-Ag, Mo_2N-Cu, etc. to enhance the tribological and mechanical performance at different temperatures as well as in vacuum (Aouadi et al. 2008; Köstenbauer et al. 2008; Ju et al. 2018; Suszko et al. 2006). Since this current chapter is basically focused on the tribological behaviour of different classes of CMCs, the discussion on such coating materials is limited.

This chapter aims to discuss the mechanical and tribological behaviour of various non-oxide CMCs in dry sliding conditions with or without the inclusion of solid lubricants and the fabrication techniques of such composites. A majority of the work is found on the evaluation of tribo-mechanical performances of CMCs without any lubricant. In addition to this, the different fabrication techniques and experimental processes are articulated. Some of the researches are carried out at elevated temperature for the CMCs and hence the discussions are made separately for room and high temperature, categorized by the base ceramic matrix. In the later stage of discussion, the key applications and advantages, as well as disadvantages are stated. The ending remark includes the future work which can be done for exploring the tribological behaviour of such classes of CMCs with the inclusion of other potential solid lubricants.

2 Fabrication and Experimental Progress

Fabrication of CMCs is the key factor for manipulating the composite overall performance when employed in wear resistance applications. In addition to this, the densification and corresponding mechanical behaviour also depend on the fabrication technique to a great extent. The traditional specimen fabrication route includes cold press (CP)/cold isostatic press (CIP) and hot press (HP)/hot isostatic press (HIP) of bulk powder materials via powder metallurgy route. In cold press technique, the mechanically ground powder mixtures are first compacted using a uniaxial pressing system inside a carbide or metallic die at room temperature followed by sintered in a tube or box furnace under an ambient or inert atmosphere. For cold isostatic press, the powder is pressed using liquid pressure and the rest of functions are similar like a cold press. On the other hand, the hot press is carried out by applying simultaneous temperature and pressure, usually inside graphite die. The operating mechanism of hot isostatic press is similar to hot press except the pressing mechanism utilized is fluid pressure. These techniques are normally carried out to fabricate products from powders if they have a high self-diffusion coefficient which improves the sintering kinetics and gives high structural stability. However, these sintering techniques take a long time to be completed and require a skilled operator. Most of the oxide based ceramics are generally fabricated by means of these conventional sintering techniques.

Non-oxide based ceramics sometimes fail to be densified properly using traditional sintering techniques because of the constraint of sintering kinetics and very low self-diffusivity. Hence some advanced technique like spark plasma sintering (SPS) is adopted. In the SPS technique, a high pulsed rate electric current is passed through the bulk powder sample to generate high localized heat which helps to bond particles strongly by melting while simultaneously pressure is applied. SPS has some advantages over traditional sintering techniques such as low running cost, easy sintering, and densification, etc. Although, the equipment cost is higher as compared to traditional sintering instruments. Most importantly, it consumes low energy and provides high manufacturing efficiency since it has a very high heating rate (~1000 °C/min) within a short period of sintering time. Besides that, the low sintering duration preserves the original microstructures to retain low grain size and hence eliminate the usage of additional sintering additives in the powder mixture.

Microwave sintering (MS) is another sintering technique that uses microwaves to heat bulk compressed materials inside a thermally insulated chamber. Bulk materials are first compacted separately followed by placed inside a microwave reactor chamber for microwave irradiation. Microwave is passed through the dielectric material leading to fast heat generation. This sintering process also takes a shorter time than traditional sintering processes. Additionally, it shows a high energy efficiency with an enhanced sintering rate. Rapid heating with almost constant heat distribution throughout the bulk material again makes it advantageous over conventional sintering.

The fabricated ceramic composites are generally characterized using a Vickers hardness tester to evaluate nano/microhardness and fracture toughness. The flexural strength and compressive strength are determined using a universal testing machine (UTM) using standardized specimen size. Tribological tests are generally carried out by means of pin-on-disc or ball-on-disc or universal tribometer by adopting reciprocating, rotary and oscillatory motion at different temperatures (ambient to elevated temperature) and wet/dry lubricated/unlubricated conditions. Heating can be done by resistive or inductive heating of pin or disc material or by the inclusion of the contact zone in a furnace.

3 Mechanical and Tribological Behaviour of Non-oxide Ceramic Matrix Composites

Numerous researches have been executed to investigate the tribological behaviour of non-oxide based ceramic composites with or without the inclusion of solid lubricants. Based on these works, the CMCs can be categorized into three broad categories i.e., (i) boride based, (ii) carbide based and (iii) nitride based composites, according to the base matrix. The fabrication process, mechanical and tribological behaviour at different operating conditions are comprehensively described. Other combinations of different ceramics and their tribo-mechanical characteristics are pointed out separately.

3.1 Tribomechanical Characteristics of Boride Based Ceramic Composites

Among the boron based CMCs, TiB_2 and ZrB_2 ceramic matrices are mostly employed for high to ultra-high temperature applications such as rocket propulsion and hypersonic flight. TiB_2 is a class of boron based ceramic matrix having high hardness, melting temperature and elastic modulus of ~25 GPa, ~2200 °C and ~500 MPa, respectively. Hence this can be easily used for high strength structural applications (Telle and Petzow 1988). But besides its sound mechanical properties, it does not have promising tribological characteristics when applied in contact situations. In fretting wear conditions, the TiB_2 itself does not show acceptable wear and friction behaviour. The coefficient of friction (COF) against bearing steel varies between 0.5 and 0.65 (Vleugels et al. 2002). The tribological properties of the monolithic TiB_2 ceramic matrix can be improved by the addition of 20 wt.% $MoSi_2$ as the second phase in fretting wear experiments (Murthy et al. 2006). Abrasive wear of monolithic TiB_2 and tribochemical wear of $MoSi_2$-TiB_2 composite vary with varying loads (2–10 N). The COFs for both combinations remain in the range of 0.5–0.6, though $MoSi_2$-TiB_2 shows better wear resistance which is in the order of 10^{-5} mm^3/N m. Abrasive wear is the predominating wear mechanism in monolithic TiB_2 against bearing steels.

On the other hand, the sinterability of pure ZrB_2 is very poor because of the kinetic constraint caused by the strong covalent bonding and low self-diffusion coefficient. Hence, sintering by conventional techniques like cold and hot sintering process at temperatures below 2000 °C is generally impossible if the grain sizes are not properly refined. To achieve a good grain size distribution, the raw powders can be crushed in a high energy ball mill and spark plasma sintering (up to 1900 °C) is a recommended sintering technique for ZrB_2 to obtain desirable densification (Zamora et al. 2012). ZrB_2 fabricated by SPS techniques at 2100 °C and 35 MPa pressure shows decent tribo-mechanical characteristics as described by Chakraborty et al. (2015). The sintered ZrB_2 via SPS technique produces a maximum relative density of ~98.65% and hardness and fracture toughness values of ~16.64 GPa and 4.69 MPa $m^{0.5}$, respectively. The tribological responses of ZrB_2 can be obtained as COF of ~0.44 and wear rate of 1.01×10^{-3} mm^3/N m against diamond under 10 N load at room temperature. In addition to this, ZrB_2 can be hot pressed at 2000 °C temperature and 30 MPa uniaxial load, with the addition of 20 vol.% SiC and 20 vol.% short carbon fibre to obtain a relative density of ~99.3%, a flexural strength of ~445 MPa and a fracture toughness of ~6.56 MPa $m^{0.5}$ (Yang et al. 2008). The addition of carbon short fibre can improve fracture toughness of the ZrB_2-SiC composite by up to 54% since debonding and pull-out of carbon short fibre resist the crack propagation through matrix. In addition to this, the fibres help in crack deflection by consuming fracture energy leading to toughening of matrix. Addition of B_4C as reinforcement into the ZrB_2 can increase the tribo-mechanical properties of ZrB_2 composite if fabricated by hot press at 2100 °C in argon atmosphere. Addition of 10 wt.% B_4C into the ZrB_2 can provide a Vickers hardness value of ~20 GPa and fracture toughness of ~3.9 MPa $m^{0.5}$. 10 wt.% B_4C added ZrB_2 composite show lower friction and wear rate value of ~0.40 and ~0.49×10^{-3} mm^3/N m, respectively, which are 0.69 and 1.97×10^{-3} mm^3/Nm for the straight ZrB_2 in dry sliding condition under 10 N normal load at room temperature (Chakraborty et al. 2014). Recently it has been found that reinforcement of B_4C, SiC and ZrC into the ZrB_2 matrix fabricated by SPS technique is meaningful in terms of tribological aspects especially for the combination of ZrC-ZrB_2. A wear rate of 6.15×10^{-6} $mm^3/(Nm)$ can be achieved with 10 wt.% ZrC-ZrB_2 composite which was lowest among all other combinations in dry sliding condition. The COF of all the composites is varied between 0.63 and 0.72 (Medveď et al. 2019). The overview of the fabrication method, mechanical and tribological characteristics of different boride based ceramic composites are given in Table 1.

3.2 *Tribomechanical Characteristics of Carbide Based Ceramic Composites*

The key carbide based ceramic matrices are B_4C, SiC and ZrC, which are conventionally used for tribological applications at a variety of operating conditions. Among them, B_4C possesses excellent abrasive wear resistance and this can be suitable for industrial applications like material removing and dressing tool, bearing, abrasive air-jet nozzle material, etc. (Sun et al. 2012; Yue et al. 2010). Apart from that, B_4C has excellent physico-mechanical properties such as low density (~2.52 g/cm^3), high melting point (~2450 °C), high hardness (~36 GPa) and Young's modulus (~445 GPa) (Sedlák et al. 2017). Nevertheless, the application of B_4C is hard as the low self-diffusion coefficient makes it difficult to achieve a fully dense matrix having complex shape. In general, SPS method is suggested to accomplish highest densified matrix producing sound mechanical properties. Grain size does not have any direct correlation with the wear rate of fully dense B_4C matrix in abrasive wear tests and this fact is in contrast with other oxide ceramics such as alumina. Furthermore, hardness is an important parameter that influences the wear rate of the B_4C ceramic matrix. If the grain size is very fine in B_4C matrix, the chance of porosity is high leading to abrasive wear resulting from pull-out of grains when slides against chrome steel ball in the presence of abrasive SiC slurry. Low porosity is advantageous for high wear resistance of B_4C matrix. Thus, porosity is an influencing factor for wear resistance of B_4C matrix other than grain size (Moshtaghioun et al. 2016). In addition to this, tribological performances of B_4C matrix fabricated by hot pressing varies with varying load during reciprocating sliding test against WC. As the load increases from 3 to 10 N at 10 Hz, the COF reduces from 0.20 to 0.15 while the wear rate remains in the order of 10^{-6} mm^3/N m (Sonber et al. 2015).

A way of introducing a self-lubricating effect into B_4C is the addition of layer structured solid lubricating elements like hBN, MoS_2 and graphene which improve the tribological behaviour of B_4C during sliding (Pan and Gao 2018). For example, graphene platelets (GPLs) can be incorporated into the B_4C matrix by hot press technique at 2100 °C and 25 MPa in argon atmosphere. Increase of GPLs reduces the hardness of overall hBN-B_4C matrix. It is noted that the COF does not vary too much with the variation of GPLs content while sliding against SiC counter surface but the wear resistance enhances significantly (Sedlák et al. 2017).

SiC ceramic matrix is well known for many structural applications till now since it caters to promising physical and mechanical properties along with its chemical and thermal stability as well as oxidation and corrosion resistance even at elevated temperatures. Because of these desirable characteristics of SiC, it is extensively used for tribological applications in different areas. Lots of efforts have been made to

Table 1 Highlight of the fabrication process, mechanical and tribological characteristics of boride based ceramic composites in dry sliding

References	Ceramics or CMCs with/without lubricating additives		Fabrication techniques and parameters	Tribological testing process, counterpart and parameters	Key findings
	Base/matrix	Reinforcements/additives			
Chakraborty et al. (2015)	ZrB_2	–	Method: SPS; Pressing: 35 MPa; Sintering: 2100 °C in Ar	Scratch tester using Vickers type diamond indenter; Load: 5 and 10 N; Speed: 0.1 mm/s; Distance: 4 mm; Environment: room temperature	• On-off pulse current pattern of SPS influences structural and consolidation properties of ZrB_2 matrix • Highest relative density of 98.65% is achieved at 50 and 5 ms time pulse on and off conditions during SPS • Maximum hardness and fracture toughness obtained as ~16.64 GPa and ~4.69 MPa·m$^{0.5}$, respectively • Wear volume and wear rate obtained at 10 N are 4.05 × 10^4 μm^3 and 1.01 × 10^{-3} mm^3/N m, respectively

(continued)

Table 1 (continued)

References	Ceramics or CMCs with/without lubricating additives		Fabrication techniques and parameters	Tribological testing process, counterpart and parameters	Key findings
	Base/matrix	Reinforcements/additives			
Vleugels et al. (2002)	Mono-lithic TiB_2	Cermet with $Ni_3(Al,Ti)$ binder; Sialon; ZrO_2	Method: HP; Pressing: 28–30 MPa; Sintering: 1450–1900 °C	Fretting wear test by ball (bearing steel) on plate; Load: 8 N; Frequency: 10 Hz; Stroke: 200 μm; Duration: 100,000 cycles; Environment: ambient temperature	• Addition of other phases lowers the hardness (~21 GPa) of monolithic TiB_2, however, TiB_2 based cermet and ZrO_2-TiB_2 show an increased fracture toughness up to ~9.7 MPa $m^{0.5}$ • Thermal conductivity (~64 W/m K) of monolithic TiB_2 decreases with addition of the secondary phases • COF varies between 0.5 and 0.65 for all composites • ZrO_2-TiB_2 shows lowest volumetric wear as compared to all composites

(continued)

Table 1 (continued)

References	Ceramics or CMCs with/without lubricating additives		Fabrication techniques and parameters	Tribological testing process, counterpart and parameters	Key findings
	Base/matrix	Reinforcements/additives			
Murthy et al. (2006)	TiB_2	20 wt.% $MoSi_2$	Method: HP; Pressing: 32 MPa; Sintering: 1700–1800 °C in vacuum (10^{-5} Pa)	Ball (steel) and flat fretting wear tester; Load: 2, 5 and 10 N; Frequency: 8 Hz; Stroke: 100 µm; Duration: 10,000 cycles; Environment: room temperature	• Hardness (~25.9 GPa) and fracture toughness (~5.3 $MPa{\cdot}m^{0.5}$) both are higher in the case of monolithic TiB_2 • Frictional behaviours (0.5–0.8) are almost similar for all the cases. At 10 N, the COF is lower than 0.5 for both the cases • Specific wear decreases with increasing. $MoSi_2$-TiB_2 composite has a better wear resistance than monolithic TiB_2 • Tribochemical wear is the predominant wear mechanism as a form of severe abrasive wear (10^{-5} mm^3/N m) for monolithic TiB_2

(continued)

Table 1 (continued)

References	Ceramics or CMCs with/without lubricating additives		Fabrication techniques and parameters	Tribological testing process, counterpart and parameters	Key findings
	Base/matrix	Reinforcements/additives			
Medveď et al. (2019)	ZrB_2	10 wt.% B_4C; 10 wt.% SiC; 10 wt.% ZrC	Method: SPS; Pressing: 50 MPa; Sintering: 1800–2050 °C in Ar	Ball (SiC)-on-flat plate; Load: 5 and 10 N; Speed: 0.1 m/s; Distance: 500 m; Environment: room temperature	• Maximum hardness of ~19 GPa achieved for ZrB_2-B_4C • Highest fracture toughness and elastic modulus obtained as ~5.26 MPa·$m^{0.5}$ and 484 GPa, respectively, for ZrB_2-ZrC • Lowest COF recorded for ZrB_2-SiC as ~0.63 at 5 N • Lowest wear rate obtained for both ZrB_2-SiC and ZrB_2-ZrC in the order of 10^{-6} mm^3/(N m) at 5 N • Hertzian crack formation occurred for all specimens • Little grain pull-out observed for all during wear test

(continued)

Table 1 (continued)

References	Ceramics or CMCs with/without lubricating additives		Fabrication techniques and parameters	Tribological testing process, counterpart and parameters	Key findings
	Base/matrix	Reinforcements/additives			
Chakraborty et al. (2014)	ZrB_2	5 wt.% B_4C; 10 wt.% B_4C; 15 wt.% B_4C	Method: HP; Pressing: 40 MPa; Sintering: 2100 °C in Ar	Scratch tester using Vickers type diamond indenter; Load: 5 and 10 N; Speed: 0.1 mm/s; Distance: 4 mm; Environment: room temperature	• Highest hardness and fracture toughness achieved for 10 wt.% B_4C addition in ZrB_2 and the values are ~20 GPa and ~3.93 $MPa{\cdot}m^{0.5}$, respectively • Scratch resistance coefficient, wear coefficient and wear rate for 10 wt.% B_4C-ZrB_2 composite are 0.40, 0.01 and 0.49×10^{-3} mm^3/N m, respectively, at 10 N load • Thermal conductivity of ZrB_2-B_4C composites lies in the range of 45.23–70.13 W/m K between 100 and 1000 °C

improve stability during sliding and erosive operations. Microstructural alteration plays a vital role in the tribological behaviour of the SiC matrix. Fine grain structure shows better wear resistance than a coarse structure. Erosion resistance of SiC matrix can be improved by refining grain size and by minimizing pores as well (Sharma et al. 2016). Sometimes, reinforcement with other hard phases can improve the wear resistance of SiC ceramic composites. For instance, WC is well accepted as a reinforcement of SiC to enhance friction and wear performances. If 30 wt.% WC is added into the SiC matrix via hot pressing at 1800 °C and 40 MPa, the overall hardness and fracture toughness can be achieved as ~26 GPa and ~6.47 $MPa \cdot m^{0.5}$, respectively. Besides that, an improved COF (0.40–0.48) and wear rate (10^{-6}–10^{-5} $mm^3/N\ m$) of WC-SiC composite can be attained as compared to straight monolithic SiC (COF as ~0.5 and wear rate in the order of 10^{-5}–10^{-2} $mm^3/N\ m$) if tested against SiC surface. Grain pull out of SiC ceramic matrix can be minimized with the addition of WC during dry sliding against SiC counterpart (Sharma et al. 2015). It is understood that counterpart plays a significant role in the tribological behaviour of monolithic SiC and WC-SiC during dry sliding conditions (Sharma et al. 2019). The wear resistance of WC-SiC is higher against a softer steel ball (hardness ~7 GPa) against a harder WC-Co (hardness ~28 GPa) counter surface. WC-SiC having a fracture toughness of ~6.7 MPa with 50 wt.% WC incorporation provides mild abrasion against steel by forming a soft tribo-layer consisting of iron oxide and iron tungsten oxide, whereas a strong penetration of WC-Co countersurface results in higher erosion of WC-SiC composite as a consequence of tribochemical changes leading to the formation of silicon oxide, cobalt tungsten oxide and tungsten oxide. The wear mechanisms causing due to mechanical fracture and tribochemical wear of SiC and WC-SiC against softer (steel) and harder (WC-Co) countersurfaces are given in Fig. 1 SiC shows intergranular fracture causing higher loss of material and this can be improved by addition of WC into the SiC matrix leading to less fracture of surface as transgranular fracture by increasing fracture toughness of the overall composite.

The tribo-mechanical characteristics of SiC matrix can be improved by reinforcing with carbon based fibres/nanotubes such as graphene and CNT (Tang et al. 2009; Llorente et al. 2016; Džunda et al. 2019). Carbon short fibre (C_f) can be added to the SiC matrix by following a powder metallurgy route. C_f-SiC composite fabricated by hot press sintering has a higher bending strength and fracture toughness than monolithic SiC. In addition, 40–55 vol.% carbon fibre can enhance the tribological performance (COF as ~0.24 and wear rate in the order of 10^{-6} $mm^3/N\ m$) of SiC matrix at ambient temperature is tested against SiC surface (Tang et al. 2009). To achieve a significant tribological performance of SiC, addition of graphene nanoparticles (GNPs) and graphene oxide (GO) can be another alternative. During SPS process for fabrication such a composite matrix, GO becomes reduced graphene oxide (rGO) and GNPs remain the same. Composites attain up to ~99.9% densification for each combination. Both graphene fillers improve the tribological properties of the SiC matrix, but GNPs are suggested as the better one since they easily come out from the matrix, then being exfoliated and crushed which in turn provides a lubricating tribo-layer during sliding (Llorente et al. 2016). Another interesting way of toughening SiC matrix is the reinforcement of carbon nanotube (CNT) which bridges cracks,

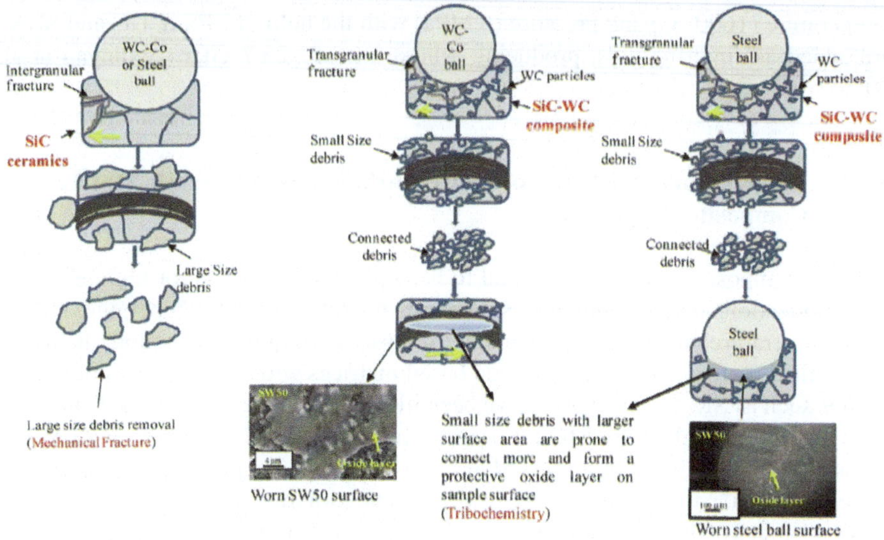

Fig. 1 Different types of wear mechanism of SiC and WC-SiC ceramics against WC-Co and steel countersurface (Sharma et al. 2019)

deflects cracks at the CNT/matrix junction and provides crack branching during fracture. However, achieving proper distribution and maximum densification has always been a challenge with CNTs. Catalytic chemical vapour deposition (CCVD) is an effective way of CNT dispersion into the ceramic matrix which controls in situ growth of CNTs into the matrix and provides a higher degree of scalability (Jourdain and Bichara 2013). Therefore, a composite powder of CNT/SiC can be developed by following this technique to achieve a maximum degree of homogeneity. By hot pressing, this powder at 1850 °C and 40 MPa a CNT-SiC composite with promising tribo-mechanical behaviour can be formed. Reinforcement of CNTs into the SiC matrix can lead to a COF and wear rate values of ~0.30 and ~2.4×10^{-6} mm^3/N m, respectively, against SiC surface (Džunda et al. 2019).

ZrC is a capable ceramics matrix for its structural and thermal stability and shows sound tribological behaviour without the addition of any other sintering additive if fabricated by SPS technique. It possesses excellent properties like high hardness, thermal conductivity and an extreme melting point of ~25 GPa, ~40 W m^{-1} $°C^{-1}$ and ~3540 °C, respectively (Shaffer and Schneider 1969). ZrC shows a very low specific wear rate (mild wear) in the order of 10^{-8} mm^3/N m when sliding against a Si_3N_4 counter surface without any lubricant. The wear resistance of this ceramic matrix is improved is small grain size and high densification are obtained, leading to high hardness and toughness as well (Bertagnoli et al. 2015). Addition of $MoSi_2$ has some influence on the ZrC matrix if fabricated via conventional hot press sintering. $MoSi_2$ addition in ZrC matrix of up to 20 vol.% fortifies the overall tribological performance, although a higher content of $MoSi_2$ shows a detrimental effect (González et al. 2011). Furthermore, almost full densification of ZrC can be achieved at low sintering temperature (1600 °C) and pressure (6 MPa) with the help of SPS if TiC and Si_3N_4 are added as sintering aids, producing a hardness of ~23.7 GPa (Aydinyan et al. 2019).

3.2.1 Progression in High Temperature Carbide Based Ceramic Matrix Composites

Carbide composites have the potential to be applied as materials for high temperature applications e.g., turbine blades, bearings for rotating high temperature shafts, high temperature nozzle, etc. Systematic exercises are required to explore the tribo-mechanical performance of the carbide based matrices with high temperature capabilities such as SiC. A few works have been illustrated on the mechanical behaviour of Si_3N_4-SiC at high temperature. For instance, a Si_3N_4-SiC composite with Y and Al as sintering additives, hot pressed at 1800–1850 °C, can provide a creep rate in the order of 10^{-5} to 10^{-6} h^{-1} at 1400 °C and oxidation rate of 0.2 to 0.4 mg/cm^2 at 1400 °C after 100 h. Creep performance depends upon the SiC content. Lowering of the SiC content can reduce the creep performance of overall composite significantly (Woetting et al. 2000). If the SiC matrix is doped with rare earth oxides such as Lu_2O_3 or Er_2O_3, room temperature flexural strength can be maintained at elevated temperatures. Lu_2O_3-SiC has a flexural strength of ~600 MPa at a temperature of 1600 °C and Er_2O_3-SiC has the same strength of ~600 MPa up to 1500 °C (Kim et al. 2007). Despite having good frictional and wear behaviour at ambient temperature, hot pressed WC-SiC does not provide sound tribological results at the elevated temperature. 30 wt.% WC-SiC has a COF of 0.7 and wear volume in the order of 10^{-3} mm^3 during sliding against SiC counterface at 500 °C temperature. WC-SiC composites have a wear rate between 1.5×10^{-6} and 4.2×10^{-5} $mm^3/N\,m$ in all temperatures and sliding conditions against SiC surface. With higher amounts of WC (>30 wt.%), chances of grain pull-out of WC from the SiC matrix are increased at high temperatures resulting severe surface fracture. A large amount of WC pull-out agglomerates at the defect zone leading to thermal stress accumulation which enhances the material loss at high temperature. Thus, high WC concentration can be detrimental to the performance of SiC matrix if this is expected to be operated at elevated temperatures (Sharma et al. 2017). On the other hand, SiC fibre reinforced SiC matrix composite performs well at a temperature of 850 °C when tested for erosion resistance. The erosion rate can be diminished by one order of magnitude at 850 °C than 25 °C (Wang and Levy 1990). A similar trend can also be found in the erosion of SiC added Si_3N_4 composite matrix up to 1400 °C. Above 800 °C, the erosion starts decreasing up to an operating temperature of 1400 °C (Li et al. 2014). Table 2 summarizes the fabrication method, mechanical and tribological characteristics of different carbide based ceramic composites.

Table 2 Highlights of the fabrication process, mechanical and tribological characteristics of carbide based ceramic composites in dry sliding/erosion

References	Ceramics or CMCs with/without lubricating additives		Fabrication techniques and parameters	Tribological testing process, counterpart and parameters	Key findings
	Base/matrix	Reinforcements/additives			
Moshtaghioun et al. (2016)	B_4C with varying grain sizes 17.2 μm, 690 nm, 370 nm and 120 nm	–	Method: SPS; Pressing: 75 MPa; Sintering: 1600–1800 °C	Ball (chrome steel)-on-flat plate tests; Load: 5 N; Speed: 0.2 m/s; Distance: 70 m; Environment: room temperature and under influence of SiC abrasive slurry	• Wear resistance of B_4C is independent of grain size but depends on hardness and porosity • Smallest grain size (688 nm) provides highest hardness as ~34 GPa • Transgranular fracture causes the mild wear of B_4C ceramic surface during abrasive wear test

(continued)

Table 2 (continued)

References	Ceramics or CMCs with/without lubricating additives		Fabrication techniques and parameters	Tribological testing process, counterpart and parameters	Key findings
	Base/matrix	Reinforcements/additives			
Sonber et al. (2015)	Mono-lithic B_4C	–	Method: HP; Pressing: 45 MPa; Sintering: 1900 °C	Ball (WC-Co)-on-flat reciprocating tribo tests; Load: 3, 5 and 10 N; Stroke: 1 mm; Frequency: 10–25 Hz; Speed: 20–50 mm/s; Distance: 200 m; Environment: ambient condition	• Hardness, fracture toughness, flexural strength and elastic modulus values are 28 GPa, 4.1 MPa $m^{0.5}$, 363 MPa and 420 GPa, respectively • Lowest static COF and specific wear rate are achieved as ~0.49 and ~1.61×10^{-6} mm^3/N m, respectively, at 3 N load and 20 mm/s sliding speed • Lowest dynamic COF is found to be 0.15 at 10 N load and 10 Hz frequency • Abrasive and tribochemical wear are observed for monolithic B_4C ceramic

(continued)

Table 2 (continued)

References	Ceramics or CMCs with/without lubricating additives		Fabrication techniques and parameters	Tribological testing process, counterpart and parameters	Key findings
	Base/matrix	Reinforcements/additives			
Pan and Gao (2018)	B_4C	10 wt.% hBN; 20 wt.% hBN; 30 wt.% hBN	Method: HP; Pressing: 30 MPa; Sintering: 1800 °C in Ar	Pin-on-disc (grey cast iron) Load: 10 N; Speed: 0.67 m/s; Distance: 850 m; Environment: ambient condition and under oil-water emulsion effects (fully immersed and drop feeding)	• Lowest COF (0.005) is achieved by 30 wt.% hBN-B_4C composite under drop feeding condition • Lowest wear rate (10^{-6} mm^3/N m) can be achieved with 30 wt.% hBN-B_4C composite under drop feeding of emulsion

(continued)

Table 2 (continued)

References	Ceramics or CMCs with/without lubricating additives		Fabrication techniques and parameters	Tribological testing process, counterpart and parameters	Key findings
	Base/matrix	Reinforcements/additives			
Sedlák et al. (2017)	B_4C	0.5 wt.% GPLs; 1 wt.% GPLs; 2 wt.% GPLs; 4 wt.% GPLs; 6 wt.% GPLs	Method: HP; Pressing: 25 MPa; Sintering: 2100 °C in Ar	Ball (SiC)-on-flat reciprocating tribo tests; Load: 5, 30 and 50 N; Stroke: 5 mm; Speed: 0.1 m/s; Distance: 500 m; Environment: room temperature	• Maximum hardness of ~30 GPa and fracture toughness of ~4.60 MPa $m^{0.5}$ are observed for 0.5 wt.% GPLs-B_4C and 6 wt.% GPLs-B_4C composites, respectively • COF (~0.34) remains low with higher GPLs content at higher load • Lowest wear rate (10^{-6} mm^3/N m) is found for 6 wt.% GPLs-B_4C composites at a load of 5 N

(continued)

Table 2 (continued)

References	Ceramics or CMCs with/without lubricating additives		Fabrication techniques and parameters	Tribological testing process, counterpart and parameters	Key findings
	Base/matrix	Reinforcements/additives			
Džunda et al. (2019)	SiC	CNT with 2.5, 5 and 10 wt.% Fe catalytic nanoparticle	Method: HP; Pressing: 40 MPa; Sintering: 1850 °C in Ar	Ball (SiC)-on-disc; Load: 5 N; Speed: 0.1 m/s; Distance: 500 m; Environment: ambient condition	• Hardness and fracture toughness for SiC lowered from ~24 GPa and ~5 MPa·$m^{0.5}$ to ~17 GPa and ~4 MPa·$m^{0.5}$ with the inclusion of CNT • With the inclusion of CNT in SiC, specific wear rate increases from 7.7 × 10^{-7} to 2.9 × 10^{-6} mm^3/N m • Lowest COF of CNT-SiC composite is ~0.30 • Wear mechanisms of tribochemical and mechanical abrasive wear are found leading to SiO_2 formation • CNT enhances the electrical conductivity of SiC significantly

(continued)

Table 2 (continued)

References	Ceramics or CMCs with/without lubricating additives		Fabrication techniques and parameters	Tribological testing process, counterpart and parameters	Key findings
	Base/matrix	Reinforcements/additives			
Sharma et al. (2019)	SiC	10 wt.% WC; 30 wt.% WC; 50 wt.% WC	Method: HP; Pressing: 40 MPa; Sintering: 1800 °C	Ball (WC-Co and AISI 52100 steel)-on-disc; Load: 5, 10 and 20 N; Speed: 0.16 m/s; Distance: 283 m; Environment: room temperature	• SiC has 5–10 times more wear against WC-Co than against steel • Lowest wear volume of 2.0 $\times 10^{-3}$ mm^3 is found by 50 wt.% WC-SiC composite against steel ball • Increase of WC content, reduces COF from 0.55 to 0.33 for WC-SiC against WC-Co, while the COF increases from 0.35 to 0.43 with the increment of WC content against steel • 50 wt.% WC-SiC having ~6.7 MPa fracture toughness, shows highest wear resistance characteristics • Mechanical and tribo-chemical are the two major wear mechanisms for WC-SiC composite against WC-Co and steel surface

(continued)

Table 2 (continued)

References	Ceramics or CMCs with/without lubricating additives		Fabrication techniques and parameters	Tribological testing process, counterpart and parameters	Key findings
	Base/matrix	Reinforcements/additives			
Sharma et al. (2017)	SiC	10 wt.% WC; 30 wt.% WC; 50 wt.% WC	Method: HP; Pressing: 40 MPa; Sintering: 1800 °C in Ar	Ball (SiC)-on-disc reciprocating tribo tests; Load: 6, 9 and 19 N; Stroke: 4.71 mm; Frequency: 8.3 Hz; Speed: 78 mm/s; Duration: 1000 cycles; Environment: room to 500 °C temperature	• In ambient condition, COF decreases from 0.5 to 0.4 with increasing loads but wear rate increases from 1.4×10^{-6} to 1.5×10^{-5} mm^3/N m • 30 wt.% WC-SiC exhibits lowest wear rate among other composites at elevated temperature • 50 wt.% WC-SiC provides highest wear resistance at ambient condition • At room temperature, both microfracture and tribochemical as material removing mechanism involve but at elevated temperature microfracture is dominating mechanism

(continued)

Table 2 (continued)

References	Ceramics or CMCs with/without lubricating additives		Fabrication techniques and parameters	Tribological testing process, counterpart and parameters	Key findings
	Base/matrix	Reinforcements/additives			
Llorente et al. (2016)	SiC	5 vol.% GNPs; 10 vol.% GNPs; 20 vol.% GNPs	Method: SPS; Pressing: 50 MPa; Sintering: 1800 °C in vacuum	Ball (Si_3N_4)-on-plate reciprocating tribo tests; Load: 5 N; Stroke: 2.5 mm; Frequency: 20 Hz; Distance: 360 m; Environment: room temperature	• 5 vol.% GNPs show best hardness (~24 GPa) among the other graphene based SiC composites • Fracture toughness value (~5.9 MPa·$m^{0.5}$) and flexural strength (~621 MPa) are maximum with 10 vol.% GNPs • 20 vol.% GNPs-SiC shows best wear resistance among all and it has almost 72% improvement as compared to monolithic SiC ceramic

(continued)

Table 2 (continued)

References	Ceramics or CMCs with/without lubricating additives		Fabrication techniques and parameters	Tribological testing process, counterpart and parameters	Key findings
	Base/matrix	Reinforcements/additives			
Sharma et al. (2015)	SiC	10 wt.% WC; 30 wt.% WC; 50 wt.% WC	Method: HP; Pressing: 40 MPa; Sintering: 1800 °C in Ar	Ball (SiC)-on-disc tribo tests; Load: 5, 10 and 20 N; Speed: 0.16 mm/s; Distance: 283 m; Environment: ambient condition. The Hertzian contact stress varies between 1.07 and 1.85 GPa	• Highest hardness (~26.33 GPa) and fracture toughness (~6.66 MPa $m^{0.5}$) are achieved with 30 wt.% WC-SiC and 50 wt% WC-SiC composites, respectively • COF and wear rate of WC-SiC composites vary between 0.40 and 0.48 and 3.3×10^{-6}–3.8×10^{-5} mm^3/N m, respectively • WC addition can reduce grain pull-out of SiC matrix resulting low wear rate • 50 wt.% WC-SiC composite show highest wear resistance

(continued)

Table 2 (continued)

References	Ceramics or CMCs with/without lubricating additives		Fabrication techniques and parameters	Tribological testing process, counterpart and parameters	Key findings
	Base/matrix	Reinforcements/additives			
Tang et al. (2009)	SiC	15 vol.% C_f; 30 vol.% C_f; 42 vol.% C_f; 53 vol.% C_f	Method: HP; Pressing: 30 MPa; Sintering: 1800 °C in Ar	Ball (SiC)-on-disc; Load: 8 N; Rotational speed: 180 rpm; Duration: 3 h; Environment: ambient condition	• Fracture toughness values are increased with C_f concentration and 53 vol.% C_f-SiC provides a maximum fracture toughness value of ~7 MPa·m$^{0.5}$ • Bending strength can be enhanced with C_f addition up to 30 vol.% in SiC • COF decreases up to a value of ~0.24 with addition of 53 vol.% C_f • Lowest wear rate in order of 10^{-6} mm^3/N m is obtained by ~42 vol.% C_f addition

3.3 *Tribomechanical Characteristics of Nitride Based Ceramic Composites*

The most commonly used matrix in the group of nitride based ceramics for the wear resistance applications in various mating parts is Si_3N_4. Si_3N_4 based composites are suitable for structural applications at an elevated temperature where factors such as brittleness, mechanical strength and toughness are important. However, the Si_3N_4 matrix by itself cannot perform well in dry fretting tests since brittle fracture of grains occurs resulting in high COF and sever wear (Novak et al. 1996). To enhance the fracture toughness of Si_3N_4, SiC whiskers can be embedded in the matrix. Si_3N_4 with 30 vol.% SiC whiskers added can reach a fracture toughness of ~5.3 MPa·m$^{0.5}$ if densified by hot press at 1750 °C and 30 MPa (Bellosi and Portu 1989). Recently, it is found that SiC nanowire (SiC NW) is a very promising material to improve the mechanical properties and thermal shock resistance of α-Si_3N_4 matrix. Addition of SiC NW is often a better solution than adding particulate SiC to a ceramic matrix if the goal is to achieve a desirable modulus of elasticity and fracture strength as well. Figure 2 describes the toughening mechanism of SiC NW into the α-Si_3N_4 composite. During sliding, the energy needed for fracture can be increased by SiC NW which restricts the propagation of cracks by bridging and by deflecting the direction of crack propagation which as a whole absorbs the fracture energy to prevent brittle fracture. In addition to this, the grain pull-out can be strongly inhibited by SiC NW which ultimately strengthens the surrounded matrix system under applied load (Wanga et al. 2019).

Another way of enhancing the matrix strength of Si_3N_4 is reinforcement by CNTs. CNTs strongly adhere to the Si_3N_4 grains and a densified composite system can be obtained if fabricated by hot isostatic process (HIP) and SPS. Interestingly, a high value of elastic modulus, hardness and fracture toughness can be obtained as ~326 GPa, ~20 GPa and ~5.2 MPa·m$^{0.5}$, respectively, when densified by SPS at a low temperature of 1500 °C and a pressure of 100 MPa (Balázsi et al. 2006). Tribological characteristics of Si_3N_4 can be enhanced with the inclusion of TiB_2

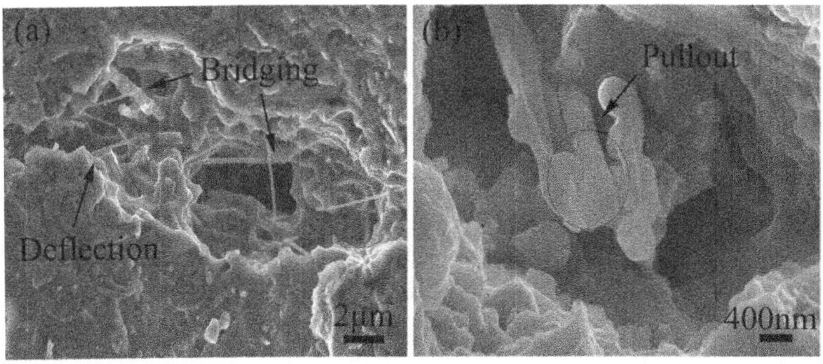

Fig. 2 Crack bridging and deflection by SiC NW during fracture of α-Si_3N_4 composite (Wanga et al. 2019)

as the second phase. 40 vol.% TiB_2 into Si_3N_4 improves the wear resistance of the composite by 5 times as compared to straight against steel balls (Jones et al. 2001). As reinforcement, SiC and TiN together lead to a better sinterability for Si_3N_4 matrix. A proper proportion of these two compounds can achieve meaningful tribo-mechanical characteristics for Si_3N_4 composite. In spite of having better hardness for the combination SiC-Si_3N_4, it fails to provide better abrasive wear resistance as compared to SiC-TiN-Si_3N_4 composite (Blugan et al. 2014). Hardness along with wear resistance of Si_3N_4 can be enhanced with the addition of TiC. Both friction and specific wear rate for TiC-Si_3N_4 composite can be improved with increasing load up to 60 N (Manzoor et al. 2019). The efficient self-lubricating characteristics of Si_3N_4 matrix composite can be obtained by introducing some class of solid lubricants into it. Hexagonal boron nitride (hBN) provides a lucrative lubricating action at the sliding interface since it gets sheared off easily. Thus, hBN added Si_3N_4 matrix composite can be used for a self-lubricating mechanical component such as metal removing tool (Kimura et al. 1999; Saito et al. 2001). The tribological performance of hBN-Si_3N_4 has been studied extensively in different operating conditions (dry and wet) (Saito et al. 2001; Chen et al. 2010, 2017, 2020; Xin et al. 2019). A very low COF of 0.03 can be obtained while sliding hBN-Si_3N_4 composite against austenitic stainless steel in dry conditions. Alongside, the wear rate can also be as low as in the order of 10^{-6} mm^3 N^{-1} m^{-1} because of the soft tribochemical layer formed by the oxidized products such as SiO_2, B_2O_3 and Fe_2O_3. The mild wear was caused by the oxidation reaction against steel (Chen et al. 2010). In addition to this, the applied load influences the tribological performances of hBN-Si_3N_4 composite against steel. At low loading conditions (10 N), abrasive wear is the dominant wear mechanism for the tribo pair. If the load is increased to 50 N, both abrasive and adhesive wear come in the picture (Chen et al. 2017). The tribological performance of Si_3N_4 can also be enhanced with the inclusion of multi-layered graphene (MLG). Spark plasma sintering (SPS) is a more efficient sintering method than the traditional hot isostatic process (HIP) to incorporate MLG into Si_3N_4. SPS can have a maximum (~98.5%) density with the inclusion of 1 wt.% MLG into Si_3N_4. The most significant benefit of using SPS is the structural degradation which can be avoided during sintering for both Si_3N_4 matrices as well as MLG. This structural stability inside the matrix can enhance the self-lubricating behaviour of MLG-Si_3N_4 composite by providing low shear strength at the contacting interface with SiC, which is the advantageous mechanism of MLG as a lubricant during sliding (Berkes et al. 2016). The wear rate is always lower against the SiC counterface irrespective of the sintering method. The wear rate is also lower for SPS fabricated composites than for those prepared by traditional HIP method.

Another significant way of reducing friction and wear rate of monolithic Si_3N_4 in dry sliding is the addition of rare earth oxide compounds viz., La_2O_3, Nd_2O_3, Y_2O_3, Yb_2O_3 and Lu_2O_3. Experiments reveal that Si_3N_4 doped with Lu_2O_3 exhibits the highest wear resistance since the higher bonding strength provided by the Lu_2O_3 having smallest ionic radius among other additives, prevents Si_3N_4 grain pull-out from the matrix during sliding. This also explains the improvement of the hardness and fracture toughness that can be obtained by doping of such additives into the Si_3N_4 matrix (Tatarko et al. 2010).

3.3.1 Progression in High Temperature Nitride Based Ceramic Matrix Composites

Ceramic composites of Si_3N_4 are widely accepted for high temperature applications because of their excellent thermal and chemical stability at elevated temperature. The use of nitride based ceramic matrices at elevated temperature has been a frequent practice since before 1990 (Tomizawa and Fischer 1986; Gee et al. 1990; Skopp et al. 1990; Park et al. 1992). Nevertheless, obtaining a desirable COF and low wear rate is a challenge for pure Si_3N_4 at high temperatures (1000 °C). For instance, a study shows that monolithic Si_3N_4 suffers severe wear while sliding at a temperature of 1000 °C. To achieve better wear resistance of Si_3N_4, its thermal diffusivity and thermal conductivity should be enhanced which in turn diminishes the thermal stress and crack propagation. Addition of TiN and BN as the second phase can improve the wear resistance of Si_3N_4 in highly demanding sliding condition such as high sliding speed, load and elevated temperature (Skopp et al. 1995). Most recently, the tribological response of β-Si_3N_4 has been characterized against laser heated Inconel 718 at a temperature of 600 °C to explore the potential of Si_3N_4 as a material removing tool for machining Ni based alloy using laser assisted machining. Interestingly, it is seen that the COF and wear rate against the high temperature laser heated Inconel is lower than at room temperature while sliding at 7.5 m/s. The wear rate of Si_3N_4 increases with increasing sliding speed above 7.5 m/s at high temperature. At higher speed, the oxidation and diffusion rate increase because of higher temperature occurring at the tribo interface and the tribo layer becomes thick and spalls away which acts detrimentally to increase the total tribochemical as well as mechanical wear of the Si_3N_4. It can be mentioned that laser heating is advantageous for Si_3N_4 ceramic during sliding since it provides fast heat distribution along the surface leading to lower material loss (Zhaoa et al. 2019).

In a different study, investigation on the tribological behaviour of Si_3N_4 with addition of rare earth oxides (Sm_2O_3, Y_2O_3, La_2O_3, Nd_2O_3, Yb_2O_3 and Lu_2O_3) and nano SiC has been carried out up to a temperature of 900 °C. A significant combined influence of rare earth oxides and nano SiC can be observed on the tribological results if fabricated by powder metallurgy and hot press sintering. Lu, Yb, and Y having smaller ionic radius have the largest influence on the wear performance of SiC-Si_3N_4 composite up to 700 °C since it provides strengthening to the SiC nanoparticles on the grain boundaries of the matrix composite Si_3N_4 during sliding (Tatarko et al. 2013). Table 3 summarizes the fabrication method, mechanical and tribological characteristics of different nitride based ceramic composites.

Table 3 Highlights of the fabrication process, mechanical and tribological characteristics of nitride based ceramic composites in dry sliding

References	Ceramics or CMCs with/without lubricating additives		Fabrication techniques and parameters	Tribological testing process, counterpart and parameters	Findings
	Base/matrix	Reinforcements/additives			
Zhaoa et al. (2019)	Commercial Si_3N_4 matrix	–	–	Pin-on-disc (Inconel 718) tribo tests; Load: 100 N; Speed: 2.5–10 m/s; Distance: 2000 m; Environment: 600 °C temperature with laser heating arrangement	• COF at high temperature is lower than room temperature with all sliding speeds • Up to 7.5 m/s, the wear rate is lower at high temperature in comparison to room temperature and beyond it wear rate increases • At high sliding speed, the tribochemical wear increases leading to higher wear rate

(continued)

Table 3 (continued)

References	Ceramics or CMCs with/without lubricating additives		Fabrication techniques and parameters	Tribological testing process, counterpart and parameters	Findings
	Base/matrix	Reinforcements/additives			
Park et al. (1992)	Si_3N_4	–	Method: HP and HIP	Ball (Si_3N_4)-on-flat reciprocating tribo tests; Load: 3.67 N; Speed: 2 cm/s; Stroke: 2.18 cm; Distance: 10,000 strokes; Environment: room to 1000 °C in Ar	• COF and wear rate increase at elevated temperature • Both COF and wear rate decreased in oxidative environment forming tribochemical products under humidity

(continued)

Table 3 (continued)

References	Ceramics or CMCs with/without lubricating additives		Fabrication techniques and parameters	Tribological testing process, counterpart and parameters	Findings
	Base/matrix	Reinforcements/additives			
Skopp et al. (1990)	Si_3N_4	–	Method: HIP	Pin (Si_3N_4)-on-disc tribo tests; Load: 10 N; Speed: 0.03–3 m/s; Distance: 1000 m; Environment: 1000 °C	• COF in between 0.5 and 1 • Wear rate increases with increase in temperature at higher speed • Tribo-oxidation accelerates the wear rate at higher speed and temperature (above 400 °C)

(continued)

Table 3 (continued)

References	Ceramics or CMCs with/without lubricating additives		Fabrication techniques and parameters	Tribological testing process, counterpart and parameters	Findings
	Base/matrix	Reinforcements/additives			
Tomizawa and Fischer (1986)	Si_3N_4	–	Method: HP	Pin (Si_3N_4)-on-plate tribo tests; Load: 0.5–5 N; Speed: 0.1–10 cm/s; Distance: 5 m; Environment: 150–800 °C	• In dry air, COF is ~0.8 at room temperature • In humid air, COFs are ~0.2 at 150 °C, ~0.3 at 650 °C and ~0.8 at 800 °C • At elevated temperature fracture increases leading to high wear

(continued)

Table 3 (continued)

References	Ceramics or CMCs with/without lubricating additives		Fabrication techniques and parameters	Tribological testing process, counterpart and parameters	Findings
	Base/matrix	Reinforcements/additives			
Manzoor et al. (2019)	Si_3N_4	1 wt.% TiC; 2 wt.% TiC	Method: CP; Sintering: 1740 °C in N_2	Ball-on-plate reciprocating tribo tests; Load: 20–60 N; Stroke: 2 mm; Frequency: 20 Hz; Duration: 45 min; Environment: room temperature	• Highest hardness is achieved as ~19.88 GPa with 2 wt.% TiC addition in Si_3N_4 • COF and wear rate both decrease with TiC addition • At higher load (60 N), 2 wt.% TiC-Si_3N_4 composite provides better tribological behaviour

(continued)

Table 3 (continued)

References	Ceramics or CMCs with/without lubricating additives		Fabrication techniques and parameters	Tribological testing process, counterpart and parameters	Findings
	Base/matrix	Reinforcements/additives			
Xin et al. (2019)	Si_3N_4	20 vol.% hBN	Method: HP; Pressing: 30 MPa; Sintering: 1800 °C in N_2	Pin-on-disc (polyethylene) tribo tests; Load: 10, 50 and 90 N; Speed: 0.1, 0.3 and 0.5 m/s; Distance: 1000 m distance Environment: room temperature	• COF and wear rate of composite decrease with increase in load and velocity • hBN provides effective lubricating film to enhance the tribological behaviour of hBN-Si_3N_4/polyethylene couple

(continued)

Table 3 (continued)

References	Ceramics or CMCs with/without lubricating additives		Fabrication techniques and parameters	Tribological testing process, counterpart and parameters	Findings
	Base/matrix	Reinforcements/additives			
Chen et al. (2017)	Si_3N_4	5 vol.% hBN; 10 vol.% hBN; 20 vol.% hBN; 30 vol.% hBN	Method: HP; Pressing: 30 MPa; Sintering: 1800 °C	Pin-on-disc (stainless steel) tribo tests; Load: 10 N; Speed: 0.52, 0.86, 1.73 m/s; Distance: 1000 m; Environment: ambient condition	• Hardness and fracture toughness and bending strength decrease with increasing hBN content in Si_3N_4 • Lowest COF (0.27) and wear rate (10^{-6} mm^3/N m) are obtained with 10 vol.% hBN-Si_3N_4 composite at 20 N load • With higher loading (30 and 50 N), the wear mechanisms are associated with abrasive and adhesive, which is only abrasive wear with lower load (10 N) particularly for hBN-Si_3N_4 composite

(continued)

Table 3 (continued)

References	Ceramics or CMCs with/without lubricating additives		Fabrication techniques and parameters	Tribological testing process, counterpart and parameters	Findings
	Base/matrix	Reinforcements/additives			
Berkes et al. (2016)	Si_3N_4	1 wt.% MLGs; 3 wt.% MLGs	Method: HIP; Pressing: 20 MPa; Sintering: 1700 °C, and Method: SPS; Pressing: 50 MPa; Sintering: 1500 °C	Ball (SiC and Si_3N_4)-on-disc tribo tests; Load: 40 N; Speed: 20 and 200 mm/s; Distance: 100 m; Environment: room temperature	• Improved physico-mechanical properties are achieved with SPS as compared to HIP • Best frictional and wear behaviour can be achieved with SPS specimen against SiC. Reducing specific wear rate by 14–71% and 22–55% while sliding speeds are 20 and 200 mm/s, respectively • 3 wt.% MLG is not beneficial for specimen by HIP, but advantageous for specimen by SPS in terms of wear resistance

(continued)

Table 3 (continued)

References	Ceramics or CMCs with/without lubricating additives		Fabrication techniques and parameters	Tribological testing process, counterpart and parameters	Findings
	Base/matrix	Reinforcements/additives			
Blugan et al. (2014)	Si_3N_4	TiN; SiC; TiN + SiC	Method: HP; Pressing: 30 MPa; Sintering: 1750–1820 °C in N_2	Ball (Si_3N_4)-on-disc reciprocating tribo tests; Load: 10 N; Stroke: 2 mm; Frequency: 10 Hz; Distance: 36 m; Environment: room temperature	• High hardness of TiN-SiC-Si_3N_4 composites provides low wear rate among other SiC-Si_3N_4 and TiN-Si_3N_4 composites • TiN-SiC-Si_3N_4 exhibits lowest COF among other combinations • TiN-SiC-Si_3N_4 is suitable for abrasive wear resistance applications

(continued)

Table 3 (continued)

References	Ceramics or CMCs with/without lubricating additives		Fabrication techniques and parameters	Tribological testing process, counterpart and parameters	Findings
	Base/matrix	Reinforcements/additives			
Tatarko et al. (2013)	Si_3N_4	5 vol.% SiC NP; 5 vol.% SiC NP + 8 vol.% Y_2O_3; 5 vol.% SiC NP + 8 vol.% La_2O_3; 5 vol.% SiC NP + 8 vol.% Nd_2O_3; 5 vol.% SiC NP + 8 vol.% Sm_2O_3; 5 vol.% SiC NP + 8 vol.% Yb_2O_3; 5 vol.% SiC NP + 8 vol.% Lu_2O_3	Method: HP; Pressing: 30 MPa; Sintering: 1750 °C in N_2	Ball (Si_3N_4)-on-disc tribo tests; Load: 5 N; Speed: 0.1 m/s; Distance: 500 m; Environment: room to 900 °C in air	• Composites having smaller ionic radius (Lu, Yb and Y) show lowest specific wear rate up to 700 °C • Presence of intergranular SiC nanoparticles provides better wear resistance of monolithic Si_3N_4 ceramics • Brittle fracture as failure mechanism is observed up to 700 °C, but above that oxidation of SiC is significant wear mechanism

(continued)

Table 3 (continued)

References	Ceramics or CMCs with/without lubricating additives		Fabrication techniques and parameters	Tribological testing process, counterpart and parameters	Findings
	Base/matrix	Reinforcements/additives			
Chen et al. (2010)	Si_3N_4	5 vol.% hBN; 10 vol.% hBN; 20 vol.% hBN; 30 vol.% hBN	Method: HP; Pressing: 30 MPa; Sintering: 1800 °C	Pin-on-disc (austenitic stainless steel) tribo tests; Load: 10 N; Speed: 1.31 m/s; Distance: 850 m; Environment: ambient condition	• Mechanical characteristics like hardness, fracture toughness and bending strength are diminished with increasing vol.% of hBN in Si_3N_4 • Lowest COF (0.03) and wear rate (10^{-6} mm^3/N m) are found with 30 vol.% hBN-Si_3N_4 composites • Tribo film consisting of SiO_2, B_2O_3 and Fe_2O_3, are responsible to reduce COF against stainless steel

(continued)

Table 3 (continued)

References	Ceramics or CMCs with/without lubricating additives		Fabrication techniques and parameters	Tribological testing process, counterpart and parameters	Findings
	Base/matrix	Reinforcements/additives			
Tatarko et al. (2010)	Si_3N_4	SiC; SiC + Y_2O_3; SiC + La_2O_3; SiC + Nd_2O_3; SiC + Sm_2O_3; SiC + Yb_2O_3; SiC + Lu_2O_3	Method: HP; Pressing: 30 MPa; Sintering: 1750 °C in N_2	Ball (Si_3N_4)-on-disc tribo tests; Load: 5 N; Speed: 0.1 m/s; Distance: 500 m; Environment: ambient condition	• SiC significantly enhances hardness of monolithic Si_3N_4 ceramic • Specific wear rate decreases with increasing hardness for SiC-Si_3N_4 composites • Smaller ionic radius of rare earth compounds provides strong bonds between grains and intergranular phases leading to high fracture toughness resulting high wear resistance of composite matrix • Composite with Lu_2O_3 provides highest wear resistance of SiC-Si_3N_4 composites

(continued)

Table 3 (continued)

References	Ceramics or CMCs with/without lubricating additives		Fabrication techniques and parameters	Tribological testing process, counterpart and parameters	Findings
	Base/matrix	Reinforcements/additives			
Jones et al. (2001)	Si_3N_4	40 vol.% TiB_2	Method: HP; Pressing: 30 MPa; Sintering: 1500–1700 °C in Ar, and Method: HIP; Pressing: 160 MPa; Sintering: 1550–1700 °C in Ar	Ball(steel/Si_3N_4)-on-disc tribo tests; Load: 5 N; Speed: 0.05 m/s; Distance: 1000 m; Environment: ambient condition	• Low temperature (1500 °C) in hot press provide full densified composite having hardness, fracture toughness and elastic modulus of ~17.4 GPa, ~4.9 MPa $m^{0.5}$ and ~365 GPa, respectively • Hot isostatically pressed 40 vol.% TiB_2-Si_3N_4 composite sintered at 1650 °C shows best hardness and fracture toughness values as ~18 GPa and ~5.8 MPa $m^{0.5}$, respectively • Wear rate is lower against steel than Si_3N_4 surface for TiB_2-Si_3N_4 composite • Lowest wear rate of ~1.3 × 10^{-6} mm^3/N m is observed with 40 vol.% TiB_2-Si_3N_4 composite against steel as compared to straight Si_3N_4 ceramic

(continued)

Table 3 (continued)

References	Ceramics or CMCs with/without lubricating additives		Fabrication techniques and parameters	Tribological testing process, counterpart and parameters	Findings
	Base/matrix	Reinforcements/additives			
Skopp et al. (1995)	Commercial Si_3N_4 ceramics	20 wt.% TiN; 5 wt.% BN; 10 wt.% BN; 20 wt.% BN	Method: HP, HIP, reaction bonded, gas pressure sintered	Ball (Si_3N_4)-on-disc tribo tests; Load: 10 N; Speed: 0.03, 0.1, 0.3, 1, 3 and 5 m/s; Distance: 1000 m; Environment: 22, 400, 800 and 1000 °C	• COF of monolithic Si_3N_4 ceramic is in the range 0.4–0.9 and wear rate in the range of 10^{-6}–10^{-2} $mm^3/N\,m$ • Addition of BN and TiN enhance the tribological performance at high temperature • Low COF (0.1) and wear rate (10^{-7} $mm^3/N\,m$) can be achieved with Si_3N_4 if thermal diffusivity and thermal conductivity is increased along with the glass phases are required to reduce with addition of solid lubricants

4 Tribomechanical Performances of Other Ceramic Composites

Apart from boride, carbide and nitride matrix composites, there are some other forms of ceramic composites which may be a combination of three or more phases together and can show great mechanical and wear resisting characteristics during dry sliding. For example, a hybrid composite made of different phases like ZrB_2, SiC, ZrC and C_f can show sound density, hardness and flexural strength values of 99.6%, 18 GPa and 565 MPa, respectively, if the composite powder mixture is densified by SPS at 1800 °C temperature and 30 MPa pressure (Adibpur et al. 2020). In another study, a composite of TiB_2-TiC-Ti_3SiC_2 is prepared by hot press technique from the in-situ preparation of the composite powder from the starting raw materials TiH_2, Si, graphite and B_4C. The chemical changes of the starting raw materials followed the reaction as given in Eq. 1. The composite shows low friction and has good wear resistance with a variation of TiB_2 (15 and 20 vol.%) content within load range 10–30 N against AISI-52 100 bearing steel (Yang et al. 2011).

$$TiH_2 + Si + C + B_4C \rightarrow Ti_3SiC_2 + TiB_2 + TiC + H_2 \uparrow \tag{1}$$

TiB_2 forms a very good wear resistance composite when combine with TiC. Generally, Ni is added into the TiC-TiB_2 composite by following different fabrication methods such as powder metallurgy, reaction sintering, etc. to enhance the toughness behaviour (Tjong and Lau 2000; Zhao and Cheng 1999; Xue et al. 2011). A good distribution of Ni, TiC and TiB_2 can be achieved in TiB_2-TiC-Ni composite if they are fabricated by field-activated pressure-assisted synthesis (FAPAS) from the starting powders Ti, B_4C and Ni by the following chemical changes given in Eq. 2:

$$3Ti + B_4C + xNi \rightarrow TiC + 2TiB_2 + xNi \tag{2}$$

This composite can be used up to a temperature level of 400 °C since it provides a low wear rate with increasing temperature (400 °C). However, the wear rate increases with a decreasing trend of COF if the load (60–120 N) and sliding speed (10–40 m/s) increase while sliding against GCr15 steel.

A Ti_3SiC_2 matrix can be produced from the proper stoichiometric mixture of TiC and Si followed by hot pressing at 1420 °C temperature and 25 MPa pressure in Ar atmosphere. Under fretting wear condition against steel, the composite shows a COF in the range of 0.5–0.6 and a wear rate in the order of 10^{-5} mm^3/N m with the increase of load from 6 to 8 N. The wear occurs due to the triboxidation of the composite resulting TiO_2, SiO_2 and Fe_2O_3 (Sarkar et al. 2006).

Recently, transition metal (Fe, Mo, Cr, Mn, W, etc.) added nanolayered MAB (M: metal, A: aluminium and B: boron) phase ceramic has become a good attraction of research since they possess superior mechanical, tribological and electrical characteristics in different working conditions (Natu et al. 2020; Benamor et al. 2019). For instance, MoAlB ceramic shows mild wear when slides against Inconel 718 at

600 °C. COF and wear rate can be 0.5 and 9.37×10^{-7} mm^3/N m, respectively, under this dry sliding operating condition (Yu et al. 2020).

5 Key Applications of Non-oxide Based Wear Resistive Ceramic Matrix Composites

Applications of such non-oxide ceramics are found in many cutting-edge industries such as aviation, manufacturing, automobile industries etc. Significant applications of these CMCs as promising wear resistive components are mentioned below.

5.1 Brakes

SiC based composites have several applications in different manufacturing industries. One of the key applications of SiC based composite is in the brake discs used in the automobile/aviation sector because of desirable frictional characteristics (Bian and Wu 2015, 2016). The cast iron used in brake disc is being replaced by carbon fibre reinforced SiC (C_f-SiC) composite. C_f is reinforced into SiC matrix to produce efficient C_f-SiC composite brake systems useful for aircraft brakes (Jiang et al. 2008). Carbon-carbon SiC matrix reinforced with ZrB_2 and ZrC (C/C-ZrB_2-ZrC-SiC) hybrid composite is another form of braking material having superior frictional behaviour in wet condition than dry condition. Even, it provides better performances than C/C-SiC composite braking material. Henceforth, this material can be used as an aircraft brake disc as well (Qian et al. 2013).

5.2 Bearing Balls and Rolling Contact Bearings

High hardness, fracture toughness, low thermal expansion coefficient and good thermal shock resistance along with thermal stability and lower density of ceramic composites balls, make them advantageous over bearing steel balls (Gal et al. 2019). Especially, Si_3N_4 as bearing balls can be advantageous over commercial grade steel balls because of having low weight, low thermal coefficient of expansion and good frictional characteristics during the rolling condition. Hence Si_3N_4 balls are used as bearing balls for rolling contact bearings (Cundill 1992). In addition to this, Si_3N_4 ceramics are used as the material of rolling contact bearing race since they show promising tribomechanical property and dimensional stability under different operating temperatures, even under wet conditions (Wang et al. 2000). High temperature sustainable ceramic composite bearings can be employed in gas turbine engines, adiabatic diesel engines, rudder bearing of aircraft etc.

5.3 *Material Removing Tools*

Si_3N_4 based composites are often used as material removing tools in a wide range of cutting conditions because of their promising mechanical characteristics and chemical stability (Guo et al. 2017; Riley 2000). TiC/Si_3N_4 composite tools with graphene reinforcement can significantly diminish the friction coefficient (0.46) and wear rate (4.29×10^{-6} mm^3/N m) while employed for cutting applications (Zhang et al. 2020). Sometimes, hard non-oxide ceramics like TiB_2/TiC/SiC etc. are used as reinforcement in an Al_2O_3 matrix to produce cutting tool components for machining of steels.

5.4 *Mechanical Seals*

Mechanical sealing materials require a low COF and high wear resistance along with high fatigue resistance, corrosion resistance and of course low weight. Ceramic composites are favourable for committing these characteristics while being applied in the form of rings or sealing elements. SiC is generally used for the fabrication of mechanical seals (Brown 1995). Apart from that, Si_3N_4 or Si_3N_4-SiC composite can be a suitable choice for sealing materials since they provide low wear rate under normal operating conditions (Carrapichano et al. 2003).

6 Merits and Demerits of Non-oxide Based CMCs

The non-oxide ceramics are potential candidates for high temperature applications. For instance, SiC and Si_3N_4 ceramics can be employed at up to ~1500 °C temperature applications which is a limitation for oxide based ceramic candidates who retain their structural property up to ~1000 °C (Klemm 2010). Non-oxide ceramics can retain their high mechanical strength at elevated temperature as well. On the other hand, non-oxide ceramics are useful for wear resisting applications in humid or wet conditions. Application of such ceramics is beneficial in producing abrasive conditions such as grinding media, grinding wheel, cutting tool etc., as they possess high hardness along with good wear resistance, even at elevated temperature. Si_3N_4 has excellent thermal shock resistance which makes it suitable in using critical areas like thermocouple shielding tubes, crucible etc.

However, ceramics like B_4C, ZrC etc. have poor sintering kinetics because of covalent bonding, and generally require the use of advanced fabrication technologies. Traditional sintering technique such as hot pressing is not recommended for such class of materials since it takes a long processing time which may lead to unnecessary grain growth. Even if they are densified using SPS, a high temperature is required. Addition of some sintering aids can improve the sinterability of these materials. Another major limitation of non-oxide based ceramics is the oxidation stability. They require some oxygen diffusion barrier to restrict the oxidation for long term applications. For example, Si_3N_4 requires the formation of silica as an oxygen diffusion barrier to restrict oxidation at elevated temperature. Other ceramics like SiC-ZrB_2, ZrB_2-SiC-C have very high oxidation instability which make them inappropriate at high temperature applications (Klemm 2010).

7 Concluding Remarks

It is found from the available literature, non-oxide ceramics such as boron-based, carbide-based and nitride-based ceramics are mostly used for tribomechanical applications in dry sliding under different operating conditions. Microhardness and fracture toughness values of boron-based (TiB_2 and ZrB_2), carbide-based (B_4C, SiC and WC) and nitride-based (BN and Si_3N_4) ceramics/composites lie in the range of ~16–26 GPa, ~3.9–9.7 MPa $m^{0.5}$; ~17–34 GPa, ~4–7 MPa $m^{0.5}$; and ~17–19.9 GPa, ~4.9–6 MPa $m^{0.5}$, respectively. Coefficient of friction and wear rate in ambient and high temperature vary between ~0.03–1, 10^{-7}–10^{-3} mm^3/N m and ~0.4–0.9, ~10^{-3}–10^{-2} mm^3/N m, respectively. Non-oxide ceramics are potential candidates for application at elevated temperatures. However, their improvement of chemical stability is a major area of further research since they get oxidized at severe contact conditions such as high speed, load and elevated temperature. Future work should be directed to the study of oxidation and tribo-corrosion resistance of non-oxide based ceramics and their further enhancement. Furthermore, very few non-oxide based ceramics have been tested for their tribomechanical characteristics at elevated temperature. In particular, it is found that nitride based ceramics, e.g. Si_3N_4 cannot perform well when employed for high temperature wear resistive applications unless some secondary phases are added. Therefore, doping or addition of other phases (reinforcement or lubricating) can be potential research to establish high temperature wear resistant suitability for such ceramic matrices.

Acknowledgements The authors are grateful for the financial support received from Faculty Research Grant of the University of Malaya, Malaysia (Grant number: GPF023A-2019).

References

Adibpur F, Tayebifard SA, Zakeri M, Asl MS (2020) Spark plasma sintering of quadruplet ZrB_2–SiC–ZrC–Cf composites. Ceram Int 46(1):156–164

Aouadi SM, Paudel Y, Luster B, Stadler S, Kohli P, Muratore C, Hager C, Voevodin AA (2008) Adaptive Mo_2N/MoS_2/Ag tribological nanocomposite coatings for aerospace applications. Tribol Lett 29:95–103

Aydinyan S, Minasyan T, Liu L, Cygan S, Hussainova I (2019) ZrC based ceramics by high pressure high temperature SPS technique. Key Eng Mater 799:125–130

Balázsi C, Wéber F, Kövér Z, Shen Z, Kónya Z, Kasztovszky Z, Vértesy Z, Biró LP, Kiricsi I, Arato P (2006) Application of carbon nanotubes to silicon nitride matrix reinforcements. Curr Appl Phys 6:124–130

Bellosi A, Portu GD (1989) Hot-pressed Si_3N_4-SiC whisker composites. Mater Sci Eng A 109:357–362

Benamor A, Kota S, Chiker N, Haddad A, Hadji Y, Natu V, Abdi S, Yahi M, Benamar MEA, Sahraoui T, Hadji M, Barsoum MW (2019) Friction and wear properties of MoAlB against Al_2O_3 and 100Cr6 steel counterparts. J Eur Ceram Soc 39:868–877

Berkes MM, Németh AK, Károly Z, Bódis E, Maros Z, Tapasztó O, Balázsi K (2016) Tribological characterisation of silicon nitride/multilayer graphene nano composites produced by HIP and SPS technology. Tribol Int 93:269–281

Bertagnoli D, López OB, Rojas FR, Guiberteau F, Ortiz AL (2015) Effect of processing conditions on the sliding-wear resistance of ZrC triboceramics fabricated by spark-plasma sintering. Ceram Int 41:15278–15282

Bian G, Wu H (2015) Friction performance of carbon/silicon carbide ceramic composite brakes in ambient air and water spray environment. Tribol Int 92:1–11

Bian G, Wu H (2016) Friction surface structure of a Cf/C–SiC composite brake disc after bedding testing on a full-scale dynamometer. Tribol Int 99:85–95

Blugan G, Hadad M, Graule T, Kuebler J (2014) Si_3N_4–TiN–SiC three particle phase composites for wear applications. Ceram Int 40:1439–1446

Brown M (1995) Seals and sealing handbook. Elsevier, Amsterdam

Carrapichano JM, Gomes JR, Oliveira FJ, Silva RF (2003) Si_3N_4 and Si_3N_4/SiC composite rings for dynamic sealing of circulating fluids. Wear 255:695–698

Chakraborty S, Debnath D, Mallick AR, Das PK (2014) Mechanical, tribological, and thermal properties of hot-pressed ZrB_2-B_4C composite. Int J Appl Ceram Technol 1–9

Chakraborty S, Mallick AR, Debnath D, Das PK (2015) Densification, mechanical and tribological properties of ZrB_2 by SPS: effect of pulsed current. Int J Refract Metal Hard Mater 48:150–156

Chen W, Gao Y, Chen C, Xing J (2010) Tribological characteristics of Si_3N_4–hBN ceramic materials sliding against stainless steel without lubrication. Wear 269:241–248

Chen W, Zhang D, Ai X (2017) Effect of load on the friction and wear characteristics of Si_3N_4-hBN ceramic composites sliding against steels. Ceram Int 43:4379–4389

Chen W, Wang Z, Liu X, Jia J, Hu Y (2020) Effect of load on the friction and wear characteristics of Si_3N_4-hBN ceramic composites sliding against PEEK in artificial seawater. Tribol Int 141:105902

Cundill RT (1992) High-precision silicon nitride balls for bearings. In: Baker LR (ed) Commercial applications of precision manufacturing at the sub-micron level. SPIE–The International Society for Optical Engineering, London, pp 75–87

Džunda R, Fides M, Hnatko M, Hvizdoš P, Múdra E, Medveď D, Kovalčíková A, Milkovič O (2019) Mechanical, physical properties and tribological behaviour of silicon carbide composites with addition of carbon nanotubes. Int J Refract Metal Hard Mater 81:272–280

Gal CW, Song GW, Baek WH, Kim HK, Lee DK, Lim KW, Park SJ (2019) Fabrication of pressureless sintered Si_3N_4 ceramic balls by powder injection molding. Ceram Int 45:6418–6424

Gee MG, Matharu CS, Almond EA, Eyre TS (1990) The measurement of sliding friction and wear of ceramics at high temperature. Wear 138(1–2):169–187

González BN, Ortiz AL, Guiberteau F, Padture NP (2011) Effect of $MoSi_2$ content on the lubricated sliding-wear resistance of ZrC–$MoSi_2$ composites. J Eur Ceram Soc 31:877–882

Guo XL, Zhu ZL, Ekevad M, Bao X, Cao PX (2017) The cutting performance of Al_2O_3 and Si_3N_4 ceramic cutting tools in the milling plywood. Adv Appl Ceram 117:1–7

Hasan M, Zhao J, Jiang Z (2019) Micromanufacturing of composite materials: a review. Int J Extreme Manuf 1:012004

Jiang G, Yang J, Xu Y, Gao J, Zhang J, Zhang L, Cheng L, Lou J (2008) Effect of graphitization on microstructure and tribological properties of C/SiC composites prepared by reactive melt infiltration. Compos Sci Technol 68:2468–2473

Jones AH, Dobedoe RS, Lewis MH (2001) Mechanical properties and tribology of Si_3N_4–TiB_2 ceramic composites produced by hot pressing and hot isostatic pressing. J Eur Ceram Soc 21(7):969–980
Jourdain V, Bichara C (2013) Current understanding of the growth of carbon nanotubes in catalytic chemical vapour deposition. Carbon 58:2–39
Ju HB, Yu D, Yu LH, Ding N, Xu JH, Zhang XD, Zheng Y, Yang L, He X (2018) The influence of Ag contents on the microstructure, mechanical and tribological properties of ZrN-Ag films. Vacuum 148:54–61
Kim YW, Chun YS, Nishimura T, Mitomo M, Lee YH (2007) High-temperature strength of silicon carbide ceramics sintered with rare-earth oxide and aluminum nitride. Acta Mater 55:727–736
Kimura Y, Wakabayashi T, Okada K, Wada T, Nishikawa H (1999) Boron nitride as a lubricant additive. Wear 232:199–206
Klemm H (2010) Silicon nitride for high-temperature applications. J Am Ceram Soc 93(6):1501–1522
Kong L, Bi Q, Zhu S, Yang J, Liu W (2012) Tribological properties of ZrO_2 (Y_2O_3) $MoBaF_2/CaF_2$ composites at high temperatures. Tribol Int 45:43–49
Köstenbauer H, Fontalvo GA, Mitterer C, Keckes J (2008) Tribological properties of TiN/Ag nanocomposite coatings. Tribol Lett 30:53–60
Li X, Ding H, Huang Z, Fang M, Liu B, Liu Y, Wu X, Chen S (2014) Solid particle erosion-wear behavior of SiC–Si_3N_4 composite ceramic at elevated temperature. Ceram Int 40:16201–16207
Llorente J, Manso BR, Miranzo P, Belmonte M (2016) Tribological performance under dry sliding conditions of graphene/silicon carbide composites. J Eur Ceram Soc 36:429–435
Manzoor S, Wani MF, Saleem SS (2019) Effect of load on the friction and wear behaviour of silicon nitride and silicon nitride titanium carbide ceramic composite. Mater Today Proc 19(2):474–477
Mazumder S, Barad BB, Show BK, Mandal N (2019a) Tribological property enhancement of 3Y-TZP ceramic by the combined effect of CaF_2 and MgO phases. Ceram Int 45:13447–13455
Mazumder S, Kumar OP, Kotnees DK, Mandal N (2019b) Tribological influences of CuO into 3Y-TZP ceramic composite in conformal contact. J Tribol 141(3):031606
Mazumder S, Kumar A, Singh BK, Roy H, Mandal N (2019c) Tribological investigation of MgO/Al_2O_3 ceramic composite with the inclusion of nano CuO in dry abrasive wear test. Mater Res Express 6:085086
Medveď D, Balko J, Sedlák R, Kovalčíková A, Shepa I, Duszová AN, Bączek E, Podsiadło M, Dusza J (2019) Wear resistance of ZrB_2 based ceramic composites. Int J Refract Metal Hard Mater 81:214–224
Moazami GM, Nemati A (2018) Tribological behavior of self lubricating Cu/MoS_2 composites fabricated by powder metallurgy. Trans Nonferrous Met Soc China 28:946–956. https://doi.org/10.1016/S1003-6326(18)64729-6
Moshtaghioun BM, Garcia DG, Rodriguez AD, Todd RI (2016) Abrasive wear rate of boron carbide ceramics: influence of microstructural and mechanical aspects on their tribological response. J Eur Ceram Soc 36:3925–3928
Murthy TSRC, Basu B, Srivastava A, Balasubramaniam R, Suri AK (2006) Tribological properties of TiB_2 and TiB_2–$MoSi_2$ ceramic composites. J Eur Ceram Soc 26:1293–1300
Natu V, Kota SS, Barsoum MW (2020) X-ray photoelectron spectroscopy of the MAB phases, MoAlB, M_2AlB_2 ($M = Cr, Fe$), Cr_3AlB_4 and their binary monoborides. J Eur Ceram Soc 40:305–314
Novak S, Drazic G, Samardzija Z, Kalin M, Vizintin J (1996) Wear of silicon nitride ceramics under fretting conditions. Mater Sci Eng A 215:125–133
Pan W, Gao Y (2018) Tribological behavior of B_4C/hBN ceramic composites coupled with grey iron under the lubrication of emulsion. Mater Res Express 5:066512
Park DS, Danyluk S, McNallan M (1992) Influence of tribochemical reaction products on friction and wear of silicon nitride at elevated temperatures in reactive environments. Am Ceram Soc 75(11):3033–3039

Qian Y, Zhang W, Ge M, Wei X (2013) Frictional response of a novel C/C–ZrB_2–ZrC–SiC composite under simulated braking. J Adv Ceram 2(2):157–161
Riley FL (2000) Silicon nitride and related materials. J Am Ceram Soc 83:245–265
Rosado L, Forster NH, Trivedi HK, King JP (2000) Solid lubrication of silicon nitride with cesium-based compounds: part I—rolling contact endurance, friction and wear. Tribol Trans 43:489–497
Saito T, Hosoe T, Honda F (2001) Chemical wear of sintered Si_3N_4, hBN and Si_3N_4–hBN composites by water lubrication. Wear 247:223–230
Sarkar D, Kumar BVM, Basu B (2006) Understanding the fretting wear of Ti_3SiC_2. J Eur Ceram Soc 26(13):2441–2452
Sedlák R, Kovalčíková A, Balko J, Rutkowski P, Dubiel A, Zientara D, Girman V, Múdra E, Dusza J (2017) Effect of graphene platelets on tribological properties of boron carbide ceramic composites. Int J Refract Metal Hard Mater 65:57–63
Shaffer PTB (1969) Engineering properties of carbides. In Schneider SJ Jr (ed) Ceramics and glasses: engineered materials handbook, vol 4. ASM International, Ohio, pp 804–811
Sharma SK, Kumar BVM, Kim YW (2015) Effect of WC addition on sliding wear behavior of SiC ceramics. Ceram Int 41:3427–3437
Sharma SK, Kumar BVM, Kim YW (2016) Tribological behavior of silicon carbide ceramics—a review. J Korean Ceram Soc 53(6):581–596
Sharma SK, Kumar BVM, Zugelj BB, Kalin M, Kim YW (2017) Room and high temperature reciprocated sliding wear behavior of SiC-WC composites. Ceram Int 43:16827–16834
Sharma SK, Kumar BVM, Kim YW (2019) Tribology of WC reinforced SiC ceramics: influence of counterbody. Friction 7(2):129–142
Skopp A, Woydt M, Habig KH (1990) Unlubricated sliding friction and wear of various Si3N4-pairs between 22 °C and 1000 °C. Tribol Int 23(3):189–199
Skopp A, Woydt M, Habig KH (1995) Tribological behavior of silicon nitride matrials under unlubricated sliding between 22 °C materials under and 1000 °C. Wear 181–183:571–580
Sonber JK, Limaye PK, Murthy TSRC, Sairam K, Nagaraj A, Soni NL, Patel RJ, Chakravartty JK (2015) Tribological properties of boron carbide in sliding against WC ball. Int J Refract Metal Hard Mater 51:110–117
Sun J, Liu C, Tian J, Feng B (2012) Erosion behavior of B_4C based ceramic nozzles by abrasive air-jet. Ceram Int 38:6599–6605
Suszko T, Gulbinski W, Jagielski J (2006) Mo_2N/Cu thin films—the structure, mechanical and tribological properties. Surf Coat Technol 200:6288–6292
Tang H, Zeng X, Xiong X, Li L, Zou J (2009) Mechanical and tribological properties of short-fiber-reinforced SiC composites. Tribol Int 42:823–827
Tatarko P, Kasiarová M, Dusza J, Morgiel J, Sajgalík P, Hvizdos P (2010) Wear resistance of hot-pressed Si_3N_4/SiC micro/nanocomposites sintered with rare-earth oxide additives. Wear 269:867–874
Tatarko P, Kasiarova M, Chlup Z, Dusza J, Sajgalik P, Vavra I (2013) Influence of rare-earth oxide additives and SiC nanoparticles on the wear behaviour of Si_3N_4-based composites at temperatures upto 900 °C. Wear 300:155–162
Telle R, Petzow G (1988) Strengthening and toughening of boride and carbide hard material composites. Mater Sci Eng A 105–106:97–104
Tjong SC, Lau KC (2000) Abrasion resistance of stainless-steel composites reinforced with hard TiB_2 particles. J Compos Sci Technol 60(8):1141
Tomizawa H, Fischer TE (1986) Friction and wear of silicon nitride at 150 °C to 800 °C. ASLE Truns 29(4):481–1187
Torres H, Rodríguez RM, Prakash B (2018) Tribological behaviour of self-lubricating materials at high temperatures. Int Mater Rev 63:309–340
Vleugels J, Kumar KCH, Vitchev RG, Biest OVD, Basu B (2002) Unlubricated fretting wear of TiB_2 containing composites against bearing steel. Metall Mater Trans A 33(12):3847–3859
Wang BQ, Levy AV (1990) Erosion behavior of SiC fiber-SiC matrix composites. Wear 138:125–136

Wang L, Snidle RW, Gu L (2000) Rolling contact silicon nitride bearing technology: a review of recent research. Wear 246:159–173

Wanga B, Shangguana D, Qiaoa R, Zhanga F, Baia Y, Wanga Z, Wanga C, Lu X (2019) Fabrication, mechanical properties and thermal shock resistance of a dense SiC NWs/α-Si_3N_4 composite coating for protecting porous Si_3N_4 ceramics. Ceram Int 45:23241–23247

Woetting G, Caspers B, Gugel E, Westerheide R (2000) High-temperature properties of SiC-Si_3N_4 particle composites. J Eng Gas Turbines Power 122(1):8–12

Xin H, Shi H, Chen W, Jia J, Yang W, Jin Z (2019) Tribological investigation of Si_3N_4-hBN on HXLPE bearing couple: effects of sliding velocity and contact load. Ceram Int 45:6296–6302

Xue P, Meng Q, Chen S, Liang L, Liu Z, Chen R (2011) Tribological property of (TiC-TiB_2)pNi ceramics prepared by field-activated and pressure-assisted synthesis. Rare Met 30:599

Yang F, Zhang X, Han J, Du S (2008) Mechanical properties of short carbon fiber reinforced ZrB_2–SiC ceramic matrix composites. Mater Lett 62:2925–2927

Yang J, Gu W, Pan LM, Song K, Chen X, Qiu T (2011) Friction and wear properties of in situ (TiB_2 + TiC)/Ti_3SiC_2 composites. Wear 271:2940–2946

Yu Z, Tan H, Wang S, Cheng J, Sun Q, Yang J, Liu W (2020) High-temperature tribological behaviors of MoAlB ceramics sliding against Al_2O_3 and Inconel 718 alloy. Ceram Int. https://doi.org/10.1016/j.ceramint.2020.02.275

Yue XY, Zhao SM, Lu P, Chang Q, Ru HQ (2010) Synthesis and properties of hot pressed B_4C-TiB_2 ceramic composite. Mater Sci Eng A 527:7215–7219

Zamora V, Ortiz AL, Guiberteau F, Nygren M (2012) Crystal-size dependence of the spark-plasma-sintering kinetics of ZrB_2 ultra-high-temperature ceramics. J Eur Ceram Soc 32:271–276

Zhang YS, Hu LT, Chen JM, Liu WM (2010) Lubrication behavior of Y-TZP/Al_2O_3/Mo nanocomposites at high temperature. Wear 268:1091–1094

Zhang J, Zhang J, Xiao G, Chen Z, Yi M, Zhang Y, Xu C (2020) Orentational effect of graphene on the friction and wear behavior of Si_3N_4/TiC based composite ceramic tool materials. Ceram Int 46(3):3550–3557

Zhao H, Cheng YB (1999) Formation of TiB_2-TiC composites by reactive sinterin. J Ceram Silik 37(1):353

Zhaoa B, Khader I, Raga R, Konrath G, Degenhardt U, Kailer A (2019) High temperature tribological properties of silicon nitride in dry sliding contact against Inconel 718 heated by laser. Wear 434–435:203000

Zhu S, Cheng J, Qiao Z, Yang J (2019) High temperature solid-lubricating materials: a review. Tribol Int 133:206–223

Tribological Properties of Composite Materials for Automotive Applications

Ram Krishna Upadhyay and Arvind Kumar

Abstract The advanced technology of composite and its allied tribological properties is now gaining importance in many industries. However, composite materials encounter few problems during the tribological testing, such as the concentration of filler particles, type of particle, compatibility of counterpart material, and running conditions. With this viewpoint, the present study provides insightful information about different composites for tribological applications. In this work, composite tribological performance of polymer, ceramic, metal-matrix composite, and the coating is studied. Among the tested composites, carbon fillers have provided better tribological properties in terms of contact friction and wear. The microstructure of graphene composites contains a single-layer graphene film, which functionalizes the polymer composite surface. The effect of particle filler orientation in normal and parallel direction is described for the ceramic composites. The friction data of ceramic composites with carbon/carbon–silicon carbide filler is less fluctuating in the parallel direction. In metal-matrix composite, the infusion of graphene particles reduces the wear rate and friction coefficient values at high-applied loads. Further, zirconium-based metal-matrix composites are tested at low, medium, and high break energy densities. The wear behavior of cubic-zirconium dioxide is found to be low at all the break energy density due to its crystalline structure. Tribological properties of polymer-based liquid/solid filler composites for coating application with their significant parameters are studied. The result shows that the wear life improves by using an equal concentration of two particles. At last, the cost associated with tribological losses and its mitigation measures is discussed.

Keywords Composite tribology · Polymer composite · Ceramic · Metal-matrix composite · Coatings

---

R. K. Upadhyay · A. Kumar (✉)
Department of Mechanical Engineering, Indian Institute of Technology Kanpur, Kanpur, India
e-mail: arvindkr@iitk.ac.in

R. K. Upadhyay
National Rail and Transportation Institute, Vadodara 390004, India
e-mail: ram.upadhyay@nrti.edu.in

M. T. Hameed Sultan et al. (eds.), *Tribological Applications of Composite Materials*, Composites Science and Technology, https://doi.org/10.1007/978-981-15-9635-3_2

1 Introduction

Composite technology is one of the fast-growing methodologies that have govern almost every industry ranging from automotive to aviation industry and electronics to the bio-medical industry (Smith 1990; Molent and Haddad 2020; Sarantinos et al. 2019; Wang and Aslani 2019). With each new developing materials, composite technology asses its suitability for immediate impact on global needs. In recent years, there uses in the most expensive growing technology of tribology are fast-forwarded due to involved high-energy losses in the tribological process (Li et al. 2019). The development related to tribological applications in the automotive industry has certainly changed the earlier technology with ease (Balakrishnan and Seidlitz 2018). Nowadays, various materials such as organic, inorganic, metal, ceramic, polymer, and wood are widely used in the composite preparation, and these materials able to resist losses resulting from tribology/material wear (Nickels 2019; Gong and Yang 2013). Among all used materials, graphene, a newly realized material with high mechanical, physical, chemical, and optical properties are extensively used in the past fifteen years (Hemanth 2019; Upadhyay and Kumar 2019a, b, c; Lu et al. 2020). Particularly, in tribological innovations, research on the graphene material is extensive due to its effective lubrication properties (Upadhyay and Kumar 2019a, b, c; Dong and Qi 2015). Graphene is used as a solid, liquid, emulsion, and composite form to reduce the friction and wear of automotive parts. Other materials, such as ceramic consisting of silicon carbide, silicon nitride, boron nitride, zinc oxide, titanium carbide, and yttrium, are also found its practical applications in tribology (Chen et al. 2020; Kumar and Srivastava 2019). The benefit of using ceramic materials in composite technology is that they have a definite high melting temperature. A practical example of ceramic in automotive application is carbon-ceramic disk brake (Bian and Wu 2015), which is resistant to brake fade at high temperatures.

Polymer composite tribology by infusing solid and liquid particles with/without additives is also researched in recent years (Upadhyay and Kumar 2019a, b, c). These polymers are used with either epoxy matrix or natural biopolymers to modify the lubrication properties. The use of polymers as corrosion-resistant coatings, protection against oxidation, cavitation, and erosion wear, is widely established (Upadhyay and Kumar 2019a, b, c; Kumar et al. 2017). However, at extremely high temperatures, the application of polymer composites in automotive tribology is still under developing stage. The application of polymer-based composites is established for automotive bearings, decorative covers, internal/external hoods, and valve design. For high-temperature applications, the metal-matrix composites are widely researched (Prasad et al. 1999; Ghodrati and Ghomashchi 2019). It consists of one phase of metal and another phase of different materials, such as ceramic and carbon materials. A hybrid metal-matrix composite usually consists of three phases of material, but primarily a metal phase. Metal-matrix composites are extensively used in automotive applications for components like disk brake, silicon–carbon fiber rotors, driveshaft, cylinder liners, and engine cylinders.

An earlier study shows the benefit of using composites in real applications of tribology. Automotive parts made by composites have improved properties of mechanical, physical, and chemical. All these properties are important to any tribological application, especially in practical wear situations. Considering the advantageous properties of composites, this work presents a complete understanding of different composites that can be helpful in tribological design. Composites based on polymer, ceramic, metal-matrix, and coatings are discussed in detail. At last, different characterization techniques that are helpful in composite tribological studies are mentioned with their physical significance.

2 Tribology of Polymer-Based Composites

Polymers are a large repetitive network of molecules or macromolecules that exists in the form of natural (deoxyribonucleic acid, DNA, and protein) or synthetic polymer (polystyrene). The application of polymers in tribology is enormous due to their unique functional (polymerization, the formation of small monomers) and physical (toughness and viscoelastic) properties. Friedrich et al. (1995) have demonstrated the sliding friction and wear properties of the extreme-temperature polymer, polyetheretherketone (PEEK), and epoxy matrix against the steel counterpart. The effect of internal lubrication by polytetrafluoroethylene (PTFE) and reinforced particles of glass, aramid, and carbon is explored. Polyetheretherketone had better wear properties than the brittle epoxy resins. The reinforcements of glass, carbon, and aramid had produced substantial decrement in the wear result. Among all tested three reinforced materials, the carbon fibres had significantly low wear properties than the other two. However, the glass fibres attain high wear, and aramid wear performance is between the carbon and glass fibres. The wear is affected by the multidirectional orientation of PEEK and the parallel sliding direction of carbon fibres in the polymer composite. For carbon fibre reinforcement, the particle orientation in the composites is parallel to the sliding direction, which provides ease in motion during the sliding test and results in low material wear. Most importantly, a woven structure is formed between the contacting surfaces due to the interlocking mechanism, which restricts wear. Results can be better realized by the antiparallel orientation opposite to the sliding direction. By considering the horizontal and the vertical plane, a suggestion of hybridization is provided. The reported wear values of composites can be further improved by reinforcement of carbon in the horizontal plane and aramid in the vertical plane.

Acrylic resin and short-wood fibres are used to prepare bio-composite material for automotive tribology applications (Akpan et al. 2018). The prepared bio-based composite (HB400, HAHO120f, HAHO50, HAHO150, HM90k, and HAHOMIX) have self-lubricating properties to provide resistance against friction. A chromium steel ball as counterpart material under the dry and lubrication mode is used for the friction and wear evaluation. Fibres aspect ratio plays an important part in specific wear measurements. An increase in the length of the fibre increases the specific wear

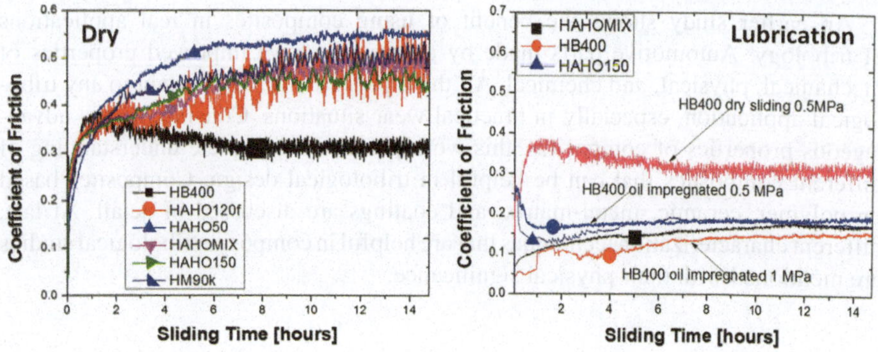

Fig. 1 Friction performance of bio-composites under dry and lubrication mode. Reprint with permission from Akpan et al. (2018)

rate of fibres due to its large aspect ratio and inferior infusion in acrodur binder. A large fibre causes surface cracking and shows the brittle behavior of composite with a high wear rate. With a decrease in fibres size, the wear tendency changes from cracking to uniform surface disruption, this behavior has diminished the effect of brittle failure. In the dry sliding, the initial stage of friction is dominated by mechanical interlocking. The abrasive particles detached from the surface are retained inside the interlocking area and increase the friction coefficient. Further, continuous sliding motion removes the abrasive wear particles and builds the transfer film to provide a stable reduction in the friction value. Another composite of HAHO150 and HAHOMIX carries more wear particles at the sliding interface and increases the friction coefficient. As mentioned in Fig. 1, among all tested composites, HB400 has achieved a low friction coefficient under the dry (solid–solid interaction) sliding tests. The specific wear rate of softwood and hardwood fibre is also compared. The softwood fibers have the highest wear rate compare to the hardwood fibres. However, the friction coefficient of softwood fibres is low compared to the hardwood fibres. Under the oil lubrication, the friction coefficient of HB400 bio-composites is substantially improved by 50% due to the oil retention capacity of these composites up to 14 h (see Fig. 1). An increase in the contact pressure under the oil medium has lowered the friction coefficient of all the prepared composites.

The tribological properties of polymer-based carbon materials are presented in recent studies (Upadhyay and Kumar 2018, 2019a, b, c). Different carbon materials such as graphene, multiwalled carbon nanotube (MWCNT), and fullerene C_{70} is used to prepare a composite with different particle concentration (see Fig. 2). Epoxy resin infused with fullerene nanoparticles has decreased the friction coefficient of polymer surfaces. Low values of friction are obtained by raising the particle concentration. More importantly, with an increase in load, the friction values improved further. However, at high particle concentration, some deviation in the friction results occurred. The wear behavior also followed the friction data trends, and it decreases with an increase in load. In comparison to the earlier study (Zhao et al. 1996), the improvement in the friction and wear result is related to the intrinsic mechanism of

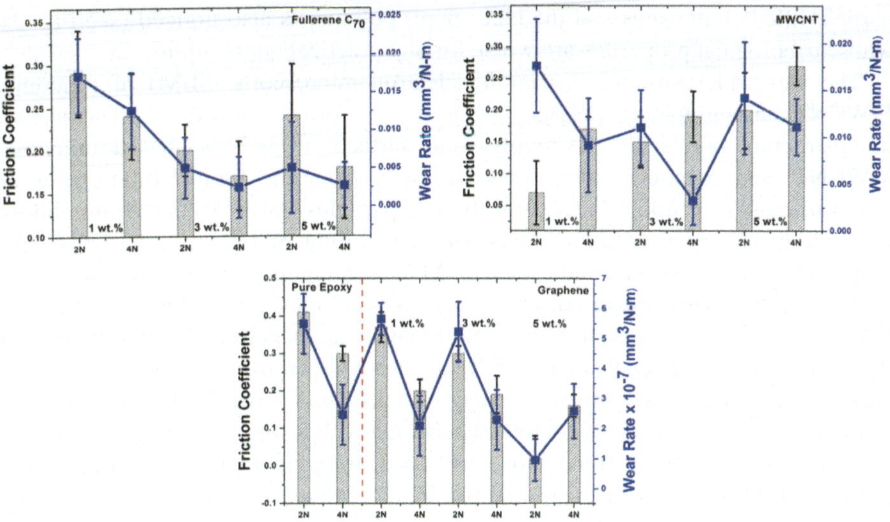

Fig. 2 Friction and wear performance of carbon-based polymer composites

the filler-polymer reaction. Rolling mechanism of a third-body particle has governed the surface friction and wear properties. The build transfer-film between the ball-and sample surface has supported the acting load. On the contrary, the use of multi-walled carbon nanotubes has provided the results opposite to the fullerene. The main influencing phenomena of a rise in friction behavior are found to be related to the accumulation of MWCNT particles. The dispersion of MWCNT particles in the epoxy-resin matrix is not controlled over stirring time, which reflects its adverse effect on the polymer surface. The continuous sliding of MWCNT particles under the ball-sample interface has detached the corresponding layer of composite material, which contributes to the high friction and wear values.

As the different quantities of graphene are infused within the polymer matrix, the friction and wear performance of composites decreased with an increase in particle concentration. Opposite to the behavior of MWCNT and fullerene, graphene has provided better surface functionalization. A build-up of more frequent tribo-film formation has resulted in surface fictionalization. Tribo-layer is achieved by the reaction between epoxide and graphene particles, which leads to forming single-layer graphene. During the sliding, this single-layer graphene exposes on the composite surface and provides effective lubrication properties. The intrinsic stresses developed in the composite surface helped to sustain the loads with minimum wear. The applied load is equally distributed in the whole contact area due to a well-dispersed (homogeneous) graphene network. During sliding, sheared particles of graphene are filled back in the wear region to provide a smooth transition between the ball and sample material. The particle sliding and rolling is found to be the most significant wear mechanism of graphene-based polymer composites. A clear distinction between the particle-filled composites and the bare (pure) polymer is also noticed (see Fig. 2), whose tribological properties are worse for any practical application.

The wear microstructure (scanning electron microscopy, SEM) of fullerene, MWCNT, and graphene polymer composite is shown in Fig. 3. The shown microstructure represents the highly wear surfaces of polymer-based fullerene, MWCNT, and graphene composites. As seen from the figure, fullerene wear microstructure is composed of several abrasive particles that are settled on the sliding wear track. However, continuous shearing and filling of fullerene particles have reduced the wear rate of composites. In MWCNT composites, an adhesion layer over the sliding wear track is observed, which is composed of polymer-matrix and MWCNT particles. These sheared adherent layers are accommodated at the top of the wear surface and result in a high wear rate. As seen from the figure, the wear surface of the graphene composite is free from detached particles. The wear surface of the graphene composite is completely covered with a single-layer graphene film. This protective graphene film acts as a self-lubricating layer and contributing to the build-up of tribo-film to achieve low friction coefficient and wear rate. The wear

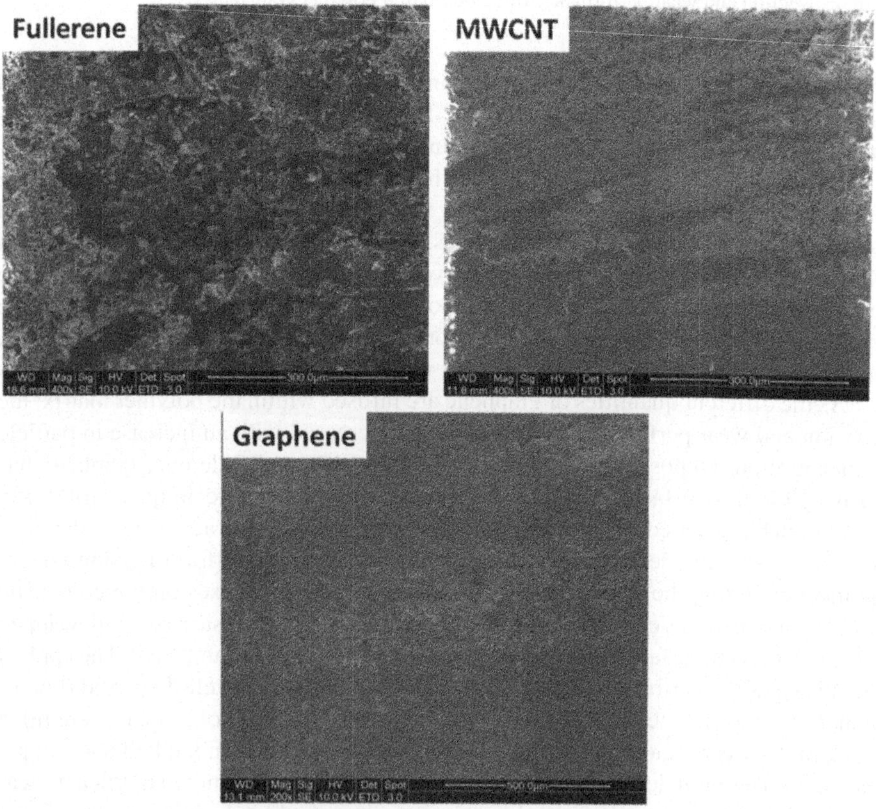

Fig. 3 SEM microstructure of carbon-based polymer composites for high wear surfaces

mechanism is controlled by the continuous shearing and filling of the wear particles in the vacant junction. This increases the contact area available for sliding and hence reduces the effect of high load at a particular area to provide a low wear rate.

3 Tribology of Ceramic-Based Composites

Ceramic composites are versatile due to its remarkable physical (crystalline structure) and heat resistance capacity. The chemical composition of ceramics makes them perfect material for marine applications (Chen et al. 2020). Ceramics application in the automotive industry is well-established (Kumar and Srivastava 2019; Bian and Wu 2015). The sliding behavior of carbon–carbon–silicon carbide and carbon–carbon composite in the presence of the brake oil is examined (Kumar and Srivastava 2019). The composite laminates are prepared in the normal and parallel orientation of particles. For the physical significance, two types of contact, such as partial and low conformity contact for the reciprocating friction and wear tests, are considered. Under low contacts in dry conditions, the carbon–carbon composite in normal orientation shows a deviation in the friction curve. While the carbon–carbon parallel orientation shows the stable behavior of friction. Carbon–carbon–silicon carbide ceramic composite has a similar tendency of sliding for both the normal and parallel orientations as of carbon–carbon composite. However, under the partial contacts, the friction coefficient of carbon–carbon–silicon carbide composites remains constant (less stress concentration compared to the low conformity contact), whereas, carbon–carbon varies with sliding time. As soon as the load is increased further, the friction performance of the ceramic composite degraded compared to the carbon–carbon composites under the low conformity contacts. Similarly, the wear loss of carbon–carbon parallel orientation and carbon–carbon–silicon carbide in normal orientation is more. However, at a later stage, it decreases for carbon–carbon composites oriented in a normal direction. Overall, the parallel orientation of ceramic composites shows the minimum fluctuation in friction and wear due to more absorption of oil between the contacts. In partial contacts, the tribological properties are enhanced for the ceramic composites under the presence of hard silicon particles, which restricts the direct material contact and lowers the wear rate. The SEM microstructure of parallel and plate tested (low conformity) composites of carbon–carbon–silicon carbide is shown in Fig. 4. From the shown microstructure, it is clear that the grains are detached at high load and provides a high friction coefficient for parallel oriented composites. Breaking of fibre in the parallel composite is witnessed under high load due to the abrasion. Silicon particles are embedded under the surface due to the high pressure of a load and lead to the abrasion process. A broken fibre in the carbon–carbon–silicon carbide plate-tested composite is also witnessed. An increase in the pressure at the surface resulted in deeper penetration of silicon particles, which covered more particles and raised the friction and wear of the composite material.

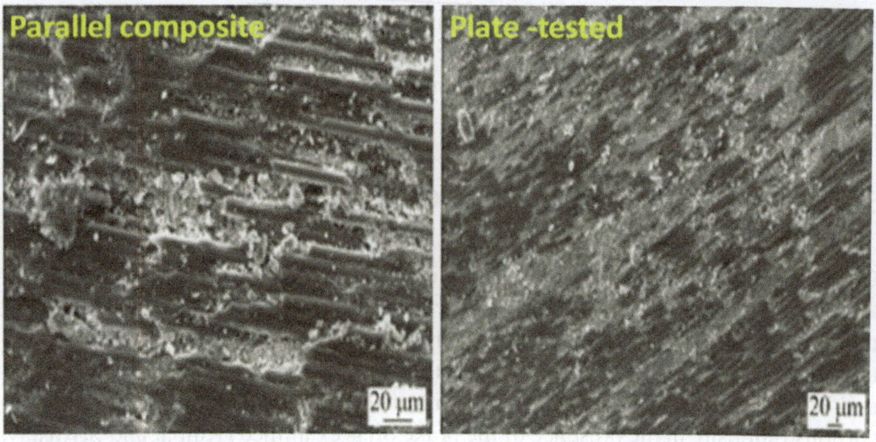

Fig. 4 SEM wear micrograph of carbon–carbon–silicon carbide composite and plate-tested samples at high loads. Reprint with permission from Kumar and Srivastava (2019)

Bian and Wu (2015) have investigated the surface fracture and friction behavior of silicon carbide ceramic disk brake against the mild steel pad. Fractography analysis of the friction surfaces under different breaking speeds (8.5, 14, and 20 m/s) is considered for the qualitative analysis. At low breaking speed, first, the friction coefficient increased to ~0.9, and after a definite number of braking stops, the friction value falls nearly at ~0.6. At the median breaking speed of ~14 m/s, a small rise in the friction coefficient is observed. However, after a few stops, the friction coefficient falls to the low value (~0.42). At the highest breaking speed ~20 m/s, friction increases rapidly from ~0.42 to 0.70, but decreases after a few stops and attain median friction values similar to earlier breaking speeds. As seen from Fig. 5, the average friction coefficient of disk brake at different breaking speeds is ~0.6. At low speed, wear debris from the mild-steel plate deposited on the silicon carbide disk material. A transmission electron microscopy (TEM) shows the thick friction transfer layer (~2 μm) on the silicon carbide disk brake (see Fig. 5).

The presence of iron (Fe), oxide (O), and silicon (Si) is validated by the energy-dispersive X-ray spectroscopy (EDX) on the silicon carbide disk brake surface. The disk brake surface is composed of cracks and plastic deformation, which builds at an early stage of braking. As the number of breaking exceeds, an iron-rich friction transfer layer is obtained. At high braking (~20 m/s) speed, fracture on the surface occurred after few braking stops, and it becomes worse after fifteen stops. The modeling analysis also confirmed the experimental data and suggested for lower initial braking speeds.

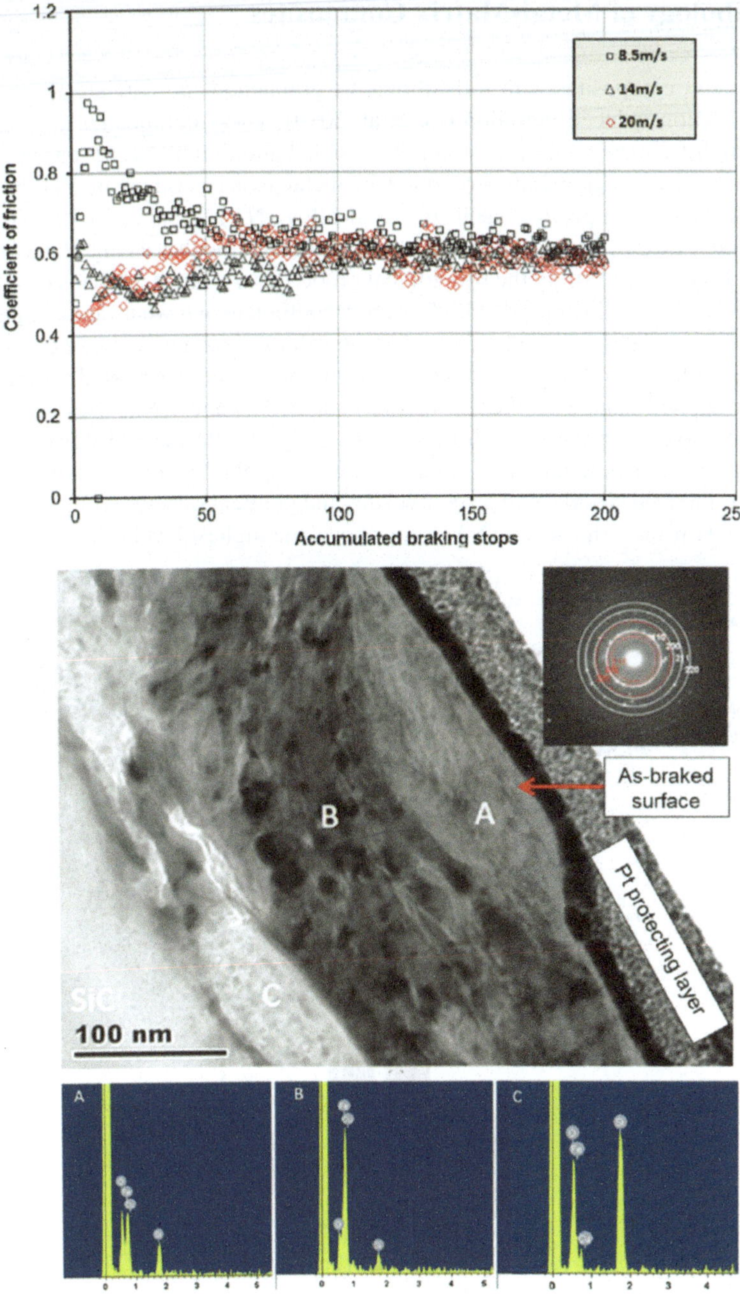

Fig. 5 Friction performance of silicon carbide brake disk under different braking speeds and TEM image of friction surface at the braking speed of 8.5 m/s with chemical composition at A, B, and C, zones respectively. Reprint with permission from Bian and Wu (2015)

4 Tribology of Metal-Matrix Composites

Metal matrix composites with self-lubricating graphene nanoplatelets are useful to control friction-induced vibration (Lu et al. 2019). These composites are meant to reduce high-frequency vibration and noise. Nickel aluminum (Ni_3Al) and graphene particles as a solid lubricant are added in metal matrix composite. Three types of metal-matrix composites, such as Ni_3Al alloy (NA), graphene-induced Ni_3Al composite (NMCs), and gradient structure (GNMMCs) with varying concentration of particles is used to study the tribological properties. At constant loads of 4.8, 9.8, and 10.8 N, Ni_3Al alloy attains a high friction coefficient and wear rate compared to the other composites (see Fig. 6). The friction values of all the prepared composites decrease with an increase in load. However, the wear rate of Ni_3Al alloy increases with the applied load. The Vickers hardness of NMCs and GNMMCs composites contributes to the better wear performance. This suggests the dependency of material wear on the composite hardness value. Under the applied loads, the noise level of graphene-induced Ni_3Al (NMCs) composites and gradient structure (GNMMCs) are the same. However, the Ni_3Al (NA) alloy shows the highest level of noise pressure.

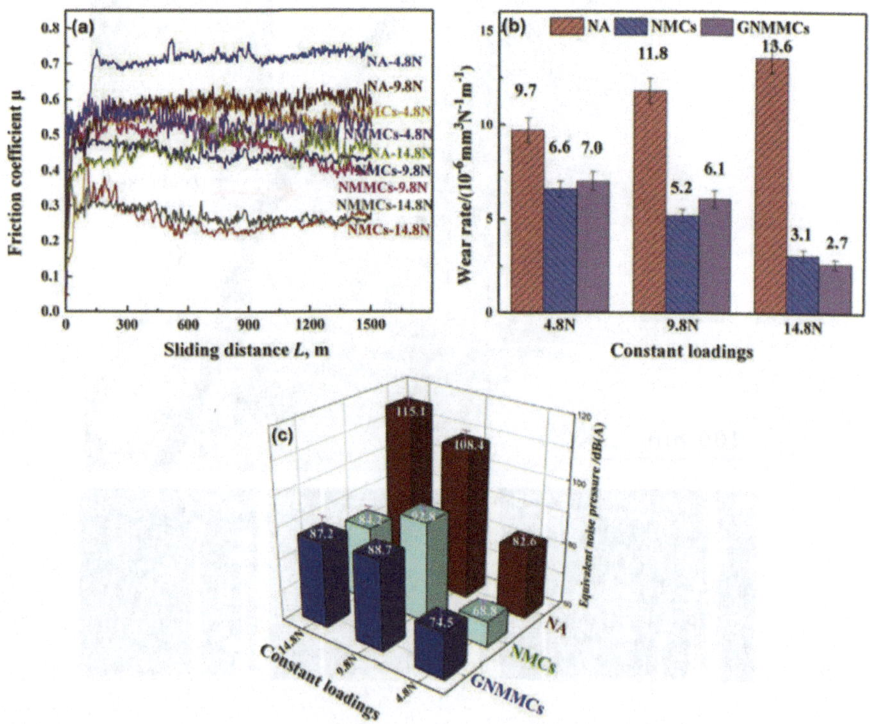

Fig. 6 **a** Friction coefficient, **b** wear, and **c** sound level of NA, NMCs, and GNMMCs composites. Reprint with permission from Lu et al. (2019)

Under the high contact loads, composite material experiences some deflection due to the weak arrangement of molecules inside the matrix. However, graphene inclusion in the composite matrix provides hardness to the material and able to sustain impact loads due to the high bonding strength. From the results, it is concluded that the graphene infusion in the metal matrix composites successfully reduces the tribological performance in terms of friction and wear response.

Zirconium abrasive, a highly stabilized (cubic structure) thermal material for tribological applications, is highly demanding. Ceramic material in the presence of included oxides, nitride, and carbide shows better tribological properties (Chen et al. 2020; Zhou et al. 2019). Zirconium dioxide (ZrO_2) in the copper metal matrix (Cu-MMCs) is introduced to study the tribological properties of composites. Two different structures of monoclinic ZrO_2 (m-ZrO_2) and cubic ZrO_2 (c-ZrO_2) are used for this purpose. The crystal structure of ZrO_2 particles affects the morphology and internal bonding of copper metal-matrix composite. Between both the structures of m-ZrO_2 and c-ZrO_2, the cubic zirconium structure is unchanged during the matrix preparation. The microstructure of the copper matrix with mixed particles is uniform throughout the direction parallel to the sliding. The presence of iron (Fe) and graphite (Gr) in the matrix increases the mechanical strength of the metal matrix composite (see Fig. 7). Graphene reduces the adhesion between contacts by forming a protective

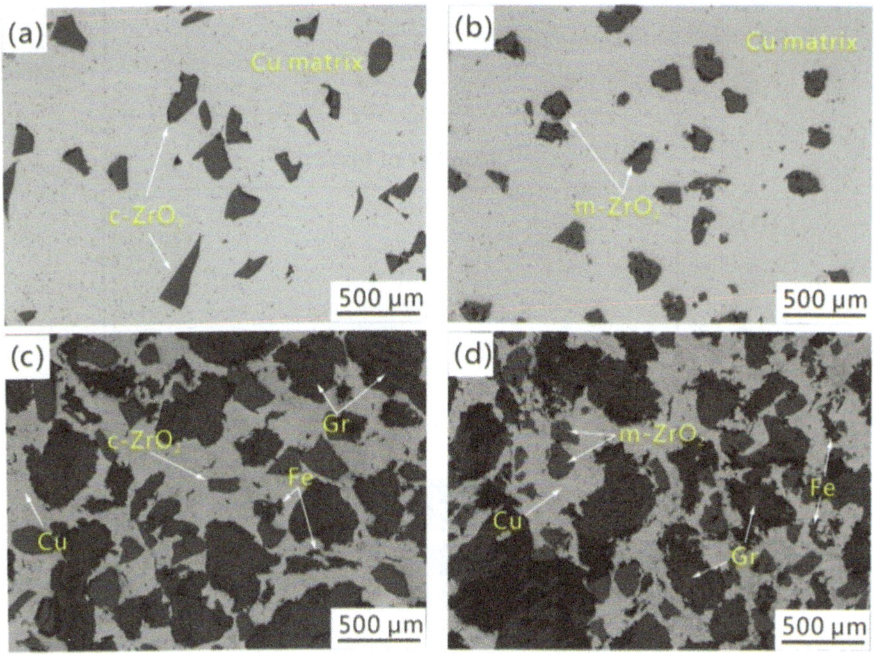

Fig. 7 Microstructure of the metal-matrix composites. **a** Cu-c-ZrO_2, **b** Cu-m-ZrO_2, **c** Cu–Fe–C-c-ZrO_2, **d** Cu–Fe–C-m-ZrO_2. Reprint with permission from Zhou et al. (2019)

layer over the wear surface. This helps in a smooth transition during the sliding wear tests.

The friction performance is studied at the three different breaking energy densities (BED), such as low, medium, and high BED (See Fig. 8). Monoclinic ZrO_2 (m-ZrO_2) in the copper material exhibits high bonding strength within the zinc matrix and gives a high friction coefficient compared to the cubic ZrO_2 (c-ZrO_2) structure at low BED. However, m-ZrO_2 achieves a low friction coefficient at high BED. The variation in friction drop at the high BED for c-ZrO_2 is high. On the other hand, the wear rate of copper-cubic zirconium metal matrix composite is low at all the BED.

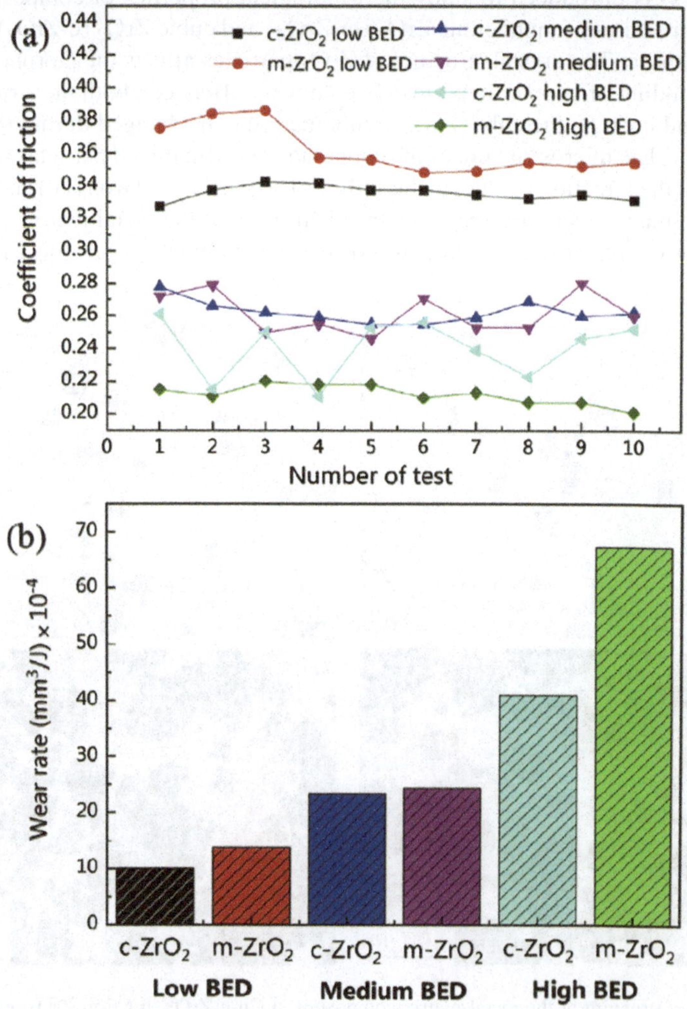

Fig. 8 **a** Friction and **b** wear performance of copper-metal matrix composites. Reprint with permission from Zhou et al. (2019)

5 Coating Tribology of Composites

Composite coating with nanofillers is a recent trend in tribology, and particularly polymer-based coatings are emerging now (Kumar et al. 2017; Katiyar et al. 2016). These polymers contain solid/liquid nanoparticles in the form of additives to prepare thin/thick composite coatings for the wear protection. Tribological properties of polymer-based graphene (10 wt%) composite coatings with liquid (SN150 oil and perfluoropolyether, PFPE) nano-lubricant is recently studied (Kumar et al. 2017). The friction behavior of coatings is studied at different sliding velocity and loads. As seen from Fig. 9, the friction values in the presence of both the additives, i.e., epoxy-graphene-SN150 (E/Gn/SN150) and epoxy-graphene-PFPE (E/Gn/PFPE) decreases

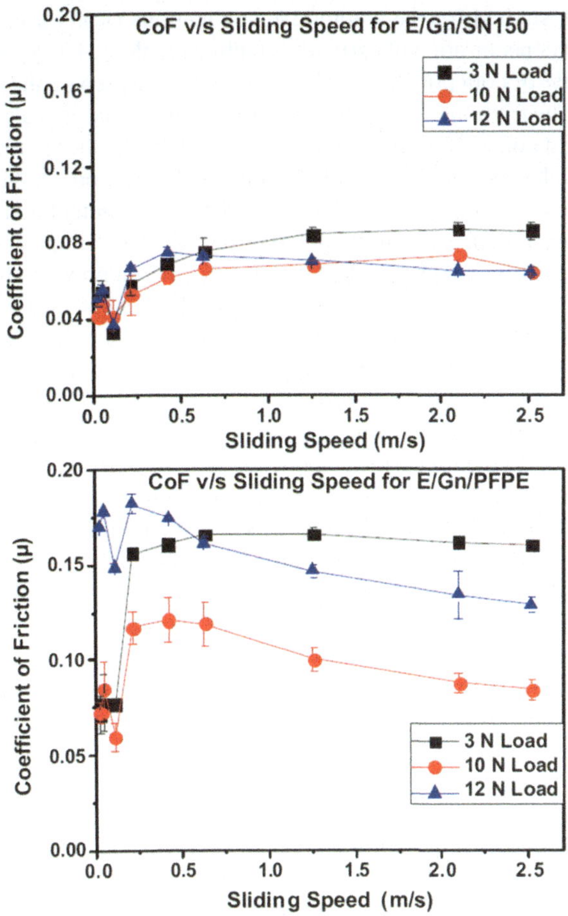

Fig. 9 Friction coefficient of epoxy-based composite coatings at different sliding speeds and loads. Reprint with permission from Kumar et al. (2017)

with sliding velocity. Most importantly, the friction performance of base-oil SN150 is significant, and it can be used as filler material for practical sliding/rolling applications. Automotive anti-friction bearings are one of the examples where all the time, minimum friction is required. Some deflection at different loads has occurred during the sliding tests. In the case of composite coating with SN150 as filler material, the friction coefficient decreases with the increase in load. The minimum value is achieved at 10 N loads. However, the difference in friction data between 10 and 12 N loads are less. The maximum deflection is achieved at the initial stage of sliding. The friction values containing PFPE composites also have similar frictional properties (first rises and then falls) but having a high friction coefficient. For PFPE composite, the friction coefficient initially ranged from 0.06 to 0.18 and attained the best friction value of nearly ~0.1 at 10 N load.

In another work, SU-8 polymer coatings prepared with graphite and talc powder filler material is studied for the automotive applications (Katiyar et al. 2016). The different weight concentration of particles (graphite or talc: 5, 10, 20, and 30 wt% and 2.5, 5, 10, 15 wt% of both powder in 1:1 ratio) are used to prepare the coatings on the glass samples. The friction of polymer composite coatings increases with an increase in particle weight concentration for individual (graphite or talc) particles. However, as both the particles of graphite and talc are mixed, the friction coefficient decreases and attains the steady-state friction. The best friction coefficient attained with 15% graphite and 15% talc powder in the polymer matrix. The composite coatings also tested for its wear life, which is calculated against the number of sliding cycles to coating failure. As similar to the friction test results, composites containing 15% graphite and 15% talc powder has performed better than the other composites and provides high wear life. This is the synergistic effect of both these powder material when infused within the polymer matrix.

Metallic and non-metallic coatings with/without deposition processes have also found their interest in the tribological applications (Paksoy et al. 2019; Endrino et al. 2002; Wood 2010). Particularly, in the severe situation of wear, the hard and thick coatings are recommended. Nevertheless, in some situations, thin coatings can also play an important part in friction reduction. Only depending on the application and load, coating materials can be recommended. Most of the refractory materials such as titanium, chromium, molybdenum, manganese, etc. are used under high-temperature applications and polymer for low-temperature applications. Process parameters and selection of material is an influential factor in determining the coating properties.

Coating thickness, roughness, load, sliding speed, substrate elastic modulus, and counterpart material affects the friction performance of coatings. The thin and soft coating may result in high friction due to poor load carrying capacity and formation of a large contact area. The friction coefficient decreases with an increase in the thickness of hard and thin coatings. However, after the predefined limit, brittle behavior occurs in the coating, and it develops tensile stresses. Under the presence of load and tensile stresses, the deposited thick coating delaminates from the surface. Substrate surface roughness highly influences the roughness of thin coatings. More surface roughness provides a high friction coefficient to the thin coatings. These coatings also have a tendency for plastic deformation. However, in the case of thick coating, the coating material penetrates the rough peaks of the substrate material and result in moderate friction behavior. Hence, smooth surface finishing is always required for the substrate material to deposit thick film coatings. The friction coefficient severely depends on the load. Thin and hard coating at low load initially attains a high friction coefficient due to the presence of several asperities. As soon as the sliding time increases, the friction value decreases. However, the friction value also depends on the sliding zone temperature, whether the surface is capable of building any oxide layer or not. For high loads, initial time to smoothen the rough surface asperities is low, but with an increase in sliding time, thin coating delaminates from the surface and provides high friction values. On the other hand, thick coating sustains the high load impact and gives low friction values. Sliding speed generally favors the friction coefficient. An increase in the sliding speed decreases the friction coefficient of applied coatings. Initially, high sliding speed attains plasticity behavior due to the shearing of asperities. After that, it starts building a transfer layer and attains self-lubrication behavior in some instances. At this stage, the friction coefficient drops expectedly. In the case of thick coatings, the friction coefficient is independent of the substrate/sample material because the coating only supports the entire load. However, in the case of thin coatings, the load is completely supported by the substrate. With an increase in the elastic modulus of substrate material, the contact area reduces, which provides low friction. Hard counterpart materials add advantage to the sliding test because it degrades slowly than the soft counterpart material and provides a low friction coefficient. Also, hard counterpart material supports a build-up of transfer-film due to continuous asperity shearing. On the other hand, soft counterpart materials wear rapidly and give rise to the friction coefficient.

6 Characterization Techniques in Composite Tribology

Composite surface material and their internal properties are vital for the technologist to prepare and install desire parts made by the composite technology. Various characterization techniques are developed to assess the surface and intrinsic (cross-section investigation, developed stresses, and density) properties. Thermal characterization of composite includes differential scanning calorimetry (DSC), thermogravimetric analysis (TGA), and dynamic mechanical analysis (DMA). DSC evaluates heat flow and temperature with respect to the reference material. TGA measures change in mass (weight loss of composite) as a function of time and temperature. DMA gives the physical property of materials under stress in the form of storage modulus and loss modulus (tan δ). Various analytical instruments like X-Ray Diffraction (XRD), scanning electron microscopy (SEM), transmission electron microscopy (TEM), electron probe microanalyzer (EPMA), and X-ray photoelectron spectroscopy (XPS) are the commonly used non-destructive imaging techniques. Some of the features of these instruments are listed in Table 1.

Table 1 Analytical techniques for composite characterization

Instrument	Energy range	Primary beam and secondary signal	Application
XRD	>1 keV	Photon–X-ray	Crystal structure/stresses
SEM	0.3–30 keV	Electron–Electron	Surface microstructure and morphology
TEM	100–400 keV	Electron–Electron	High resolution structure
EPMA	1–30 keV	Electron–X-ray	Surface region composition
XPS	>1 keV	Photon–Electron	Chemical composition at surface

7 Composite Tribology Significance

The tribology significance of composite materials for practical uses in the automotive industry is highly rated. Except for the decorative and covering pats in the automotive industry, other important composite parts installed in the system either function in sliding/rolling motion. These parts tend to deteriorate over time, with an increase in pressure and temperature. The surrounding environment and operating parameters also affect the performance of parts. Composite parts made by the infusion of a polymer-particle matrix could not sustain very high temperatures. Hence, such parts need to be made by the metallic infusion. Also, the load sustaining capacity is another challenge for these composites. However, in some cases of low-load conditions (ball joints), polymer composites can outperform other materials. Sometimes, hard materials in sliding/rotating action wear faster than the softer material due to external factors, such as induced corrosion/erosion. For a particular situation, lightweight structure material can be employed to restrict material wear. This could be achieved by making the composite with at least one hard material and other softer material. Hard material provides suitable toughness to the composite part for longer operation, and other material with self-lubrication properties provides a minimum effect of surface adhesion and wear. Similarly, for composite coatings, it must possess sufficient hardness and strong adhesion so that the coating material should not encounter surface spallation. However, before coating, the deposition process, material quantity, and thickness should be optimized for the overall weight-reduction of the prepared part. Multicomponent composite fabrications improve the strength and tribological properties of prepared parts. In this, each material provides its physical characteristics to the system for achieving high performance in terms of strength, friction, and wear. Without better tribological properties, the system may receive continuous maintenance, depending on the failure severity (see Fig. 10). The most dominant expenditure is associated with maintenance costs (see Fig. 10). At present, worldwide, the total cost of maintenance is almost about 50% in a calendar year. Other economic losses are linked to friction and wear, catastrophic failure, incorrect lubrication supply, and human error. With improved strength, the sliding/rolling behavior of composite material could be enhanced for the better service life of components.

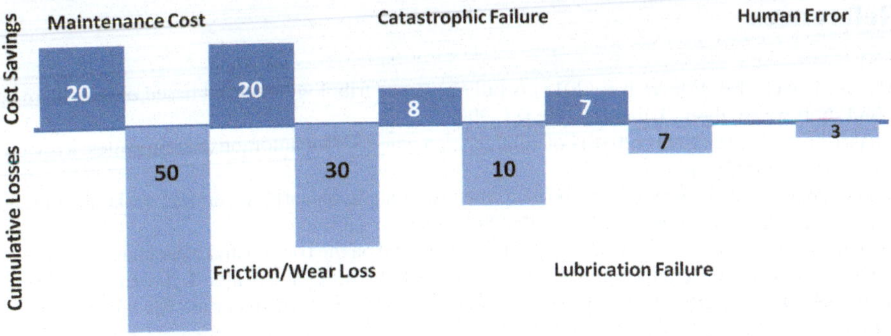

Fig. 10 Targeted cost savings by ternary composites (upward column) and generalized losses (downward column) associated at present in the automotive industry

8 Conclusion and Future Perspective

Composite tribology of polymer, ceramic, metal-matrix, and coating technology is studied. All the work related to the composite tribology shows surface modification behavior in the presence of two materials having different properties. Comparing all the composites used in this study, the matrix containing carbon nanoparticles, such as graphene, shows better performance than other non-carbon fillers. This increases the application of carbon-based fillers in composite tribology. Graphene, a self-lubricating material is currently in high demand due to their surface functionalization properties and high hardness. Epoxy-graphene composites provide low friction as similar to the sliding under the lubrication environment. Metal-matrix composites with ceramic as a filler material have high resistance against friction and wear. The formation of carbide and oxide during the composite sliding test provides a lubrication mechanism within the contact pairs and reduces surface flaws. The normalization of tribological performance increases with an increase in sliding interface temperature. This is valid for any composite (solid form) material and applied coatings made/deposited with oxide/carbide fillers of ceramics. Without lubrication and wear-resistant properties, the sustainability of industrial parts is difficult. Hence, the main realization to overcome frequent maintenance and losses incurred by the friction and wear processes, application of novel ternary composites will be explored in the near future. These composites will contain a gradient structure that will be able to reduce material wear and increase the service life of components without frequent maintenance. Implementation of ternary composites in the practical automotive application will increase vehicle performance at a comparatively low investment cost. Less wear and tear will allow reducing downtime of an industry, which can result in high production.

References

Akpan EI, Wetzel B, Friedrich K (2018) A fully biobased tribology material based on acrylic resin and short wood fibres. Tribol Int 120:381–390

Balakrishnan VS, Seidlitz H (2018) Potential repair techniques for automotive composites: a review. Compos B 145:28–38

Bian G, Wu H (2015) Friction and surface fracture of a silicon carbide ceramic brake disc tested against a steel pad. J Eur Ceram Soc 35:3797–3807

Chen W, Wang Z, Liu X, Jia J, Hua Y (2020) Effect of load on the friction and wear characteristics of Si3N4-hBN ceramic composites sliding against PEEK in artificial seawater. Tribol Int 141:105902

Dong HS, Qi SJ (2015) Realising the potential of graphene-based materials for biosurfaces—a future perspective. Biosurf Biotribol 1:229–248

Endrino JL, Nainaparampil JJ, Krzanowski JE (2002) Microstructure and vacuum tribology of TiC–Ag composite coatings deposited by magnetron sputtering-pulsed laser deposition. Surf Coat Technol 157:95–101

Friedrich K, Lu Z, Hager AM (1995) Recent advances in polymer composites' tribology. Wear 190:139–144

Ghodrati H, Ghomashchi R (2019) Effect of graphene dispersion and interfacial bonding on the mechanical properties of metal matrix composites: an overview. FlatChem 16:100113

Gong Y, Yang Z-G (2013) Fracture failure analysis of automotive accelerator pedal arms with polymer matrix composite material. Compos B 53:103–111

Hemanth J (2019) Quartz (SiO_2p) reinforced chilled metal matrix composite (CMMC) for automotive applications. Mater Des 30:323–329

Katiyar JK, Sinha SK, Kumar A (2016) Friction and wear durability study of epoxy-based polymer (SU-8) composite coatings with talc and graphite as fillers. Wear 362–363:199–208

Kumar P, Srivastava VK (2019) Reciprocating sliding tribology of brake oil treated carbon fiber reinforced ceramic matrix composites. Trans Nonferrous Met Soc China 29:1903–1913

Kumar V, Sinha SK, Agarwal AK (2017) Tribological studies of epoxy composites with solid and liquid fillers. Tribol Int 105:27–36

Li Y, Wang Q, Wang S (2019) A review on enhancement of mechanical and tribological properties of polymer composites reinforced by carbon nanotubes and graphene sheet: molecular dynamics simulations. Compos B 160:348–361

Lu G, Shi X, Liu X, Zhou H, Chen Y (2019) Effects of functionally gradient structure of Ni_3Al metal matrix self-lubrication composites on friction-induced vibration and noise and wear behaviors. Tribol Int 135:75–88

Lu N, He G, Yang Z, Li Y, Li J (2020) Fabrication and densification mechanism of MgO/graphene composites with LiF as additive. Scr Mater 174:91–94

Molent L, Haddad A (2020) A critical review of available composite damage growth test data under fatigue loading and implications for aircraft sustainment. Compos Struct 232:111568

Nickels L (2019) New innovations in automotive thermoplastics. Reinf Plast 63:185–188

Paksoy AH, Deprem O, Tazegul O, Cimenoglu H (2019) Tribology of SiCp reinforced Al-12Si matrix composite coatings in water. Tribol Int 110:392–400

Prasad SV, McDevitt NT, Zabinski JS (1999) Tribology of tungsten disulfide films in humid environments: the role of a tailored metal-matrix composite substrate. Wear 230:24–34

Sarantinos N, Tsantzalis S, Ucsnik S, Kostopoulos V (2019) Review of through-the-thickness reinforced composites in joints. Compos Struct 229:111404

Smith GF (1990) Design and production of composites in the automotive industry. Compos Manuf 1:112–116

Upadhyay RK, Kumar A (2018) A novel approach to minimize dry sliding friction and wear behavior of epoxy by infusing fullerene C_{70} and multiwalled carbon nanotubes. Tribol Int 120:455–464

Upadhyay RK, Kumar A (2019a) Effect of particle weight concentration on the lubrication properties of graphene based epoxy composites. Colloid Interface Sci Commun 33:100206

Upadhyay RK, Kumar A (2019b) Effect of humidity on the synergy of friction and wear properties in ternary epoxy-graphene-MoS_2 composites. Carbon 146:717–727

Upadhyay RK, Kumar A (2019c) Epoxy-graphene-MoS_2 composites with improved tribological behavior under dry sliding contact. Tribol Int 130:106–118

Wang L, Aslani F (2019) A review on material design, performance, and practical application of electrically conductive cementitious composites. Constr Build Mater 229:116892

Wood RJK (2010) Tribology of thermal sprayed WC–Co coatings. Int J Refract Met Hard Mater 28:82–94

Zhao W, Tang J, Puri A, Sweany RL, Li Y, Chen L (1996) Tribological properties of fullerenes C_{60} and C_{70} microparticles. J Mater Res Technol 11:2749–2756

Zhou H, Yao P, Gong T, Xiao Y, Zhang Z, Zhao L, Fan K, Deng M (2019) Effects of ZrO_2 crystal structure on the tribological properties of copper metal matrix composites. Tribol Int 138:380–391

Tribological Test of Composites Material Lubricated with Various Solid-Liquid Lubricating System

Y. Aiman, N. F. Azman, and S. Syahrullail

Abstract Nowadays, the use of lubricants in the industrial sector is rapidly increasing, particularly in applications that are subject to high loads and high speeds. Lubricant is a material, generally synthetic, applied to reduce friction between surfaces in reciprocal interaction, which essentially decreases the heat generated by moving surfaces. It may also function as transporting foreign particle or carry any debris to prevent clogging. The solid lubricant is present as a thin film, paste or powder that serves as a friction reducer and also avoids wear on both surfaces in order to reduce the scar on the side. Solid lubricant is a crucial technology for increasing the efficiency of jet and car engines, including engine running at significantly higher temperatures, and for improving the durability of spacecraft. Solid lubricants act as a special application to minimize harm in situations where the use of liquid lubricants is inefficient or insufficient, for example in space technology, automotive or also in vacuum areas. There are a few types of solid lubricants such as transition metal dichalcogenide compounds, carbon-based materials; soft metals and polymers. They can be divided into two specific groups for convenience: soft (hardness < 10 GPa) and hard (hardness > 10 GPa) solid lubricants. Many research has been done in solid-liquid lubricant and mostly of the solid-liquid lubricant is not capable lubricate properly without any modification or additives. In this chapter the different modification, additives and also tribology testing is been reviewed and discussed to get better understanding in solid-liquid lubricants in composite material.

Keywords Tribological composites · Solid-liquid lubricating system · Materials

---

Y. Aiman (✉) · N. F. Azman · S. Syahrullail
School of Mechanical Engineering, Faculty of Engineering, Universiti Teknologi Malaysia, 81310 UTM Skudai, Johor, Malaysia
e-mail: wmaiman91@gmail.com

N. F. Azman
e-mail: nurulfarhanahazman@gmail.com

S. Syahrullail
e-mail: syahruls@mail.fkm.utm.my

M. T. Hameed Sultan et al. (eds.), *Tribological Applications of Composite Materials*, Composites Science and Technology, https://doi.org/10.1007/978-981-15-9635-3_3

1 Introduction

Lubricant is used in most equipment and engine as a wear and tear reducer (Golshokouh et al. 2013). Without lubricant, spinning or sliding machine or engine component cannot function properly and may cause the machine or engine to experience catastrophic failure. Load capabilities without failure, temperature stability, wear resistance capability and their flash point are the most important parameter to be considered in oil lubricant research (Jabal et al. 2014; Syahrullail et al. 2013). Today, lubricant use is rapidly increasing in the industrial sector, especially in applications that are exposed to high loads and high speeds. According to Hassan et al. (2016) lubricant plays an important role in reducing the damage to the surface of the device caused by the small particles. It also acting as a transporting medium for the contaminant and debris as stated by Razak et al. (2015).

Over the past 10 years, there has been a revival in strong lubricant research and development. This resurgence was motivated due to the advanced technological applications that need lubricants to be extremely good under different conditions (Sapawe et al. 2016). There are a few types of solid lubricant that been research and develop in the past 10 years to working under very high temperature with longer lifespan. Solid lubricant is a key technology in increasing the performance of jet and car engines, which includes engine operation at significantly higher temperatures, and increasing the endurance of space vehicles (Booser 1984; Hasannuddin et al. 2018a, b). Most of the solid lubrication, their production has given benefit to what is learned about its action and also the mechanism. However, the research with solid lubricant awareness is still incomplete, especially under extremely high temperatures, considerable progress has been made in material performance (Sliney 1982).

The solid lubricant is presence as a thin film, paste or powder that function as friction reducer and also provide wear prevention on both surface contact to reduce the scar on the part. Different from liquid friction models, solid lubricant friction has a specific view for of friction on its mobility, shape, crystallography and size of particles. Constructing a rubbing surface with continuous adherent soft or hard film is the main purpose of solid lubricants. The film distribution can be implemented by physical, electro-chemical or mechanical, processes. The standard test is according to ASTM D2714 test method to assess a solid lubricant's friction coefficient (Bart et al. 2013).

Solid lubricants serve a special niche to reduce damage in cases with ineffective or insufficient use of liquid lubricants for example in space technology, automobile or even in vacuum space. For lubrication under extreme pressure where the bearing surfaces still need to be effective, solid lubricants are required, where the lubricant needs to remain without moving.

Usually the solid lubricants is used at to the sliding surface under high load conditions that going through the boundary and mixed regimes frictional, extremely low hydro-dynamically efficient speeds, or to work under very high temperature and also very extreme condition such as in aviation. The solid lubrication is also needed when the part of the components cannot be contaminated with liquid lubricant such as in nuclear reactor and high vacuum system. In this chapter, the author would to address the solid-liquid lubricant types, properties, modification, and additive in composite material with various tribological test.

The use of solid-liquid lubricant is very famous nowadays and many modifications has been made to improve the tribological performance of the lubricants by using additive, coating, or modifying (Zeng et al. 2013; Mistery et al. 2011; Vengudusamy et al. 2012). However, there is some certain cases the solid-lubricant has shown poor tribological performance in composite material even after the modification (Okubo and Sasaki 2017; Zhang et al. 2019). This chapter is addressing the study the tribological performance of solid-liquid lubricant on composite material that have done by previous researcher.

2 Solid Lubricant Classes

To provide low friction, non-organic solid lubricants need to possess small shear resistance, for one crystallographic direction at least. There are a few types of solid lubricants such as transition metal dichalcogenide compounds, carbon-based materials; soft metals and polymers. We can classify them into two broad categories for in the pursuit of convenience: soft (hardness < 10 GPa) and hard (hardness > 10 GPa) strong lubricants (Holmberg et al. 1998). Table 1 shows the lists of category for the solid lubricants. In addition to reduced friction relative to soft lubricants,

Table 1 Classification of the tribological coatings according to the quality of the substance in consideration (Donnet and Erdemir 2004)

Hard Coatings Hardness higher than 10 Gpa	Soft coating Hardness lower than 10 Gpa
Nitrides TiN, CrN, ZrN, BN $BaSO_4$	**Soft metals** Ag, Pb, Au, In, Sn, Cr, Ni, Cu
Carbides Tic, WC, CrC	**Lamellar solids** MoS_2, WS_2, Graphite H_3BO_3, HBN, GaS, GaSe
Oxides Al_2O_3, Cr_2O_3, TiO_2, ZnO, CdO, Cs_2O, PbO, Re_2O_7	**Halides Sulfates, sulfure** CoF_2, BaF_2, PbS, $CaSO_4$, $BaSO_4$
Borides TiB_2	**Polymers** PTFE, PE, Polymide Polymer like DLC
DLC Coating a-C, ta-C. a-C:H, ta-C:H, CN_x a-C:X(:H), (nc-)diamond	

Polytetrafluorethylene
Polyethylene
Dia amorphous carbon mond-like carbon
Tetrahedral amorphous

the hard-solid lubricants have greater in resistance formation of high friction, which can provide low friction but not always high wear resistance. Many carbon-based coatings (DLC and diamond) and some oxides are address as a hard durable lubricant coatings. Soft solid lubricant coatings include soft metal, halides, polymers, and sulfates of alkaline earth metals, and the well-known lamellar solids, including graphite, transition-metal dichalcogenides, and boric acid. In recent years, many of self-lubricating materials have been researched to study the development of it, and some of them has modified to produce better lubricity when used under different type conditions or in situations such as high loading (Sutor 1991; Donnet and Erdemir 2004).

Figure 1a shows the molybdenum disulphide MoS_2 of layered lattice structure that under class of graphite (Fig. 1b). Figure 1c shows the new material of that is boric acid, H_3BO_3, it is a layered lattice composite that has been carefully studied only

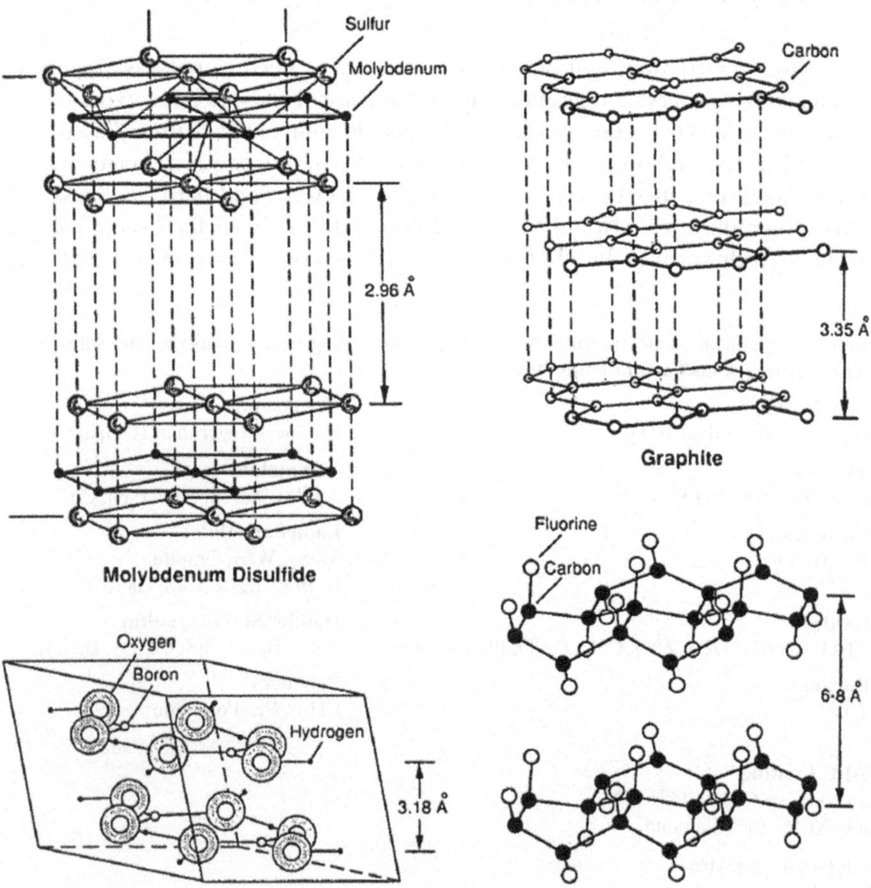

Fig. 1 Layered lattice structures of several lamellar solid lubricants (Sutor 1991)

recently. Triclinic is the most common form of boric acid, orthoboric acid. Figure 1d shows the lamellar structure for graphite fluoride, where fluorination of graphite is to provide sub stoichiometric graphite fluoride.

2.1 Carbon-Based Materials

Graphite is one of the carbon-based substances renowned for their role in solid lubricating. Of greater significance are the various types of DLC coatings which were specifically synthesized for friction and wear reduction. The DLCs are amorphous in nature that is completely different to the graphite. The first to studied and research DLC coatings were by Schmellenmeier (1953) and later Eisenberg and Chabot (1971). Initially diamond is not considered as a strong lubricant material; but, in the presence of water an oxygen vapor, solid lubricity is found when the grains become nanocrystalline under certain operating conditions.

2.1.1 Graphite

Graphite is a low-density, crystalline and soft carbon allotrope. Graphite is one of the solid lubricant that has a lamellar form and inorganic lubricants with also contains boron nitride, molybdenum disulphide (see Fig. 1a), and some other molybdenum, tantalum, selenides and tellurides (chalcogenides), niobium, tungsten, and titanium sulphides. Its crystal lattice consists of parallel thin planes (graphenes) forming hexagonal rings. For every carbon atom in the plate was bonded covalently to 3 other atoms (120° angle of bond). Soft Van der Waals powers bind the graphenes to each other.

The staggered structure allows the parallel planes to slip. Poor bonding between the planes produces weak shear strength in sliding motion direction but high compression strength in perpendicular direction to sliding motion (see Fig. 2a). It is believed that the combination of oxygen and water vapor in the atmosphere would facilitate graphite crystal interlamellar shearing. Such close basic planes have small energies surface and low adhesion at the bot surfaces (Deacon and Goodman 1958). This result is supported by Buckley (1981) that found when smooth, metallic materials, such as tantalum and iron, are under ultra-high vacuum contact sliding at basic pyrolytic graphite planes, there is no adhesive metal transfer is detected, that generate in smaller coefficient of friction.

2.1.2 Diamond-Like Carbon (DLC)

According to Robertson (1986) DLC is thin film substance with a greater fraction of carbon bonding metastable sp^3, and amorphous-carbon (a-C) or hydrogenated-amorphous (a-C: H). This is usually processed via a deposition cycle which contains

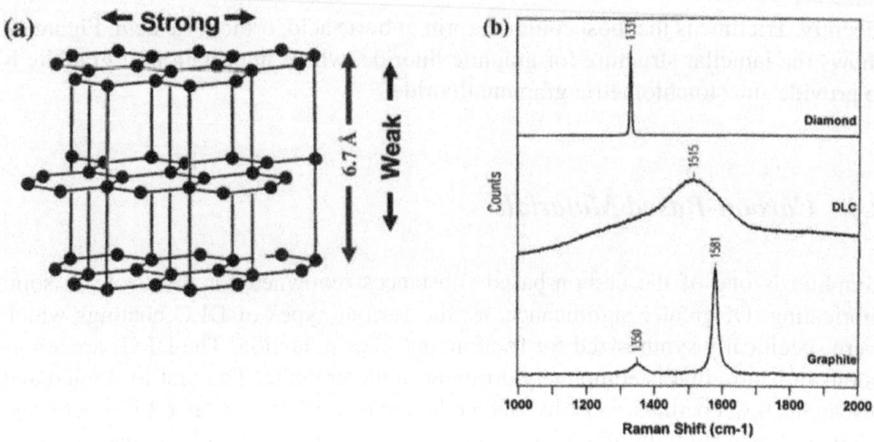

Fig. 2 **a** Graphite Lamellar crystal structure. **b** Typical Raman spectra taken with $k = 458$ nm of graphite, DLC coating and diamond (Scharf and Prasad 2013)

energetic ions. Such ions has increase to the bonding of sp^3 that is actually metastable relative to the bonding of sp^2 when stabilized by C–H bonds. The ion-induced cycle is distinct from plasma polymerisation, where the C sp^3 bond emerges from a C–H group condensation of molecular H_2 evolution. Some of elements for example B, N and F or Si, metals can be alloyed with DLC.

The DLC forms can be shown on a ternary phase diagram, as shown in Fig. 3. It indicates the proportion of C sp^3, C sp^2 or hydrogen (H) sites in the alloy. Jacob and Moller (1993) derived the diagram for the first time. Remember that the total fraction of atoms here is up to 1. Cases that quote the sp^3 fraction (of only C) or the H/C ratio instead of the H/(C + H) ratio should be careful decide. This network's mechanical

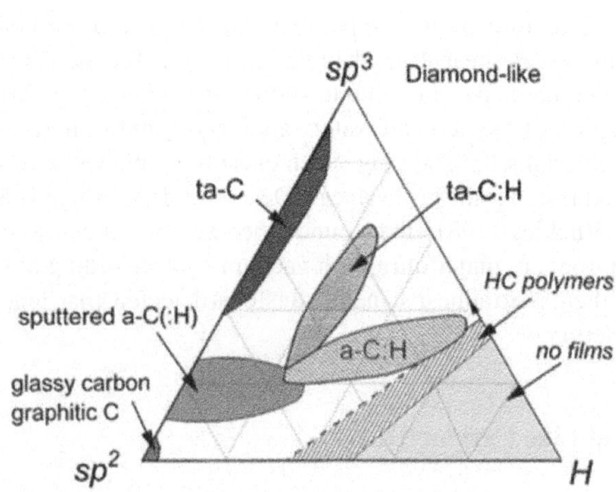

Fig. 3 Ternary phase diagram of bonding in amorphous carbon-hydrogen alloys (Robertson 1986)

properties can be seen as a combination of multiple bonding elements, C–H, C–C sp^2 and C–C sp^3 bonds. The longevity, elastic module and general diamond-like consistency of the C–C sp^3 bonds are increased. C–C sp^2 bonds make no meaningful contribution. The C–H bonds do not bind the network, so they are like a hanging bond mechanically, contributing nothing to the mechanical properties (Robertson 1986). According to Ferrari et al. (1999), therefore, it is found that the Elastic modulus is monotonically dependent on the mean number of C–C coordinates, or in other words the fraction of C–C sp^3.

2.1.3 Nanocrystalline Diamond

Diamond films and diamond are not considered low-interface shear stable lubricant materials until smooth nanocrystalline diamond coatings (Erdemir et al. 1999) and ultranocrystalline (3–5 nm) of the grain shape showed fascinating tribological performance. The UNCD films show low friction coefficients and low surface roughness (*13 nm RMS) in both dry nitrogen and moist air (0.03–0.1) when slipping against a UNCD-coated Si_3N_4 counter face. Diamond coatings coated with chemical-vapor (CVD) have gained significant interest over the past decade, this is partially due to the high level of attention because of the extremely good tribological properties of diamond (Spear et al. 1994; Braza and Sudarshan 1992; Tsai and Bogy 1987). In addition to all known essential parameters influencing the tribological properties, it is clear that the sum of H and the sp^3/sp^2 bonding ratio influence the behavior of the nanotribology.

2.2 Transition Metal Dichalcogenide Compounds (TMD)

Transition metal dichalcogenides (TMD) are in many ways a gift of design to mechanical engineers who seek to minimize friction. TMDs occur in 2 quartz types: hexagonal and rhomboedral. The hexagonal structure is the most common and most critical for low-friction applications as shown in Fig. 4 (Polcar and Cavaleiro 2011). The hexagonal crystal structure with two molecules per unit cell, exhibits a laminar or layer-lattice structure, six-fold symmetry. TMD is often used as an oil additive or as a coating to reduce friction. WS_2 and MoS_2 are most famous due to the strong lubricating performance among the other types of the compound TMD group and usually it used in several applications. WS_2 and MoS_2 lubricating activity derives from the inter-mechanical instability in their crystal structure. MoS_2 crystallizes in the hexagonal form, identical to graphite, where a plane of molybdenum atoms is interlayer with two hexagonally that packed sulphur particles with a strong c/a ratio ($c = 12.29$ Å, $a = 3.16$ Å).

There is a laminar or layer-lattice arrangement in the hexagonal crystal structure with six-fold symmetry, two molecules per unit cell. Every chalcogenide atom is

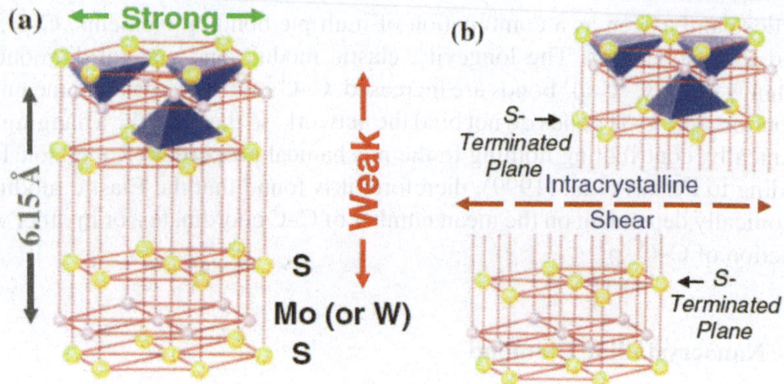

Fig. 4 **a** LMS of Mo (or W). **b** Weak VDW slip planes to interlamellar

equidistant from three metal atoms, stopping at six dichalcogenide atoms equidistant from each metal atom. According to Lansdown (1999) the attraction between dichalcogenide atoms and metal is because of strong covalent bonding, but between lattice layers there is only a mild attraction of Van der Waals. TMD family consists of disulphides and diselenides of molybdenum, tungsten and niobium. TMD is also used as an additive to oil or also as a friction reduction coating. Waghray et al. (1995) says the latter could be prepared as a thick film, and preparation of electrochemical processes or thin films deposited primarily by physical vapor deposition (PVD) is suggested by Cardinal et al. (2009). TMD can also be produced by tribochemical reactions on contact surfaces during the sliding process (Grossiord et al. 1998). Figure 4 shows the Lamellar crystal structure (LMS) of Mo and the weak van der walls forces slip. Figure 5 shows the schematic views of previous film design for nanocomposite, nanograns and the nanolayer. Nanocomposite coatings with strong WC nanoparticles (red) embedded in amorphous carbon matrix (grey) along with WS2 nanograins (blue).

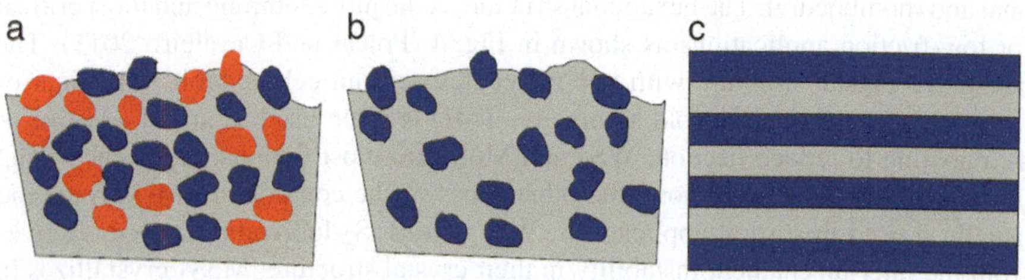

Fig. 5 **a** Nanocomposite coatings, **b** nanograins and **c** nanolayered (Polcar and Cavaleiro 2011)

Fig. 6 Chemical illustration of Polytetrafluoroethylene (Franklin 2017)

2.3 Polymers

PTFE is a plastic fluoropolymer called polytetrafluoroethylene and it is one of the well-known stable lubricant polymers. PTFE has many desirable features and was influential in the lubrication of colloids. Water and water-containing liquids do not humidify PTFE, thereby inhibiting adhesion to PTFE surfaces. According to Franklin (2017), it is extremely non-reactive, and usually used in pipework and container for corrosive and reactive chemicals. The anti-friction property of PTFE is well known (Mortimer 1991), this is due to the result of its low intermolecular cohesion and molecular profile (Scharf et al. 2009). PTFE does not have unsaturated bonds and does not polarize quickly. This forms a thin transfer film on the counter face during sliding contact with small contact pressure (7.3–73 MPa), but uneven films thickness can generate low and not stable friction behaviour if the films are not distributed thinly afterwards (Fig. 6).

PTFE colloids are highly appealing as oil additives at first glance. The fact that PTFE colloids do not like water or oil, though, has made their use far from easy in this use. To make a suitable lubricant, PTFE cannot just be rubbed into a liquid. It is crucial to choose the dispersion chemistry and additives used with carefully and the dispersion skill is extremely important. There are powerful synergistic interactions with certain additives. Extraordinary lubricants produce when properly done and a new class of lubricants has been achieved with remarkably low friction and tear. Four-ball tests show low wear and a friction coefficient dependent on this technique for a professionally formulated lubricant (Riceo et al. 2007).

2.4 Solid-Liquid Lubricant

Recently, development of solid-liquid lubricating system has becoming an attractive research topic especially for space and automotive applications since it is a promising

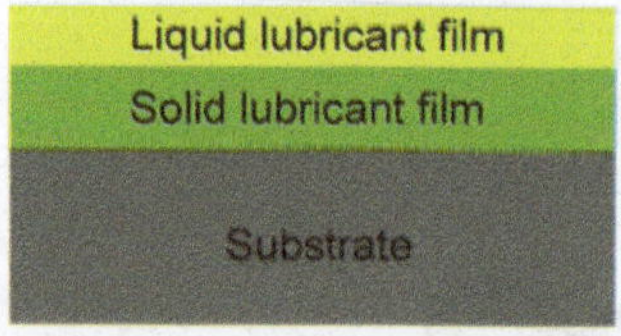

Fig. 7 The schematic diagram for solid-liquid lubricating system

way to achieve desired tribological performance (Liu et al. 2011a; Lv et al. 2015). solid-liquid lubricating system was composed of a protective layer of solid lubricant film deposit to any substrate material and upper layer of liquid lubricants film (Liu et al. 2011a). The schematic of the solid-liquid lubricating system can be found in Fig. 7. An advanced tribological performance of solid-liquid lubricating system was important to improve energy efficiency, service reliability and durability of mechanical assemblies of an engine and space drive mechanisms (Fan et al. 2015; Guo et al. 2018; Liu et al. 2019). Previous researchers have made comprehensive efforts to develop and explore the synergy between solid and liquid lubricants of a solid-liquid lubricating system to achieve optimal tribological efficiency (Liu et al. 2011b; Fan et al. 2015; Zhang et al. 2019).

Multiple factors can affect the tribological performance of solid-liquid lubricating system, these includes:

- Intrinsic condition of solid lubricants: microstructure, hardness, thickness, interlayers, elasticity, roughness, bonding strength, mechanical strength, thermal and chemical properties (Liu et al. 2011a; Fan et al. 2015)
- Intrinsic condition of liquid lubricants: structure and composition, viscosity-temperature performance, physico-chemical properties (Fan et al. 2015)
- Tribo-test parameters: temperature, applied load, sliding speed, counter materials
- Environment conditions: relative humidity, inert atmosphere and space conditions (vacuum, irradiation, atomic oxygen) (Liu et al. 2012, 2013, 2017b).

3 Tribological Properties of Solid-Liquid Lubricating System

Diamond-like carbon (DLC) films are widely use as solid lubricants in automotive and space industry due to their outstanding properties (Zeng et al. 2013). Numerous studies were conducted by researchers to evaluate the tribological performance of DLC films as protective layer of substrates and with the presence of liquid lubricants. In an earlier study, Liu et al. (2011a) explored the tribological performance of DLC-based coating composed of TiC/a-C:H solid coating lubricated with multi-alkylated cyclopentanes (MACs) in high vacuum condition. It was observed that DLC/MACs solid-liquid lubricating system enhanced the friction reduction performance, anti-wear performance and load capacity compared to that of bare TiC/a-C:H

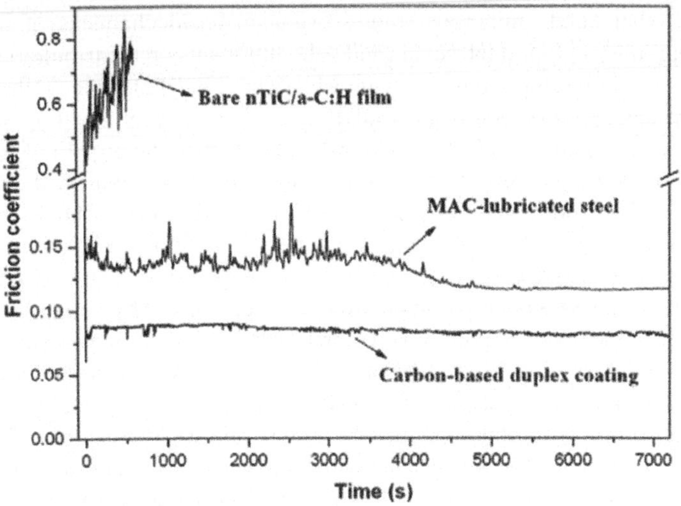

Fig. 8 The difference of the friction process between bare TiC/a-C:H film, the MACs lubricated steel and the DLC-based solid-liquid lubricating system (Liu et al. 2011a)

solid coating and MACs-lubricated steel (see Fig. 8). The significant improvement in the tribological performance was attributed to the synergistic interaction between solid and liquid lubricant to form tribofilm on the counter surface. As reported by Zhiqiang et al. (2013), the presence of lubricants on the DLC coating lead to the formation of a layer of lubricant films on the contact surface. These lubricant films have lower shear strength than that of transfer layer formed by DLC coatings, thereby decreases friction and wear. Guo et al. (2018) also reported the presence of poly-alpha-olefin synthetic oil (PAO) lubricant had improved the tribological performance of DLC coatings due to the formation of lubricant film in the sliding interface. Additionally, the presence of lubricant prevents the formation of transfer film and prevents the titanium (II) oxide (TiO) and titanium carbide (TiC) hard particles from entering the sliding interface.

On top of that, the effect of doping content in DLC films on the tribological performance of dry and lubricated condition of DLC coating was studied by Fu et al. (2013). They found that increased in tungsten (W) content in DLC coatings had increased the friction coefficient while the wear rate show a minimum value under dry DLC coatings. Whereas under PAO lubricated condition, the friction coefficient decreased as the W content in DLC coatings increased. However, presence of PAO on DLC coatings had antagonist effect on wear, showing an increment with increase in W content. The addition of thiophosphoric acid amine (T307) salt as additive in PAO resulted in low wear rates in both undoped DLC and W-doped DLC. Their finding proved that the intrinsic condition of DLC coatings will affect the tribological performance of solid-liquid lubricating system.

On the other hand, numerous studies evaluate the mechanisms of interaction between additives in liquid lubricants and solid lubricants on the tribological performance of solid-liquid lubricating system. Mistry et al. (2011) evaluate the tribological performance of two additives (ZDDP and thiadiazole) interacting with tungsten carbide doped DLC (WC–DLC) in order to elucidate the growth of tribochemical films. Tribological performance turned up to be dependent on the chemistry of tribofilm. The tribofilm composition of WC–DLC coated and ZDDP–thiadiazole interaction were consisted of zinc sulphide, tungsten sulphide (WS_2), tungsten oxide (WO_3), tungsten carbide (WC) and organic compounds. They confirmed the tribofilm composition using XPS analysis. In previous work conducted by Vengudusamy et al. (2013), they found that the hydrogen (H) and tungsten (W) (doping element) concentration affect the tribological performance of DLC coatings lubricated with ZDDP solutions. They also concluded that this intrinsic condition of DLC coatings markedly effect on adhesion, tribofilm formation by ZDDP and wear resistance. This shows that intrinsic condition of solid lubricant will influence the DLC/additive interactions.

Moreover, Okubo and his team (2016) formulated PAO based lubricant with several primary- and secondary-alkyl groups ZDDPs in order to study the effect of structure of ZDDPs on the tribological performance of the tetrahedral amorphous carbon (ta-C) film. Their results revealed that friction coefficient of the ta-C lubricated with PAO + ZDDP is independent with length of alkyl chains. Under DLC/DLC contact, primary-alkyl ZDDP exhibited excellence friction reduction performance, while secondary-alkyl ZDDP exhibited excellence anti-wear performance than the other lubricants. However, primary-alkyl ZDDP shows antagonisms effect on wear volume due to the occurrence of chemical wear caused by interaction between primary-alkyl ZDDP and ta-C films. They concluded that the chemical composition of ZDDP tribofilms affect the tribological performance of the ta-C films. The low friction coefficient of the ta-C film was attributed to sacrificial sulphur-rich tribofilm, while the low wear volume attributed to the excellence anti-wear performance of phosphorus-rich tribofilms themselves. This proved that intrinsic factor of liquid lubricant (i.e. composition) will influence the DLC/additive interactions (Fig. 9).

Yue and his team (2013) investigated the interaction mechanism between sulfurized W-DLC coating and PAO containing molybdenum dithiocarbamate (MoDTC). The improvement of friction reduction performance of sulfurized W-DLC/MoDTC solid-liquid lubricating system compared with dry W-DLC coating was attributed to the formation of metallic sulphide (WS_x) layer on the surface coating. Tribochemical reactions between MoDTC and sulfurized W-DLC coating results in the formation of tribofilm rich in Mo sulfide/Mo oxide and sp^2/sp^3. Surprisingly, Yoshida and Kunitsugu (2018) found that amorphous hydrocarbon (a-C:H)/MoDTC solid-liquid lubricating system leads to an increment of the degree of wear. The wear loss of a-C:H/MoDTC solid-liquid lubricating system was significantly increased as the sp^2/sp^3 ratios of DLC coating increased, compared to DLC coating without MoDTC which shows no significant change in wear loss as the sp^2/sp^3 ratios increases. During friction and wear of a-C:H/MoDTC solid-liquid lubricating system, molybdenum disulphide (MoS_2) tribofilm will form on the tribological contact area. The degree

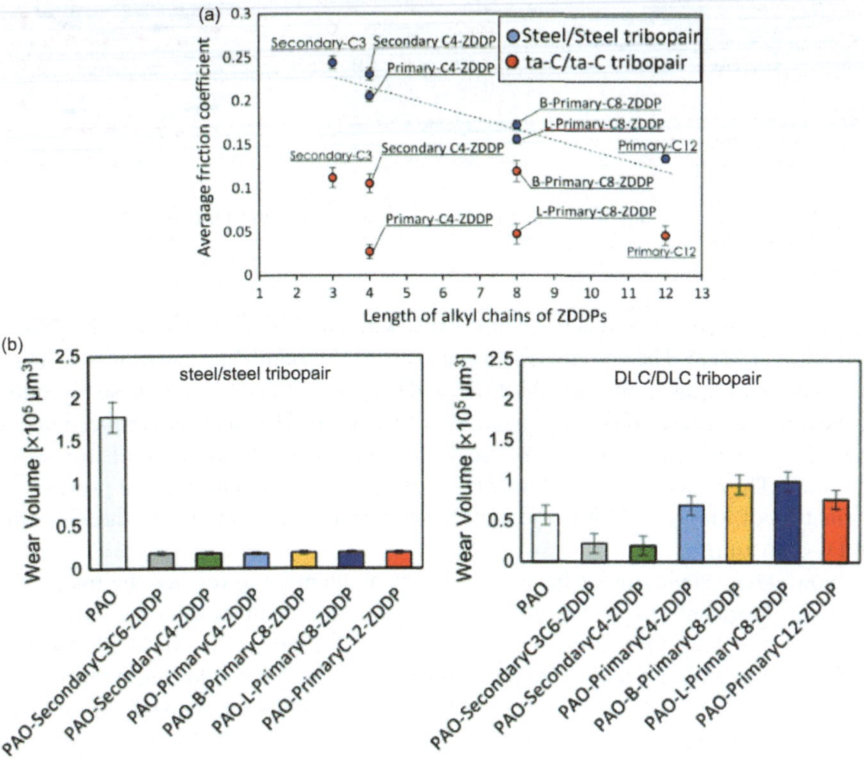

Fig. 9 Effect of length of alkyl chains of ZDDPs on **a** friction coefficient and **b** wear volume of steel/steel and ta-C/ta-C tribopair lubricated with PAO + ZDDP (Okubo et al. 2016)

of wear increases either by mechanical wear (abrasive wear) or chemical wear when DLC films oxidized by MoO_3 (Vengudusamy et al. 2012). Okubo and Sasaki (2017) investigated the mechanism of wear acceleration of DLC/MoDTC solid-liquid lubricating system using in-situ Raman tribotester. Their results revealed that the antagonism of MoDTC in wear acceleration is occurred when ferrous-based counterface material was used. Based on their findings, they proposed the mechanism of wear acceleration as follows: (1) Mo compounds (MoO_3 and MoS_2) chemically reacts with ferrous-based counterface material surfaces, (2) formation of Mo-carbides (Mo_2C) through tribochemical reactions between product of Mo and DLC films caused chemical wear, and (3) Mo_2C particles caused mechanical wear (abrasive wear). The proposed mechanism was schematically shown in Fig. 10.

Some researchers added ZDDP additive in lubricant solution to counter the antagonisms of MoDTC in terms of wear acceleration (Kosarieh et al. 2013; Yang et al. 2014; Gorbatchev et al. 2016). Kosarieh and his team (2013) stated that the formation of MoS_2 tribofilm produced by MoDTC and DLC wear debris are responsible to provide low friction of a-C:15H/cast iron tribocouple. While the antagonism effect of MoDTC in wear acceleration was attributed to tribochemistry of rubbing surface

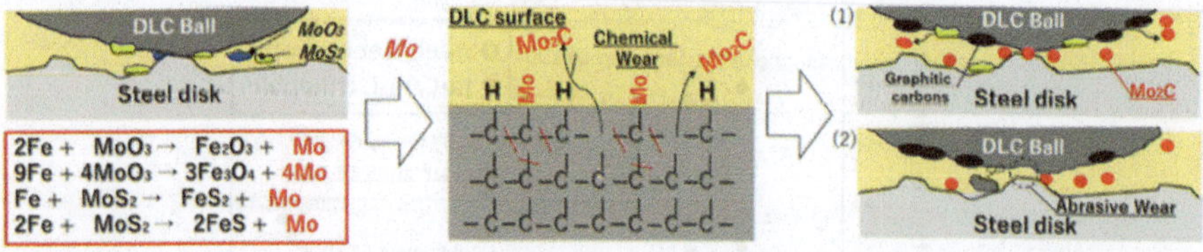

Fig. 10 Schematic diagram of the mechanism of wear acceleration of DLC films lubricated with MoDTC solution (Okubo and Sasaki 2017)

rather than the chemical reactions of the oils with DLC films, as reported by Okubo and Sasaki (2017). The authors (Kosarieh et al. 2013) found that ZDDP was successfully counter the antagonism of MoDTC and the interaction between these two additive exhibited synergistic effect on the wear resistance. The wear was reduced due to the formation of glassy phosphate species tribofilm or/and formation of iron oxide particles. They also suggested that ZDDP offers oxidation inhibitors to prevent the formation of MoO_3, which is the main character in contributing wear acceleration of DLC coating.

Nowadays, nanoparticles have attracted great attention to replace the traditional additives lubricant, due to the problem in finding the compatibility between traditional additives and DLC coatings (Zeng et al. 2013; Zhang et al. 2019). Zhang et al. (2019) compared the tribological performance of DLC/PAO solid-liquid lubricating systems. Two types of additives was added into PAO lubricant, namely Cu nanoparticles modified with diisooctyl dithiophosphoric acid (NPCuDDP) and ZDDP. Four different types of DLC coatings were used such as a-C, a-C(Si), a-C(Al) and a-C(H) to further explore the effects of intrinsic conditions of solid lubricants (i.e. hardness and composition of DLC coatings) to the mechanism and conformability of NPCuDDP and ZDDP additives. PAO/NPCuDDP lubricants exhibit excellent tribological performance under all DLC coatings compared to that of PAO/ZDDP lubricants (Fig. 11). The friction reduction performance of PAO/NPCuDDP was improved due to the formation of tribofilms derived from NPCuDDP, where Cu will released

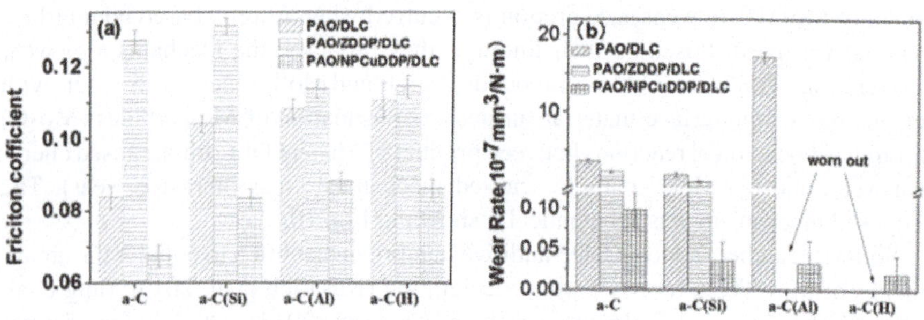

Fig. 11 COF (**a**) and Wear DLC films (**b**) (Zhang et al. 2019)

from NPCuDDP and partially turned into oxides and sulphates through tribochemistry reaction. The formation of the tribofilm is disregarding with composition and hardness of DLC coatings. Their results revealed that a-C(H) and a-C(Al) coatings have bettter anti-wear performance than that of a-C(Si) and a-C coatings due to the participation of H and Al elements in tribochemical reaction. It revealed that different doping elements on DLC films will produce different chemical reaction with lubricants, thereby results in different tribological performances.

In an experimental study, Zeng et al. (2013) use PAO oil with boron nitride nanoparticles to lubricate the silicon nitride (Si_3N_4)/DLC tribocouple. The results indicate that boron nitride nanoparticles in PAO oil improve the tribological performance of DLC films. The reported mechanism of the improved tribological performance was the reduction in the contact area due to the ball bearing effect and interlayer sliding of the weak van der Waals interaction force between boron nitride nanoparticles at contact surface. It is widely known that 2D-layered materials exhibit excellent mechanical and tribological properties due to weak van der Waals bonds between the molecular layers and strong in-plane covalent bonds characteristics. Liu et al. (2019) explored the effect of 2D-layered materials namely graphitic carbon nitrogen (g-C_3N_4) as lubricant additives on the tribological performance of Ti-DLC films. The authors proposed the synergisms between g-C_3N_4 nanosheets and Ti-DLC films contributing to the tribofilm formation on the contact surface, thereby improved tribological performance. The same mechanism was also reported by Kogov and Kalin (2019). Their study also revealed that the quantity of tribofilm formed on the contact surface were depends on the concentration of the nanoparticles and contact velocities or changing of the lubrication regime. Based on these reported literatures, researchers needed to identify the compatibility between liquid lubricants and solid lubricants as well as gain a comprehensive understanding on the interaction mechanisms between additives in lubricating oil and solid lubricants to design an effective solid-liquid lubricating system for practical application (Table 2).

Another point needed to be considered to design an effective solid-liquid lubricating system for practical application is controlling the tribology-test variable (for example temperature, applied load, sliding speed and counter materials) and environment conditions. The effects of these operating parameters can be termed as extrinsic conditions. Liu and his team (2013) studied the effect of environment conditions (atomic oxygen (AO) and ultraviolet irradiations (UV)) on the tribological performance of DLC-based solid-liquid lubricating system for space applications. Three different types of liquid lubricants were used in their studies such as poly(tetrafluoroethylene oxide-co-difluoromethylene (Zdol), ionic liquid (IL) and MACs. They found that AO irradiation was more pronounced in changing the structure of DLC films including oxidation, bond breaking and crosslinking reactions and liquid lubricants, compared to UV irradiation. Fortunately, DLC coatings presented unique advantage on its high radiation resistance which make them able to provide better tribological performances than that of ferrous-based solid-liquid lubricating system even after irradiations.

Table 2 Summary of tribological performance of solid-liquid lubricant

Type of lubricant		Performance	References
TiC/a-C:H solid coating lubricated with multialkylated cyclopentanes (MACs)		DLC/MACs solid-liquid lubricating system enhanced the friction reduction performance, anti-wear performance and load capacity compared to that of bare TiC/a-C:H solid coating and MACs-lubricated steel the presence of poly-alpha-olefin synthetic oil (PAO) lubricant had improved the tribological performance of DLC coatings due to the formation of lubricant film in the sliding interface	Liu et al. (2011a) Zhiqiang et al. (2013) Guo et al. (2018)
Additives	ZDDP and thiadiazole	Tribological performance turned up to be dependent on the chemistry of tribofilm. The tribofilm composition of WC–DLC coated and ZDDP–thiadiazole interaction were consisted of zinc sulphide, tungsten sulphide (WS_2), tungsten oxide (WO_3), tungsten carbide (WC) and organic compounds	Mistry et al. (2011)
	PAO and ZDDP	The results revealed that friction coefficient of the ta-C lubricated with PAO + ZDDP is independent with length of alkyl chains. They concluded that the chemical composition of ZDDP tribofilms affect the tribological performance of the ta-C films	Okubo et al. (2016)

(continued)

Table 2 (continued)

Type of lubricant		Performance	References
	ZDDP as counter of antofonism of MoDtc	The antagonism effect of MoDTC in wear acceleration was attributed to tribochemistry of rubbing surface rather than the chemical reactions of the oils with DLC films. The authors found that ZDDP was successfully counter the antagonism of MoDTC and the interaction between these two-additive exhibited synergistic effect on the wear resistance	Kosarieh et al. (2013), Yang et al. (2014), Gorbatchev et al. (2016)
	sulfurized W-DLC coating and PAO containing molybdenum dithiocarbamate (MoDTC)	The improvement of friction reduction performance of sulfurized W-DLC/MoDTC solid-liquid lubricating system compared with dry W-DLC coating was attributed to the formation of metallic sulphide (WSx) layer on the surface coating	Yue et al. (2013)
	DLC/PAO with Cu nanoparticles modified with diisooctyl dithiophosphoric acid (NPCuDDP) and ZDDP	The friction reduction performance of PAO/NPCuDDP was improved due to the formation of tribofilms derived from NPCuDDP, where Cu will released from NPCuDDP and partially turned into oxides and sulphates through tribochemistry reaction	Zeng et al. (2013), Zhang et al. (2019)

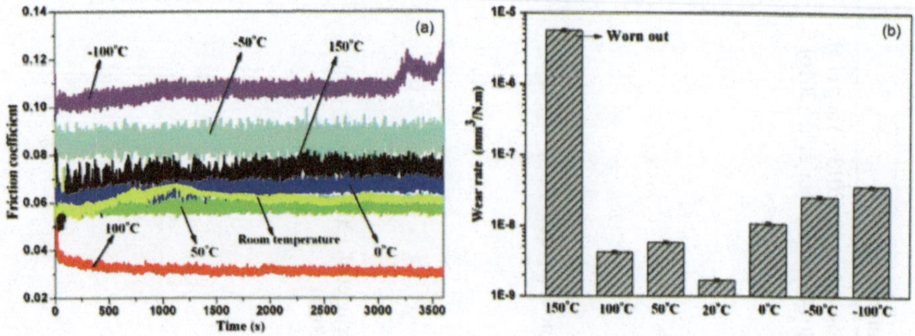

Fig. 12 COF and wear of DLC/IL solid-liquid lubricating system at different temperatures in high vacuum condition (Wang and Liu 2013)

Wang and Liu (2013) evaluated the effects of temperature on the tribological performance of DLC/IL solid-liquid lubricating system in high vacuum condition under a wide range of temperature between −100 and 150 °C. The tribological performance of the DLC/IL solid-liquid lubricating system and temperatures can be found in Fig. 12. Below room temperature (20 °C), the tribological performances decreased as the temperature decreased. This is due to the degradation of wetting, spreading and self-repairing capacity of IL that hindered tribochemistry reaction between DLC films and IL lubricant to occur. Friction coefficient was lowest at 100 °C due to slight surface graphitization and at this temperature DLC/IL lubricating system exhibited excellent synergistic lubrication mechanism. Moreover, wear rate under a high- or low- temperature condition were higher than that of room temperature due to surface graphitization. Under high-temperature condition, an increase in temperature resulted in higher degree of graphitization, whereas it oppositely occurs under low-temperature condition. It has been reported that graphitization will results in wear acceleration (Zahid et al. 2016), and the graphitization was temperature dependent.

Al Mahmud's group (2014) confirmed that the high degree of graphitization resulted in high wear rate. They investigated the effect of temperature (50, 100 and 150 °C) on the tribological performance of a-C:H and ta-C DLC coatings lubricated with SAE 40. The results revealed that the wear rate increases with increasing temperature, however friction coefficient decreases as the temperature increases for both tribopairs (ta-C/steel and a-C:H/steel) in the absence of delamination. At higher temperature (150 °C), the wear rate of ta-C coating surface was lower than that of a-C:H coating surface. The tribochemical interactions between ZDDP additive in SAE40 and ta-C DLC coating resulted in formation of polyphosphate glass to protect against high wear rate. However, full delamination occurs and no signs of polyphosphate glass were observed on the a-C:H coating, thereby resulted in high wear rate.

On the other hand, some researchers focused on the effects of sliding speed and applied load for specific applications. Yamaguchi et al. (2011) studied the effects of sliding velocity (0.1–2.0 m/s) on the friction performance of Si-containing diamond-like carbon (DLC-Si)/automatic transmission fluid (ATF) solid-liquid lubricating system at varying contact pressures, P_{max} from 0.42 GPa to 3.61 GPa for electromagnetic clutch applications. The relationship between friction coefficient (μ) and sliding velocity (ν) were studied to identify the stick–slip motion of the tribopairs, where a positive μ-ν slope (values near to 0 m/s) means no stick–slip motion. Their findings revealed that DLC-Si film can suppress stick–slip motion compared to that of nitride steel, and the reason is due to adsorption film of the succinimide on the sliding contacts. Moreover, DLC-Si film had fulfilled another two required characteristics for the electromagnetic clutch, including high wear resistance and low aggression. However, hydrodynamic effect of DLC-Si film degraded due to high initial surface roughness and poor wettability with the lubricant. Wettability can be determined by measuring the contact angle where excellence wettability between solid and liquid lubricants is essential for a lubricant so that it will be able to maintain contact on the film surface (Quan et al. 2016).

Kim and Kim (2014) found that the friction reduction and anti-wear performances of DLC-steel tribopairs lubricated with SAE 5 W-20 containing glycerol-mono-oleate (GMO) additive were improved with increasing sliding velocity. They evaluate the tribological performance under boundary lubrication to represent machine components applications, including valve train tappets, valve caps and piston rings. Their studies postulated a decrease in surface asperities contact between two sliding contacts was responsible for the reduction for both friction and wear, which confirmed by the measurement of electrical contact between the tribopairs. However, different phenomenon was reported by Liu et al. (2017a) where the friction coefficient of H-DLC/Ti-6Al-4 V tribopairs lubricated with SAE 15W40 was first reduce to a minimum value and then it start to increase again after the reciprocating frequency increases from 2.54 to 5 Hz. Three mechanisms were reported in the reduction of friction coefficient as the frequency increased: (1) increment in oil film thickness, (2) transformation from sp^3 to sp^2 carbon site on DLC coating and (3) the formation of transfer layer. While increases in friction coefficient from the increment of frequency 2.54–5 Hz was due to the increment of surface roughness. Besides that, the wear depth shows similar trend with friction coefficient results. At first, the wear depth reduced due to graphitization of DLC coating and the formation of transfer layer on the counter surface. Then, wear depth increased due to the increased in asperity contact. They also reported the effects of load (30 N and 60 N) on friction and wear, where friction coefficient and wear depth increases with normal load. Under large normal load, the durability of H-DLC films was reduced and the wear debris were carried away by the lubricant, thereby increase friction and wear.

Different phenomenon was found by Ren et al. (2015), where the friction coefficient of a-C and a-C/WC coatings lubricated with PAO and PAO + ZDDP decreases with increasing applied loads from 20 to 140 N, and vice versa for wear rate results. Since a-C coatings had relative high hardness than that of a-C/WC coatings, it results in better wear resistance when lubricated with PAO. Surprisingly, the wear rate of a-C/WC coatings was lower than that of a-C coatings when lubricated with PAO + ZDDP due to the formation of ZDDP-derived tribofilms. The improvement of wear resistance also was attributed to the formation of WS_2/WO_x tribofilms derived from the tribochemical reaction between a-C/WC coatings and ZDDP additive, and no tribochemical product was found on a-C coatings. This shows that doped DLC coatings can increase the reactivity with lubricant additive. However, excessive applied load (140 N) weakened the ZDDP tribofilms and triggered the tribochemical reactions between a-C/WC coatings and ZDDP to form hard particles (WO_3), which accelerates the abrasive wear. Their findings revealed that multiple factors, including additive, doping and operating parameter will affect the mechanism of interaction between solid and liquid lubricant of a solid-liquid lubricating system.

Abdullah Tasdemir et al. (2014) also found the tribological performance of solid-liquid lubricating system was depends on multiple factors, including concentration of lubricant additive, test temperature and counter materials. They reported that ta-C DLC coatings had poor durability causing wear acceleration at higher temperature (80 and 110 °C), but the additions of GMO and ZDDP additive in PAO base oil cancelled this effect. The passivation of dangling bonds of carbon atoms on the ta-C coatings through GMO have resulted in ultralow friction, depending on the additive concentration and temperature. The formation of ZDDP tribofilm on the ta-C films resulted in excellent wear reduction performance, but it negatively affect friction. Moreover, ta-C/steel tribopairs lubricated with PAO exhibited limited lifetime and have higher wear rate than that of ta-C/germanium tribopairs under high temperature since germanium have low solubility of carbon. The wear mechanism of ta-C/steel tribopairs is polishing mechanical wear associated with tribochemical wear, whereas ta-C/germanium tribopairs is polishing wear.

Mannan et al. (2018) used DLC and steel as counter materials to investigate the effects of counter materials on the tribological properties of ta-C DLC coating lubricated with and without ZDDP additive in canola oil. Their findings is in agreement with Abdullah Tasdemir et al. (2014) where the addition of ZDDP in base oil have improved wear performance of both DLC/DLC and steel/DLC tribopairs, but it resulted in an increment of friction coefficient. Under additivated oil, passivation in DLC/DLC tribopairs and prevention of thermo-chemical interaction in steel/DLC tribopair was responsible in the wear reduction compared to that of pure lubricant. While the increased in friction coefficient were attributed to the increment of surface roughness and increased in shear strength. It appears that tribological performance of solid-liquid lubricating system significantly depends on the extrinsic conditions (Table 3).

Table 3 Extrinsic conditions effects on tribological performance test of solid-liquid lubricant

Test condition	Analysis	References
Effect of environment conditions (atomic oxygen (AO) and ultraviolet irradiations (UV))	Three different types of liquid lubricants were used in their studies such as poly(tetrafluoroethylene oxide-co-difluoromethylene (Zdol), ionic liquid (IL) and MACs. They found that AO irradiation was more pronounced in changing the structure of DLC films including oxidation, bond breaking and crosslinking reactions and liquid lubricants, compared to UV irradiation. After irradiation, the coefficient of friction of DLC-based solid-liquid lubricating coatings was lower than before irradiation (except for AO irradiation), but the rate of wear was higher due to irradiation destruction	Liu et al. (2013)
The effects of temperature on the tribological performance of DLC/IL solid-liquid lubricating system in high vacuum condition	The tribological performances decreased as the temperature decreased. This is due to the degradation of wetting, spreading and self-repairing capacity of IL that hindered tribochemistry reaction between DLC films and IL lubricant to occur. It has been reported that *graphitization will results in wear acceleration*	Wang and Liu (2013), Zahid et al. (2016)
The effect of temperature (50, 100 and 150 °C) on the tribological performance of a-C:H and ta-C DLC coatings lubricated with SAE 40	The results revealed that the wear rate increases with increasing temperature, however friction coefficient decreases as the temperature increases for both tribopairs (ta-C/steel and a-C:H/steel) in the absence of delamination	Al Mahmud et al. (2014)

(continued)

Table 3 (continued)

Test condition	Analysis	References
Effects of sliding velocity (0.1–2.0 m/s) on the friction performance of Si-containing diamond-like carbon (DLC-Si)/automatic transmission fluid (ATF) solid-liquid lubricating system at varying contact pressures	Their findings revealed that DLC-Si film can suppress stick–slip motion compared to that of nitride steel, and the reason is due to adsorption film of the succinimide on the sliding contacts. Moreover, DLC-Si film had fulfilled another two required characteristics for the electromagnetic clutch, including high wear resistance and low aggression. However, hydrodynamic effect of DLC-Si film degraded due to high initial surface roughness and poor wettability with the lubricant	Yamaguchi et al. (2011)
Effect of applied load using coatings lubricated with PAO and PAO + ZDDP	The result shows the friction coefficient of a-C and a-C/WC coatings lubricated with PAO and PAO + ZDDP decreases with increasing applied loads from 20 to 140 N, and vice versa for wear rate results. Surprisingly, the wear rate of a-C/WC coatings was lower than that of a-C coatings when lubricated with PAO + ZDDP due to the formation of ZDDP-derived tribofilms. The improvement of wear resistance also was attributed to the formation of WS2/WOx tribofilms derived from the tribochemical reaction between a-C/WC coatings and ZDDP additive, and no tribochemical product was found on a-C coatings	Ren et al. (2015), Abdullah Tasdemir et al. (2014)

4 Conclusion

Based on the reported literature above, the main findings that can be summarized are as follows:

- Without additives, the lubricant may shape a low shear resistance film and prevent the formation of DLC coating transfer film
- Doping of DLC coating will affect the synergism between solid and liquid lubricant, where it can give negative effect on wear
- The addition of traditional additive such as ZDDP, MoDTC and GMO in base lubricant can act either protagonist or antagonist effect on the tribological performance of solid-liquid lubricating system, but the combination of additives can counter the antagonist effect
- The improved tribological performance of solid-liquid lubricating system were attributed to tribochemical reaction between solid and liquid lubricants and the formation of additive-derived tribofilms
- Different intrinsic condition of solid and liquid lubricants (such as doping concentration, type of DLC, types of additive, additives concentration and additive structure) will affect the tribochemical interaction between solid and liquid lubricants thereby resulting in different tribological performance of solid-liquid lubricating system
- Friction reduction performance was improved with increasing temperature, but anti-wear performance was reduced due to graphitization. The degree of graphitization of DLC coating increased with temperature, resulted in wear acceleration
- Under low speed and load condition, friction and wear decrease with increasing speed and load due to increment in oil film thickness, transformation from sp^3 to sp^2 carbon site on DLC coating, formation of transfer layer and decrease in surface asperities contact
- Under high speed and load condition, friction and wear increase with increasing speed and load due to high initial surface roughness and poor wettability with the lubricants
- Proper selection of counter materials can improved the tribological performance of solid-liquid lubricating system
- Tribological performance of solid-liquid lubricating system depends on the intrinsic conditions of solid and liquid lubricants and on the operating parameters.

The major concern in designing solid-liquid lubricating system is synergistic interaction between solid and liquid lubricants. Further efforts are needed to study the interrelationship between the intrinsic condition of solid and liquid lubricants and operating parameters on the interaction mechanism between solid and liquid lubricants as well as tribological performance of solid-liquid lubricating system. In depth study on the underlying mechanism of solid-liquid lubricating system need to be accomplished by researchers in order to design an effective solid-liquid lubricating system for practical applications.

References

Abdullah Tasdemir H, Wakayama M, Tokoroyama T et al (2014) The effect of oil temperature and additive concentration on the wear of non-hydrogenated DLC coating. Tribol Int 77:65–71

Aisenberg S, Chabot R (1971) Ion-beam deposition of thin films of diamondlike carbon. J Appl Phys 42(7):2953–2958

Al Mahmud KAH, Varman M, Kalam MA et al (2014) Tribological characteristics of amorphous hydrogenated (a-C : H) and tetrahedral (ta-C) diamond-like carbon coating at different test temperatures in the presence of commercial lubricating oil. Surface Coat Technol 245:133–147

Bart JCJ, Gucciardi E, Cavallaro S (2013) Environmental life-cycle assessment (LCA) of lubricants. Biolubricants: Sci Technol 527–564

Booser ER (1984) CRC handbook of lubrication. Theory and practice of tribology: Volume II: Theory and design

Braza JF, Sudarshan TS (1992) Tribological behaviour of diamond and diamondlike carbon films: status and prospects. Mater Sci Technol 8(7):574–581

Buckley DH (1981) Surface effects in adhesion, friction, wear, and lubrication, vol 5. Elsevier

Cardinal MF, Castro PA, Baxi J, Liang H, Williams FJ (2009) Characterization and frictional behavior of nanostructured Ni–W–MoS_2 composite coatings. Surf Coat Technol 204(1–2):85–90

Deacon RF, Goodman JF (1958) Lubrication by lamellar solids. Proc R Soc Lond A Math Phys Sci 243(1235):464–482

Donnet C, Erdemir A (2004) Solid lubricant coatings: recent developments and future trends. Tribol Lett 17(3):389–397

Erdemir AGRARDMTRT, Fenske GR, Krauss AR, Gruen DM, McCauley T, Csencsits RT (1999) Tribological properties of nanocrystalline diamond films. Surf Coat Technol 120:565–572

Fan X, Xue Q, Wang L (2015) Carbon-based solid-liquid lubricating coatings for space applications—a review. Friction 3:191–207

Ferrari AC, Robertson J, Beghi MG, Bottani CE, Ferulano R, Pastorelli R (1999) Elastic constants of tetrahedral amorphous carbon films by surface Brillouin scattering. Appl Phys Lett 75(13):1893–1895

Reick FG (2017) https://www.fluoramics.com/wp-content/uploads/EPQ217_52-54.pdf

Fu ZQ, Wang C biao, Zhang W et al (2013) Influence of W content on tribological performance of W-doped diamond-like carbon coatings under dry friction and polyalpha olefin lubrication conditions. Mater Des 51:775–779

Golshokouh I, Golshokouh M, Ani FN, Kianpour E, Syahrullail S (2013) Investigation of physical properties for jatropha oil in different temperature as lubricant oil. Life Sci J 10(8s)

Gorbatchev O, De Barros Bouchet MI, Martin JM et al (2016) Friction reduction efficiency of organic Mo-containing FM additives associated to ZDDP for steel and carbon-based contacts. Tribol Int 99:278–288

Grossiord CKJMTCK, Varlot K, Martin JM, Le Mogne T, Esnouf C, Inoue K (1998) MoS_2 single sheet lubrication by molybdenum dithiocarbamate. Tribol Int 31(12):737–743

Guo Y, Guo P, Sun L et al (2018) Tribological properties of Ti -doped diamond-like carbon coatings under dry friction and PAO oil lubrication, pp 1–10

Hasannuddin AK, Yahya WJ, Sarah S, Ithnin AM, Syahrullail S, Sidik NAC, Abu Kassim KA, Ahmad Y, Hirofumi N, Ahmad MA, Sugeng DA, Zuber MA, Ramlan NA (2018) Nano-additives incorporated water in diesel emulsion fuel: fuel properties, performance and emission characteristics assessment. Energy Convers Manage 169:291–314

Hasannuddin AK, Yahya WJ, Sarah S, Ithnin AM, Syahrullail S, Sugeng DA, Razak IFA, Abd Fatah AY, Aqma WS, Rahman AHA, Ramlan NA (2018) Performance, emissions and carbon deposit characteristics of diesel engine operating on emulsion fuel. Energy 142:496–506

Hassan M, Ani FN, Syahrullail S (2016) Tribological performance of refined, bleached and deodorised palm olein blends bio-lubricants. J Oil Palm Res 28(4):510–519

Holmberg K, Matthews A, Ronkainen H (1998) Coatings tribology—contact mechanisms and surface design. Tribol Int 31(1–3):107–120
Jabal MH, Ani FN, Syahrullail S (2014) The tribological characteristic of the blends of Rbd palm olein with mineral oil using four-ball tribotester. Jurnal Teknologi 69(6)
Jacob W, Möller W (1993) On the structure of thin hydrocarbon films. Appl Phys Lett 63(13):1771–1773
Kim DW, Kim KW (2014) Effects of sliding velocity and ambient temperature on the friction and wear of a boundary-lubricated, multi-layered DLC coating. Wear 315:95–102
Kogov J, Kalin M (2019) Lubrication performance of graphene-containing oil on steel and DLC-coated surfaces. Tribol Int 138:59–67
Kosarieh S, Morina A, Laine E, et al (2013) The effect of MoDTC-type friction modifier on the wear performance of a hydrogenated DLC coating. Wear 302:890–898
Lansdown AR (1999) Molybdenum disulphide lubrication, vol 35. Elsevier
Liu J, Zhang Z, Ji Z, Xie Y (2017) Friction and wear behavior of hydrogenated diamond-like carbon coating against titanium alloys under large normal load and variable velocity. Ind Lubr Tribol 69:199–207
Liu W, Li W, Li R, et al (2019) Green oil additive g-C 3 N 4: a feasible strategy to enhance the tribological properties of DLC film Green oil additive g-C 3 N 4: a feasible strategy to enhance the tribological properties of DLC film
Liu X, Wang L, Pu J, Xue Q (2012) Surface composition variation and high-vacuum performance of DLC/ILs solid—liquid lubricating coatings: Influence of space irradiation. Appl Surf Sci 258:8289–8297
Liu X, Wang L, Xue Q (2011a) A novel carbon-based solid—liquid duplex lubricating coating with super-high tribological performance for space applications. Surf Coat Technol 205:2738–2746
Liu X, Wang L, Xue Q (2011b) DLC-based solid—liquid synergetic lubricating coatings for improving tribological behavior of boundary lubricated surfaces under high vacuum condition. Wear 271:889–898
Liu X, Wang L, Xue Q (2013) High vacuum tribological performance of DLC-based solid-liquid lubricating coatings: Influence of atomic oxygen and ultraviolet irradiation. Tribiology Int 60:36–44
Liu Y, Xin L, Zhang Y et al (2017) The effect of Ni nanoparticles on the lubrication of a DLC-Based solid-liquid synergetic system in all lubrication regimes. Tribol Lett 65:1–9
Lv M, Yang L, Wang Q et al (2015) Tribological performance and lubrication mechanism of solid-liquid lubricating materials in high-vacuum and irradiation environments. Tribol Lett 59:1–10
Mannan A, Sabri MFM, Kalam A, Hassan MH (2018) Tribological performance of DLC/DLC and steel/DLC contacts in the presence of additivated oil. Int J Surf Sci Eng 12:60–75
Mistry KK, Morina A, Neville A (2011) A tribochemical evaluation of a WC-DLC coating in EP lubrication conditions. Wear 271:1739–1744
Mortimer Jr WP (1991) U.S. Patent No. 4,985,296. U.S. Patent and Trademark Office, Washington, DC
Okubo H, Sasaki S (2017) In situ Raman observation of structural transformation of diamond-like carbon films lubricated with MoDTC solution : mechanism of wear acceleration of DLC films lubricated with MoDTC solution. Tribol Int 113:399–410
Okubo H, Watanabe S, Tadokoro C, Sasaki S (2016) Effects of structure of zinc dialkyldithiophosphates on tribological properties of tetrahedral amorphous carbon film under boundary lubrication. Tribol Int 98:26–40
Polcar T, Cavaleiro A (2011) Review on self-lubricant transition metal dichalcogenide nanocomposite coatings alloyed with carbon. Surf Coat Technol 206(4):686–695
Quan X, Hu M, Gao X et al (2016) Friction and wear performance of dual lubrication systems combining WS_2–MoS_2 composite fi lm and low volatility oils under vacuum condition. Tribiol Int 99:57–66

Razak DM, Syahrullail S, Sapawe N, Azli Y, Nuraliza N (2015) A new approach using palm olein, palm kernel oil, and palm fatty acid distillate as alternative biolubricants: improving tribology in metal-on-metal contact. Tribol Trans 58(3):511–517

Ren S, Zheng S, Pu J, Zhang G (2015) Study of tribological mechanisms of carbon-based coatings in antiwear additive containing lubricants under high temperature. RSC Adv, 66426–66437

Rico EF, Minondo I, Cuervo DG (2007) The effectiveness of PTFE nanoparticle powder as an EP additive to mineral base oils. Wear 262(11–12):1399–1406

Robertson J (1986) Amorphous carbon. Adv Phys 35(4):317–374

Sapawe N, Samion S, Zulhanafi P, Nor Azwadi CS, Hanafi MF (2016) Effect of addition of tertiary-butyl hydroquinone into palm oil to reduce wear and friction using four-ball tribotester. Tribol Trans 59(5):883–888

Scharf TW, Prasad SV (2013) Solid lubricants: a review. J Mater Sci 48(2):511–531

Scharf TW, Diercks DR, Gorman BP, Prasad SV, Dugger MT (2009) Atomic layer deposition of tungsten disulphide solid lubricant nanocomposite coatings on rolling element bearings. Tribol Trans 52(3):284–292

Schmellenmeier H (1953) Die Beeinflussung von festen Oberflachen durch eine ionisierte. Exp Tech Phys 1:49–68

Sliney HE (1982) Solid lubricant materials for high temperatures—a review. Tribol Int 15(5):303–315

Spear KE, Dismukes JP (eds) (1994) Synthetic diamond: emerging CVD science and technology, vol 25. Wiley

Sutor P (1991) Solid lubricants: overview and recent developments. MRS Bull 16(5):24–30

Syahrullail S, Wira JY, Wan Nik WB, Fawwaz WN (2013) Friction characteristics of RBD palm olein using four-ball tribotester. In: Applied mechanics and MATERIALS, vol 315, pp 936–940. Trans Tech Publications Ltd.

Tsai HC, Bogy DB (1987) Characterization of diamondlike carbon films and their application as overcoats on thin-film media for magnetic recording. J Vacuum Sci Technol A Vacuum Surfaces Films 5(6):3287–3312

Vengudusamy B, Green JH, Lamb GD, Spikes HA (2012) Behaviour of MoDTC in DLC/DLC and DLC/steel contacts. Tribol Int 54:68–76

Vengudusamy B, Green JH, Lamb GD, Spikes HA (2013) Influence of hydrogen and tungsten concentration on the tribological properties of DLC/DLC contacts with ZDDP. Wear 298–299:109–119

Waghray H, Lee TS, Tatarchuk BJ (1995) A study of the tribological and electrical properties of sputtered and burnished transition metal dichalcogenide films. Surf Coat Technol 76:415–420

Wang L, Liu X (2013) Tribological behavior of DLC/IL solid-liquid lubricating coatings in a high-vacuum condition with alternating high and low temperatures. Wear 304:13–19

Yamaguchi T, Ando J, Tsuda T et al (2011) Sliding velocity dependency of the friction coefficient of Si-containing diamond-like carbon film under oil lubricated condition. Tribol Int 44:1296–1303

Yang L, Neville A, Brown A et al (2014) Friction reduction mechanisms in boundary lubricated W-doped DLC coatings. Tribol Int 70:26–33

Yoshida Y, Kunitsugu S (2018) Friction wear characteristics of diamond-like carbon coatings in oils containing molybdenum dialkyldithiocarbamate additive. Wear 414–415:118–125

Yue W, Liu C, Fu Z et al (2013) Synergistic effects between sulfurized W-DLC coating and MoDTC lubricating additive for improvement of tribological performance. Tribiology Int 62:117–123

Zahid R, Masjuki HH, Varman M et al (2016) Influence of intrinsic and extrinsic conditions on the tribological characteristics of diamond-like carbon coatings: a review

Zeng Q, Yu F, Dong G (2013) Superlubricity behaviors of Si_3N_4/DLC Films under PAO oil with nano boron nitride additive lubrication. Surface Interface Anal, 1283–1290

Zhang Y, Zhang S, Sun D et al (2019) Wide adaptability of Cu nano-additives to the hardness and composition of DLC coatings in DLC/PAO solid-liquid composite lubricating system. Tribiology Int 138:184–195

Zhiqiang F, Jian S, Chengbiao W et al (2013) Tribological performance of DLC coatings deposited by ion beam deposition under dry friction and oil lubricated conditions. Vaccum 94:14–18

Tribology of Fiber Reinforced Polymer Composites: Effect of Fiber Length, Fiber Orientation, and Fiber Size

P. S. Sarath, Rakesh Reghunath, Józef T. Haponiuk, Sabu Thomas, and Soney C. George

Abstract This chapter presents a brief account of the current state-of-the-art in the area of the tribology of fiber reinforced polymer composites. The important factors which determine the friction and wear properties of fibers from the surface modification are mentioned here. Tribological trends for fiber reinforced polymer composites, both traditional and nanocomposites, are presented using data currently available in the literature. Variation in fiber length, fiber orientation, type of treatments and physical characteristics are significantly influence the tribological properties. Finally, based on our current understanding of this field, we have speculated upon some future trends and directions in the area of polymer tribology.

Keywords Tribology · Polymer composite · Length · Orientation · Size

P. S. Sarath · S. C. George (✉)
Centre for Nanoscience and Technology, Amal Jyothi College of Engineering, Kanjirappally, Kottayam, Kerala, India
e-mail: soneygeo@gmail.com

P. S. Sarath
e-mail: sarathps005@gmail.com

R. Reghunath
Department of Mechanical Engineering, N.S.S. Engineering College, Palakkad, Kerala, India
e-mail: rreghunath11@gmail.com

J. T. Haponiuk
Department of Polymer Technology, Chemical Faculty, University of Technology, G. Narutowicza Str. 11/12. Gdansk, 80-233 Gdansk, Poland
e-mail: jhp@urethan.chem.pg.gda.pl

S. Thomas
Center for Energy Materials, Mahatma Gandhi University, Kottayam, Kerala, India
e-mail: sabuthomas@mgu.ac.in

M. T. Hameed Sultan et al. (eds.), *Tribological Applications of Composite Materials*, Composites Science and Technology, https://doi.org/10.1007/978-981-15-9635-3_4

1 Introduction

Today composite materials are considered as the most promising candidates for replacing conventional metals in aerospace industries because of its high strength to weight ratio and low density (Friedrich 1986). Most of the machine parts are exposed to tribological loadings such as adhesive, abrasive etc. in their service. Therefore, tribological studies of materials have an important role in design mechanical parts. Studies confirm that the friction and wear behavior of polymeric materials can be improved by a lower adhesion and a higher stiffness and strength (Czichos and Habig 1992; Friedrich 1997; Reinicke et al. 1998). Application window of these polymer composites can be widened by using different kinds of multifunctional fiber reinforcements. Fibers in these materials are primary load carrying members and provide strength and rigidity while the polymer matrices maintain the alignment of fibers. By reinforcing with these fibers, dramatic improvement in mechanical properties can be achieved, along with more dimensional stability. It is also possible to tailor made the properties according to the end user requirements like enhanced wear resistance, corrosion resistance, moisture absorption etc. by using a variety of fibers in different polymeric matrices (Schwartz and Bahadur 2000; Zhang et al. 2006; Haupert et al. 2004; Werner et al. 2004). During initial stages of research and development in the field of polymer composites, studies were more concentrated on synthetic fibers like Glass, Carbon and Kevlar composites. These composites have high strength, stiffness along with excellent fatigue resistance. These enhanced properties make it a suitable candidate for many aerospace applications. But poor recycling and non-biodegradable properties limits the usage of synthetic fibers. Since the early 1990s researchers are trying to replace synthetic fibers with natural fibers for developing polymeric composites due to increasing demand for eco- friendly materials for sustainability. Beyond the concern of ecological consideration, the properties of natural fibers are quite good, relatively low density which makes its suitable for light weight application. These fibers also offer significant cost advantages over synthetic fibers. Coir, Jute, hemp, sisal and abaca fibers are commonly used natural fibers. But still the properties of natural fibers are inferior to synthetic fiber composites which make it not suitable for many specific applications. For the last few decades amble research work are progressing for improving the mechanical properties of the natural fiber polymeric systems. The sliding wear behavior of polymer composite against a steel counter surface is essential to categorize according to their appearance, performance and other characteristics. Studies have been emphasized that the tribology behavior of natural fiber in composite is not an intrinsic behaviour and it largely depend on many other parameters such as operating parameters, characteristics of polymer material, physical and interfacial adhesion properties of fibers, additives and contact conditions.

This chapter is mainly focusing to give a limelight in the field of different kinds of fiber reinforcement generally used in polymeric systems for enhanced tribological properties. The basic mechanical and tribological properties of fiber reinforced composites, role of fiber length, fiber size and fiber orientation on optimizing the tribological properties will be discussed.

1.1 Overview of Polymer Composites for Tribological Application

Today in many industries rolling and sliding components such as bearings, rollers, seals gears etc. are manufactured by polymers and its composites. In these applications sustain friction and wear loading in services. When a polymer material comes in contact with any counter surfaces, there are chances of wear and friction. By suitable selection of polymer matrix and fibers, wear resistance can be improved to a greater extent. It was well proved that the friction and wear rate between polymer surfaces depends on roughness of rubbing surfaces, relative motion, temperature, vibration and relative humidity. The parameters that affect the tribological performance of polymer and its composites also include polymer molecular structure, processing and treatment, properties, viscoelastic behavior, and surface texture. There have been also a number of investigations exploring the influence of test conditions, contact geometry and environment on the friction and wear behavior of polymers and composites (Mathew 2007).

Polymer tribology is based on the analysis of abrasion, adhesion, and fatigue of polymer materials in a friction contact. The coefficient of friction (COF) (μ) is largely depend on the mechanical load carrying capacity and the wear rate (Ws) that determine their acceptability in industrial applications. The wear of material is not a simple material property, it largely depends on the two surfaces which comes in contact. The structural features of polymers provide a variety of tribological applications of basic polymers mostly as matrices and fillers of composite materials. Friction is greatly influenced by the class of polymers viz. elastomers, thermosets and thermoplastics (semi-crystalline and amorphous). Semi-crystalline linear thermoplastic would give lowest coefficient of friction whereas elastomers and rubbers show large values. This is because of the molecular architecture of the linear polymers that helps molecules stretch easily in the direction of shear giving least frictional resistance. Table 1 provides some typical values of the coefficient of friction for pristine or virgin polymers (Fig. 1).

1.2 Fiber Reinforcements—Synthetic and Natural Fibers

Scientists and engineers have been actively exploring to find the materials that will be used as replacement of conventional materials and this leads to the development of features of new design units and innovations. It is very important to select proper fibers based on the application of composite materials because of laminate density,

Table 1 Friction coefficient of few polymers when slide against a steel disk counter face (surface roughness, Ra = 1.34 μm). Corresponding specific wear rates and the pressure (P) × velocity (V) values are also presented

Polymer	Coefficient of friction	Specific wear rate ($\times 10^{-6}$ mm^3/Nm)
PMMA	0.48	1315.90
PEEK	0.32	31.72
UHMWPE	0.19	15.54
POM	0.32	168.24
Epoxy	0.45	3506.65

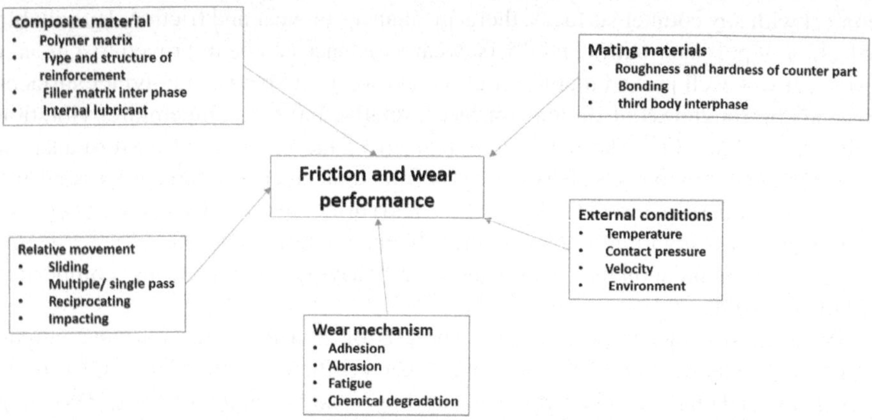

Fig. 1 Areas of influence on the tribological performance of composite materials. Ref -Friedrich et al. (1993)

tensile and compressive strength, conductivity and fatigue strength which depends on the properties of fibers. Fiber reinforced composite material consist of major volume fraction and takes up the major portion load acting on the composite structure (Fig. 2).

Fibers can generally be categorized into three types: synthetic fibers, natural fibers and mineral fibers. Synthetic fibers are made from raw materials such as petroleum, based on chemicals or petrochemicals. These materials are polymerized into a long, linear chemical with different chemical compounds and are used to produce various types of fibers. There are several methods of manufacturing synthetic fibers, but the most common is the melt-spinning process. It involves heating the fiber until it begins to melt, then fiber must be drawn out of the melt with tweezers as quickly as possible. The next step would be to align the molecules in a parallel arrangement. This brings the fibers closer together, and allows them to crystallize and orient. Synthetic fibers are more durable than most natural fibers, and will readily pick up different dyes. In addition, many synthetic fibers offer consumer-friendly functions, such as stretching, waterproofing, and stain resistance. Glass fibers are the most commonly used synthetic fibers for reinforcing polymer matrices because of its low cost, high

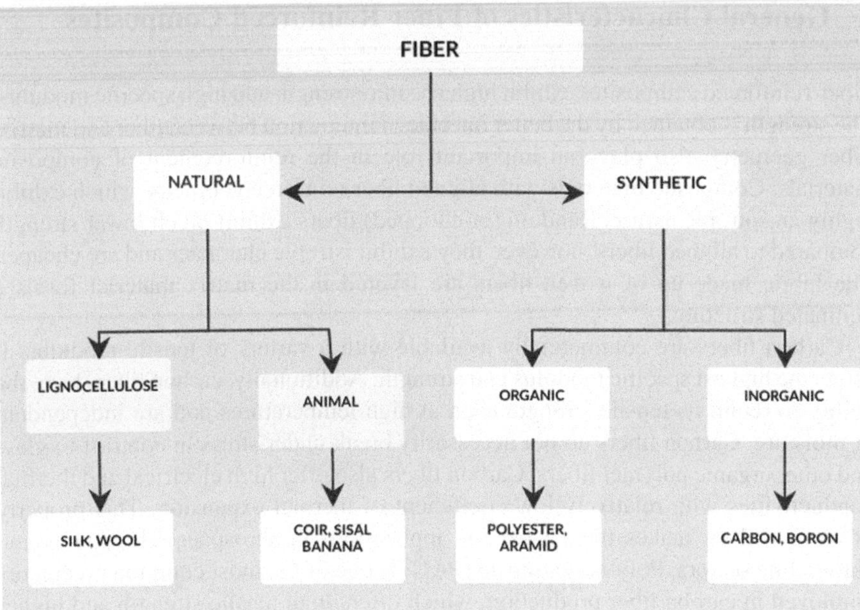

Fig. 2 General classification of fibers

tensile strength and chemical resistance. Two types of glass fibers are quite popular E-glass and S Glass fibers. Glass fibers (GFs) have been employed in various forms such as longitudinal, woven mat, chopped fiber (distinct) and chopped mats to enhance the mechanical and tribological properties of the fiber reinforced composites. Figure 3 shows that properties of such composites were however dependent on the nature and orientation of the fibers laid during composite preparation (Alam et al. 2010). The mechanical behavior of a fiber-reinforced composite basically depends on the fiber strength and modulus, the chemical stability, matrix strength and the interface bonding between the fiber/matrix to enable stress transfer.

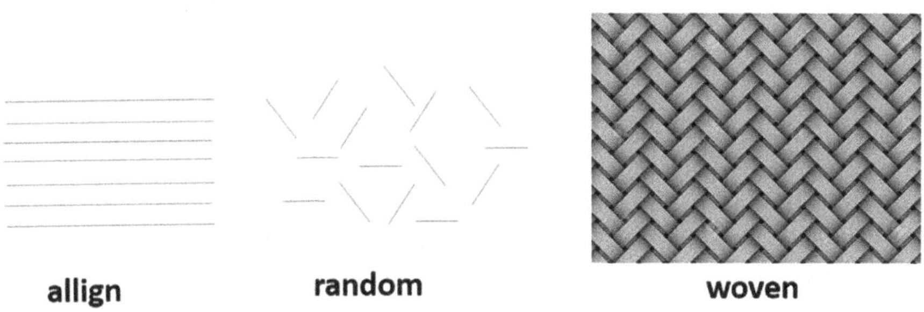

Fig. 3 Different types of fiber alignment

2 General Characteristics of Fiber Reinforced Composites

Fiber-reinforced composites exhibit high specific strength and high specific modulus. The strength is obtained by the better interfacial interaction between fiber and matrix. Fiber geometry also plays an important role in the reinforcement of composite materials. Composite materials with aligned fiber reinforce geometry which exhibit highly anisotropic nature. Random (or chopped) fibers exhibit much lower strength compared to aligned fibers, however, they exhibit istropic character and are cheaper. The fabric made up of woven fibers are layered in the matrix material forms a laminated structure.

Carbon fibers are commercially available with a variety of tensile modulus. It offers the highest specific modulus and strength. Additionally, carbon fibers have the ability to retain its tensile strength even at high temperatures and are independent of moisture. Carbon fibers do not necessarily break under stress in contrast to glass and other organic polymer fibers. Carbon fibers also offer high electrical and thermal conductivities with relatively low coefficient of thermal expansion. This property of carbon fibers makes them ideal for applications in aerospace, electronics and automobile sectors. Poly-acrylonitrile (PAN) is one of the most common precursors employed in carbon fiber production, which offers high tensile strength and higher elastic modulus, extensively applied for structural material composites in aerospace and sporting/recreational goods. Depending on the final curing temperature, different classes of carbon fibers namely high tenacity (HT) fibers, intermediate modulus (IM) fibers, high modulus (HM) fibers and ultra-high modulus (UHM) fibers are formed with PAN precursors (Prashanth et al. 2017).

2.1 Natural Fibers

During the last few years, research has been conducted to replace the conventional synthetic fibers with natural fibers (Mahir et al. 2019). For instance, fibers of sisal, jute, coir, oil palm, bamboo, wheat and banana have been found to be an effective reinforcement in the polymer matrices. The advantages of natural fibers over traditional reinforcing materials such as glass and carbon fiber are their strength, toughness, corrosion resistance, thermal properties, wear resistance etc. Natural fibers reinforcement, have attracted the attention of researchers because they are: (a) environmentally friendly, (b) fully biodegradable, (c) abundantly available, (d) renewable, (e) inexpensive, and (f) lightweight. Fibers in general play a crucial role in deciding the end property of the fiber reinforced composite systems. Thus, appropriate selection of fiber material and their relative orientation can lead to composite to composites with tailor made properties to suit specific application requirements. But the growth in environmental consciousness, community interest, the new environmental regulations and unsustainable consumption of man-made materials, led to thinking of the use of environmentally friendly materials. Owing that view natural

Table 2 Properties of a few natural fibers

Fiber	Density	Young's Modulus (GPa)	Tensile strength (MPa)	Elongation at break (%)
Coconut	1.15	4–6	131–175	15–40
Cotton	1.5–1.6	5.5–12.6	287–587	7–8
Bamboo	0.6–1.1	11–17	140–230	–
Jute	1.44	10–30	393–773	1.5–1.8
Hemp	1.47	17–70	368–800	1.6

fiber is considered as one of the best environmentally friendly materials which have good properties compared to synthetic fiber (Chandramohan and Marimuthu 2011). Table 2 shows the properties of natural fibers. Production of natural fibers causes less severe environmental impacts as compared to that of synthetic fibers. The applicability of tribological testing setup in various natural fibers reinforced polymer composites and its tribological applications have been summarized and illustrated in Table 3.

2.2 *Tribology of Fiber Reinforced Composites*

Tribology is the science that deals with design, friction, wear and lubrication of interacting surfaces in relative motion. Composites have diverse range of mechanical and tribological properties that can be obtained using different types of reinforcements in different orientations with different volume fractions. Today conventional metals have been replaced by composite materials in most of structural applications in aerospace and automotive industries. Comparing with conventional metals composites have high strength to weight ratio and low density and even it can with stand high temperature, high load, high fatigue resistance, high corrosion resistance and less noisy operating condition. Frictional properties of polymer composites are different from that of metals. Like metals polymers deform with higher loads. Adhesion and deformation are characteristic of the friction on the matting surfaces. As a result, the coefficient of friction decreases as the load increases. Among the different classes of composite materials such as polymer matrix composites (PMCs), metal matrix composites, ceramic matrix composites and carbon–carbon composites, PMCs are quite popular in tribological applications because of its self-lubrication capacity.

Today natural fiber composites are gaining a lot of attraction comparing with synthetic fiber mainly due to ease of accessibility, renewability, lower weight, less price low density and biodegradability (El-Tayeb 2008). Reinforcement is a process by which tribological properties of fibers or polymers are altered (positively or negatively) (El-Tayeb et al. 2006). Chin and Yousif (2009) used kenaf fibers reinforced with epoxy composite for a kind of bearing application in which they reported 85 percent increase in wear efficiency and standard composite orientation. Friction and wear rate of the composites depend on the materials selected for reinforcement

Table 3 Role of various natural fiber reinforced polymer composites in tribological applications

Fiber	Matrix	Conditions	Test conducted	Manufacturing Method	Tribological application	References
Rice straw dust/Rise husk	Phenolic	Untreated	Wear test	Hot pressing	Brake pad	El-Sayed et al. (1995)
Sisal	Phenolic	Silane coupling	Adhesive friction and wear	Hot compression	Brake pad	Mutlu (2009)
Betelnut	Polyester	Untreated	Constant speedtester	Hand lay-up		El-Tayeb (2008)
Sugarcane/Glass	Polyester	Untreated	Friction assessment	Hand lay-up	Bearing	Srivastava et al. (2015)
Sea shell nano Powder	Poly - Methyl methacrylate	Untreated	Wear	Micro-hardness Mold	Dental	Xin et al. (2007)
Banana and Kenaf	Polyester	NaOH	Mechanical	Hand lay-up	Clutch	Franklin and de Kraker (2003)
Grewiaoptiva fibers	PLA	Untreated	Wear test	Hot compression		Sabeel Ahmed et al. (2012)

and resin, manufacturing process, operating parameters, fiber volume fraction, fiber orientation, fiber length, and surface treatments. No material is perfect for all types of wear modes. Tribo properties of composites can be predicted only by evaluating them in the laboratory under the different operating conditions. Tribological and mechanical characteristics of fiber reinforced composite material not only depend on the properties of fiber but it also depends on the interfacial interactions of fiber and counter surface. In the case of fiber reinforced composite material there is some critical fiber length is necessary for effective load transfer in addition to the strengthening and stiffening of the composites. Critical fiber length dependence on fiber strength and diameter, and fiber-matrix bond strength/matrix shear yield strength of the composite material.

Bijwe et al. studied the friction and wear analyses under varying fiber percentage, of polyetherimide glass fiber composites were conducted (Bijwe et al. 2001). The authors have revealed that rate of wear resistance of composites is different for different types of wear modes and fiber percentage.

In the case of phenolic compounds it is proved that irrespective of graphite filler size and loading condition, with increasing temperature, friction and wear rate were increased (Kolluri et al. 2018). It was also well proved that nano particles like clay and silicon carbide when mixed with polymer matrices improves the tribological performance of composites (Nguong et al. 2013). Rubber dust was identified as suitable filler for improving anti wear performance. Study was conducted by varying the volume fraction of rubber and it was found that corresponding to 10% of rubber tribological performance was optimum (Mishra 2012). Basavarajappa and Ellangovan (2012) has done studies on glass fiber reinforced epoxy composites with SiO_2 fillers, all these studies indicates that fiber contributes to a key role in controlling the wear rate of the composites.

Friction is the opposing force which is generated when a surface is slide across another surface. Wear on the other hand is the progressive loss of material on a surface, caused by rubbing by another surface. In tribology, friction and wear depend on factors such as rubbing surface roughness, relative motion, and type of material, temperature, normal force, stick slip, relative humidity, lubrication and vibration (Ravikumar and Murali 2018). Abrasive wear of polymer matrix composite is a serious issue because repeated abrasion between two layers of polymer composites (two-body abrasion) or when loose particles are embedded in between two layers (three-body abrasion) causes loss of material. For effective working of fiber-reinforced polymer composites (FRPC's), abrasion wear is to be minimized (Taylor et al. 2014). It was well proved that incorporation of micro and nano-ceramic fillers into fiber reinforced polymer composites have improved their tribo-performance (Friedrich et al. 2005). Studies were conducted to analyze the effect of various fillers like graphite and SiC and it was found that by adding 5 wt% of these fillers wear resistance was improved. It was found that there exist an optimum wt% of these fillers for optimum performance. Normal load was found to have great influence on wear rate of epoxy/glass/SiC/Gr composites followed by sliding velocity and sliding distance were of least significance. Adding 3 wt% graphite into epoxy glass multilayered laminates has led to enhancement of both mechanical and dry sliding performance

Table 4 Tribological properties of different fiber reinforced polymer composites

Materials	Specific wear rate (mm^3/Nm)	Coefficient of friction	References
Polyester	16–22	0.9–0.95	Yousif (2009)
Chopped glass/polyester	2–3.7	0.23-0.7	Yousif and El-Tayeb (2007)
Coir/polyester	1.4–2	0.57–0.8	Hashmi et al. (2007)
Sisal/polyester	0.84–1.12	0.6–0.65	(Yousif et al. 2010)
Cotton/polyester	1.5–3.5	<1	Chin and Yousif (2009)
Un treated oil palm/polyester	4.2–5.5	0.2–0.65	Nirmal et al. (2012)
Betel nut/polyester	2–2.2	0.22–0.55	Prasad et al. (2014)
Kenaf/epoxy	1–1.9	0.36–0.42	Callister (2007)
Bamboo/epoxy	5.5-7.5	0.57–0.64	Rasheva et al. (2010)

of laminates. By adding beyond 3 wt%, it was interesting to note that agglomeration of these fillers took place which adversely affect the wear performance (Shivamurthy et al. 2013).

From Table 4 it is clear that natural fiber reinforced polymer material obtained tribological properties similar to that of synthetic fibers. The table also provides details of the conditioning of natural fibers and their corresponding consequence on friction and wear characteristics.

2.2.1 Effect of Fiber Length

The mechanical characteristics of fiber reinforced polymer composites not only depends on the properties of fiber but also the degree to which load is transmitted to the fiber by the matrix phase. The extent of this load transmittance is the magnitude of the interfacial bond between the fiber and matrix phases. Under an applied stress, this fiber–matrix bond ceases at the fiber ends, yielding a matrix deformation pattern as shown schematically in Fig. 4.

Some critical fiber length is necessary for effective strengthening and stiffening of the composite material. This critical length is dependent on the fiber diameter d and its ultimate or tensile) strength σ_f and on the fiber–matrix bond strength (or the shear yield strength of the matrix, whichever is smaller) τ_c according to

$$l_c = \frac{\sigma\sigma_f d}{\tau c} \tag{1}$$

- When the fiber length $l < l_c$; the reinforcement is particulate in nature.
- When the fiber length $l >$ lc but $l < 1.5l_c$ the reinforcement is short fiber types
- When the fiber length $l > 15l_c$; the reinforcement is continuous in nature.

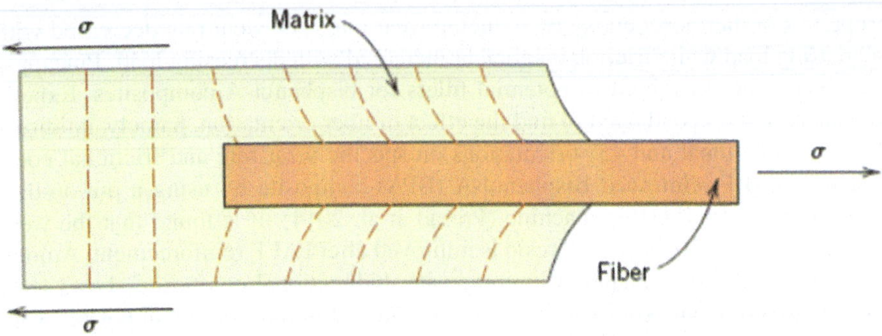

Fig. 4 Deformation pattern in the matrix surrounding a fiber that is subjected to an applied tensile load (Callister et al. 2012)

Compared to continuous fiber composites, short fiber reinforced polymers (SFRP) combine easier process ability with low manufacturing cost. Therefore, in recent years the use of SFRP composites grows rapidly in many engineering applications, in particular in automobile and mechanical engineering industry (Callister 2007). For the last few years numerous amounts of research work has been progressing to find the effect of fiber length in determining the mechanical properties of the composites. All these investigations indicate that the fiber length is a crucial parameter in determining the mechanical performance of SFRP. It is known fact that fiber length plays a main role to create interfacial bonding between fiber and the matrix. In the theoretical sense too, short length reduces the load carrying capacity of fiber and responsible for high wear rate and excessive length results in easy pull out of the fibers and produces more wear rate.

2.2.2 Effect of Fiber Orientation

There are various parameters which influence the performance of composites such as amount of matrix and fibers, their alignment with respect to loading direction, fiber–matrix interface, processing technique etc. Different orientation of the fiber with respect to the sliding direction of the counter face was considered; fiber orientations classified as normal, parallel, anti-parallel and random. From the literature it is very clear that the orientation of fiber is a very important critical parameter in composite design. Fiber reinforced composites generally behave in different way for different modes of loading conditions, for example fibers aligned in parallel direction to load gives excellent tensile properties, but in tribological loading it can lead to fiber pullouts. Therefore, at each loading conditions, the orientation of the fibers should be comprehensively studied about the performance of the composites. Friction and wear behaviour of unidirectional carbon fiber-reinforced epoxy composites containing various unidirectional carbon fibers was investigated on a pin-on-flat plate configuration (Zhang and Friedrich 2007). The carbon fiber improved the tribological properties of thermoset epoxy by reducing wear rate. The wear rate decreased with decreasing load while friction coefficient increased with decreasing load. Pineapple leaf fibers were identified as potential fillers for bisphenol-A composites. Experimental study was conducted to find the effect of fiber orientation, namely, unidirectional, bidirectional and 45° orientations on specific wear rate and frictional coefficient of PALF reinforced Bisphenol-A (BPA) composite by using a pin-on-disc wear and frictional testing machine (Prasad et al. 2014). It is found that the wear resistance of pure Bisphenol-A resin is improved after PALF reinforcement. Among three types of fiber orientation in composite, bidirectional composite shows least specific wear rate and coefficient of friction. Three different set of composites were prepared perpendicular, anisotropic and isotropic orientation were taken for the study. It was found the polyester/kenaf sandwich composite with kenaf fiber in anisotropic orientation design will form a strong bridge over the cracks, thus increasing the breakage resistance of the kenaf fiber. Polyester/kenaf composite with kenaf fiber in anisotropic arrangement achieved the highest tensile, flexural, and impact properties. This was followed by a sandwich composite with kenaf fiber in isotropic and perpendicular orientations (Chin and Yousif 2009). PTFE and graphite powder filled SCF reinforced PEEK composites were prepared and effect of fiber orientation on the tribological properties were studied and a correlation between mechanical and tribological properties with fiber orientation was developed (Rasheva et al. 2010) (Table 5).

Test results shows that the tribological performance of the composite MA (a total of 20 vol.% solid lubricants and 10 vol.% SCFs) is significantly dependent on the fiber orientation with an advantage for the perpendicular fiber orientation. Furthermore, a low solid lubricants content and high SCF-content (MC—a total of 10 vol.% solid lubricants and 15 vol.% SCFs) lead to a tendency for higher wear rates, especially for the perpendicular fiber orientation, since a transfer film is hard to be built. The worn surfaces of this material combination is very rough with a lot of grooves and wear debris. The composite MC presents low wear resistance. The composition MB (a total of 20 vol.% solid lubricants and 15 vol.% SCFs), presents the most stable wear behavior with a stable worn surface, independent from the fiber orientation and the applied load (Table 6).

Table 5 The composition and specific wear rate parallel and perpendicular to the fiber direction

Material code PEEK/PTFE/Graphite/SCF	Apparent pressure	Specific wear rate parallel	Relative error %	Specific wear rate perpendicular	Relative error
MA 70/10/10/10	1	0.554	8.68	0.410	5.85
	4	0.494	6.88	0.367	2.39
MB 65/10/10/15	1	0.413	15.74	0.424	17.68
	4	0.420	5.00	0.468	8.33
MC 75/05/05/15	1	0.501	13.77	0.397	10.30
	4	0.454	11.23	0.525	15.12

Table 6 Effect of sliding parameter on synthetic fiber reinforced Polymer composites

Polymer	Reinforcement	Load (N)	Sliding speed (m/s)	Environment	Observation	References
Epoxy	Glass fiber	40–120	2.51–3.14	Oil	Weight loss, wear rate, and COF increases with increase in load and sliding speed	Sarkar et al. (2017)
UHMWP	Glass fiber		0.2–1	Dry Water	Friction coefficient decreases	Vadivel et al. (2018)
Polyetheri-mide	Glass fiber MoS_2 Graphite PTFE	70–100			Addition of filler material improve performance of composite	Bijwe et al. (2001)
Polypropylene	MWCNTs (0 wt% to 7 wt%)	10–50	1–5	Dry	Increase in weight % of CNTs reduces weight loss and friction coefficient	Gandhi et al. (2013)

Friedrich et al. studied tribological anisotropy of different fiber orientations in continuous carbon fibers (CCF) reinforced polymers. Study shows that tribological sliding in direction parallel to the fibers axis leads to a higher wear resistance than in perpendicular direction. Effect of fiber orientation on Lyocell reinforced polypropylene composites show that mechanical properties are strongly depending on the fiber orientation (Cordin et al. 2018).

3 Surface Modification of Fibers—Physical and Chemical Methods

Now a days, there has been several attempts to replace the synthetic fibers with natural fibers for reinforcing the composite materials due to increasing environmental awareness. There are some limitations for using natural fibers as a reinforcing material such as poor compatibility with different matrices, high moisture absorption, swelling property etc. These limitations can be successfully overcome by modifying the surface of natural fibers using various techniques. Interfacial adhesion between fiber and matrix play a significant role in controlling the tribological properties of polymeric composite.

3.1 *Physical Techniques*

3.1.1 Plasma Treatment

Plasma treatment can be successfully utilized to improve surface properties of fibers in many applications. The process can be utilized to introduce new functional groups into the surface of natural fibers, which can form strong covalent bond with matrix leads to strong fiber/matrix interaction. This process also can be used for surface etching which improve in surface roughness and results in better interfacial interactions. Sarkar et al. (2017) studied the tribological behavior of PEEK reinforced with and without plasma treated carbon fiber. The mean friction coefficient dropped from 0.42 to 0.23 and the mean specific wear rate 5% dropped from 10^{-5} to 10^{-6} ($mm^3\ N^{-1}\ m^{-1}$). The improvement of tribological behaviour of PEEK and its composites after plasma surface treatment is due to cross-linking of PEEK and improvement of the interface strength of carbon fiber reinforced PEEK composites. They also studied the effect of different ratios of carbon fiber (CF) reinforcing polyimide (PI) and surface treatment of CF on the microstructure and wear resistance of surface layers. The friction coefficients of the composite increased with the increase of CF content. The reason may be the improvement of the microhardness and wear resistance of the CF/PI composite aroused by the addition of CF in the composite. Basalt woven fabric was surface-treated by atmospheric oxygen plasma to improve adhesive force at the fiber/matrix interface and the wear volume of the basalt/epoxy woven composite was reduced from 2.95 to 0.65 mm^3 (Kim et al. 2011). The surface treatment improves the interfacial adhesion between the fiber and resin, thus reducing the debonding of basalt fibers on the wear surface.

3.1.2 Laser Method

Laser treatment method modifies the polymersurface without any changes in its bulk properties. This method creates morphological changes on the smooth surface of synthetic fibers, that further changes its physical (roughness) and chemical properties (water absorption, dyeing. Advantage of laser treatment is that the small area can be treated and depending on the level of power chosen, chemical and physical changes can occur (Abdolahifard et al. 2011).

3.1.3 Electron-Beam Modification

Electron beam is a way of radiation that can produce polymer free radicals. These free radicals combine with each other to form cross links resulting in the formation of a three-dimensional network structure. Polypropylene fabric showed improved wet ability and dye ability due to formation of (O–H) and (C=O) groups on the surface of samples after electron beam irradiation (Ibrahim et al. 2005)

3.2 *Chemical Techniques*

In chemical techniques the natural fibers have been treated with different chemicals such as silane, alkali, permanganates, peroxides etc. It has been found that some of these chemical treatments successfully improve the properties of natural fibers. This type of modifications removes the weak components of lignin from fibers and modifying their crystalline structure. Main objectives of this treatment to improve the fiber strength and adhesion between fiber surface and polymer matrix.

3.2.1 Enzymatic Modification

The use of enzyme in the field of textile and natural fiber modification is rapidly increasing. Enzymatic treatment is ecofriendly method of fiber surface modification as it do not discharge harsh effluents to the environment and use milder conditions (Ibrahim et al. 2005). Other benefits of this treatment are cost reduction, energy and water saving, improved product quality and potential process integration.

Ozone gas treatment

Ozone is excellent oxidizing agent and is used for fiber modification. In this treatment hydrophilic groups are incorporated on fiber surface which results in change in fiber surface chemistry.

Oxidation of wool fiber by ozone gas leads to increase in polymer adsorption by increasing the polarity (Bradley et al. 1993).

3.2.2 Sol-Gel Technique

The sol-gel technique is of particular importance for textile materials. The principles of the sol-gel process include hydrolyzation, application and curing (Textor 2009). The deposition of coating on fibre surface in the form of sol gel is applied to improve some fiber properties such as abrasion resistance, UV protection and attains water repellency.

Adhesion between the fiber and the matrix can be improved by various surface treatment methods viz. electrochemical, chemical, thermal, discharge plasma etc. It is improved by various means such as,

(a) Wettability of the fiber surface improved by using the matrix resin.
(b) Removing the weak boundary layer on the fiber surface. This would provide a more intimate contact between the fiber and the polymer.
(c) Promoting mechanical interlocking between the fiber and the matrix by which creating surface porosity, and resin molecules can penetrate into porosity of fiber.
(d) Chemical bonding between the fiber and unreacted species in the matrix resin can be improved by increasing the number of active sites on the fiber surface.

4 Tribology—Future Aspects in Fiber Reinforced Composites

Now a days polymer composite materials replace metals from almost all area and their usage increases steadily. Matrix materials and wide variety of different fiber permits the design of composites with unique properties for different kinds of application. Polyether ether ketone (PEEK) is reinforced with carbon fiber can be used for the development of an artificial hip joint. In thermoplastics matrix short fibers can be used to process complex geometries. There is remarkable improvement in tribological properties observed with combination of SCF (short carbon fibers) with micro- and nano-filler combination. Nano fillers get freely movable in the contact region between the mating surfaces, nano rolling effect which smoothen the topographies and reduce the coefficient of friction and temperature of the contact region. Load bearing capacity of the thermosetting polymers improved by the addition of fiber, now a days these materials are used in automobile for composite break materials. In future, definitely there is a high demand for high performance materials. i.e., materials which can operate high pressure, velocity and temperature conditions. In this aspect, functionalized fiber reinforced composite material and new emerging area of polymer nanocomposites could be a great promise. The main advantage of certain fiber (carbon) reinforced polymer composites over traditional composites is its ability to improve both strength and tribological properties simultaneously. Polymers reinforced with fibers possess low specific wear rates and significant improvement in mechanical properties. Synthetic fibers such as glass, carbon, graphite and aramid are all commonly used fibers in thermosets as well as thermoplastics. Among all these fibers, continuous carbon fibers reinforced polymer exhibits excellent tribological properties. Studies of tribological properties of nano-fiber reinforced composites are still at a relatively early stage. There are lot of innovative research work that will happen in this area for coming years. There are some research papers available in the farea of high performance polymeric material mixed with nanoparticle that enhance the tribological property to a great extent. Hence this could be another major area of growth that occur in polymer tribology field.

5 Conclusions

A detailed study of different types of fiber reinforcements generally used for developing polymer composites was discussed in this chapter. The physical and mechanical properties of both natural and synthetic fibers were compared. It was found that fiber length and fiber orientation are a critical factor that effects the mechanical and tribological properties of the composites. Surfaces of the fiber can be easily modified using different techniques it will enhance the interaction between fiber and the matrix. Fiber orientation has very significant influence of the wear and frictional performance of fiber reinforced materials. Studies of tribological properties of polymer nanocomposites are still in the early stage. In future, there is a great demand for polymer composite material prepared with nano fibers which can be easily moulded in any shape and high strength and toughness properties simultaneously and isotopically.

Acknowledgements The authors are grateful to Defence Research and Development Organization-(Order No. ERIP/ER/1504758/M/01/1667) New Delhi, India for providing the financial assistance. The authors are also acknowledging the support from Department of Science and Technology, GOI [(DST-FIST- SR/FST/COLLEGE-346/2018).

References

Abdolahifard M, Bahrami SH, Malek RMA (2011) Surface modification of PET fabric by graft copolymerization with acrylic acid and its antibacterial properties. ISRN Organic Chem

Alam S et al (2010) Effect of glass fiber orientation on mechanical properties of GRP composites, vol 32, no 3

Basavarajappa S, Ellangovan S (2012) Dry sliding wear characteristics of glass—epoxy composite filled with silicon carbide and graphite particles. Wear 296(1–2):491–496

Bijwe J, Indumathi J, Rajesh JJ, Fahim M (2001) Friction and wear behavior of polyetherimide composites in various wear modes. Wear 249:715–726

Bradley RH, Clackson IL, Sykes DE (1993) UV ozone modification of wool fibre surfaces. Appl Surf Sci 72(2):143–147

Callister WD (2007) Material science & engineering—an introduction. Wiley, USA, pp 578–579, 702–706

Callister Jr WD, Rethwisch DG (2012) Materials science and engineering: an introduction. Wiley

Chandramohan D, Marimuthu K (2011) A review on natural fibers, vol 8, no August, pp 194–206

Chin CW, Yousif BF (2009a) Potential of kenaf fibres as reinforcement for tribological applications. Wear 267(9–10):1550–1557

Chin CW, Yousif BF (2009b) Potential of kenaf fibres as reinforcement for tribological applications. Wear 267:1550–1557

Cordin M, Bechtold T, Pham T (2018) Effect of fibre orientation on the mechanical properties of polypropylene—lyocell composites. Cellulose 25(12):7197–7210

Czichos H, Habig KH (1992) Tribologie Handbuch Reibung und Verschleiss. Vieweg, Braunschweig
El-Sayed AA, El-Sherbiny MG, Abo-El-Ezz AS et al (1995) Friction and wear properties of polymeric composite materials for bearing applications. Wear 184:45–53
El-Tayeb NSM (2008a) A study on the potential of sugarcane fibres/polyester composite for tribological applications. Wear 265:223–235
El-Tayeb NSM (2008) A study on the potential of sugarcane fibers/polyester composite for tribological applications. Wear 265:223–235
El-Tayeb NSM, Yousif BF, Yap TC (2006) Tribological studies of polyester reinforced with CSM 450-R-glass fiber sliding against smooth stainless steel counterface. Wear 261(3-4):443–452
Franklin SE, de Kraker A (2003) Investigation of counter-face surface topography effects on the wear and transfer behaviour of a POM-20% PTFE composite. Wear 255:766–773
Friedrich K (ed) (1986) Friction and wear of polymer composites. In: Pipes RD (ed) Composite materials series, vol 1. Amsterdam, Elsevier
Friedrich K (1997) Wear performance of high temperature polymers and their composites. In: Luise RR (ed) Application of high temperature polymers. CRC Press, Boca Raton
Friedrich K, Zhang Z, Schlarb AK (2005) Science and effects of various fillers on the sliding wear of polymer composites. Compos Sci Technol 65:2329–2343
Gandhi RA, Palanikumar K, Ragunath BK, Davim JP (2013) Role of carbon nanotubes (CNTs) in improving wear properties of polypropylene (PP) in dry sliding condition. Mater Des, 52–57
Häger AM, Davies M (1993) Short-fiber reinforced, high temperature resistant polymers for a wide field of tribological applications. In: Friedrich K (ed) Advances in composite tribology. Elsevier, Amsterdam
Hashmi SAR, Dwivedi UK, Chand N (2007) Graphite modified cotton fibre reinforced polyester composites under sliding wear conditions. Wear 262:1426–1432
Haupert F, Xian G, Oster F, Walter R, Friedrich K (2004) Tribological behaviour of nanoparticle reinforced polymeric coatings. In: Bartz WJ (ed) Proceedings of 14th international colloquium tribology, Stuttgart
Ibrahim MS, El Salmawi KM, Ibrahim SM (2005a) Electron-beam modification of textile fabrics for hydrophilic finishing. Appl Surf Sci 241(3–4):309–320
Ibrahim MS, El Salmawi KM, Ibrahim SM (2005) Electron-beam modification of textile fabrics for hydrophilic finishing. Appl. Surface Sci. 241(3–4):309–320
Kim MH, Rhee KY, Park SJ (2011) Plasma treatment and its effects on the tribological behaviour of basalt/epoxy woven composites in a marine environment. Polym Polym Compos 19(1):29–34
Kolluri DK, Satapathy BK, Bijwe J, Ghosh AK (2018) Analysis of load and temperature dependence of tribo-performance of graphite filled phenolic composites Analysis of load and temperature dependence of tribo-performance of graphite filled phenolic composites, no July
Mahir FI, Keya KN, Sarker B, Nahiun KM, Khan RA (2019) A brief review on natural fiber used as a replacement of synthetic fiber in polymer composites
Mathew MT et al (2007) Tribological properties of the directionally oriented wrap knit GFRP composites. Wear, 1–9
Mishra A (2012) Dry sliding wear behavior of epoxy-rubber dust composites. World Acad Sci Eng Technol 6(7):1218–1223
Mutlu I (2009) Investigation of tribological properties of brake pads by using rice straw and rice husk dust. J Appl Sci 9:377–381
Nguong CW, Lee SNB, Sujan D (2013) A review on natural fibre reinforced polymer composites. Int J Mater Metall Eng 7(1):52–59
Nirmal U, Hashim J, Low KO (2012) Adhesive wear and frictional performance of bamboom fibres reinforced epoxy composite. Tribol Int 47:122–133
Prasad MP, Vinod B, Sudev DL (2014) Effect of fiber orientations on tribological behaviour of PALF reinforced Bisphenol-A composite. Int J Eng Res General Sci 2(4):809–814
Prashanth S, Km S, Nithin K, Sachhidananda S (2017) J Mater Sci Eng Fiber Reinforced Compos Rev 6(3)

Rasheva ZÃ, Zhang G, Burkhart T (2010a) Tribology International A correlation between the tribological and mechanical properties of short carbon fibers reinforced PEEK materials with different fiber orientations. Tribiol Int 43(8):1430–1437

Rasheva Z, Zhang G, Burkhart Th (2010) A correlation between the tribological and mechanical properties of short carbon fibers reinforced PEEK materials with different fiber orientations. Tribol Int 43(8):1430–1437.

Ravikumar DCL, Murali CSG (2018) Review on tribological performance of natural fibre-reinforced polymer composites. J Bio-Tribo-Corrosion

Reinicke R, Haupert F, Friedrich K (1998) On the tribological behaviour of selected, injection moulded thermoplastic composites. Compos A 29:763–771

Sabeel Ahmed K, Khalid SS, Mallinatha V et al (2012) Dry sliding wear behavior of SiC/Al_2O_3 filled jute/epoxy composites. Mater Des 36:306–315

Sarkar P, Modak N, Sahoo P (2017) Effect of normal load and velocity on continuous sliding friction and wear behavior of woven glass fiber reinforced epoxy composite. Mater Today Proc 4:3082–3092

Schwartz CJ, Bahadur S (2000) Studies on the tribological behavior and transfer filmcounterface bond strength for polyphenylene sulfide filled with nanoscale alumina particles. Wear 237:261–273

Shivamurthy B, Bhat KU, Anandhan S (2013) Mechanical and sliding wear properties of multi-layered laminates from glass fabric/graphite/epoxy composites. Mater Des 44:136–143

Srivastava A, Jana KK, Maiti P, Kumar D, OmParkash (2015) Investigations on structural, mechanical, and dielectric properties of PVDF/ ceramic composites. J Eng 2015, Article ID 205490, 9 p

Taylor P, Agarwal G, Patnaik A, Sharma RK (2014) Comparative investigations on three- body abrasive wear behavior of long and short glass fiber-reinforced epoxy composites, no July, pp 37–41

Textor T (2009) Modification of textile surfaces using the sol-gel technique. In: Surface modification of textiles. Woodhead Publishing, pp 185–213

Vadivel HS, Golchin A, Emami N (2018) Tribological behaviour of carbon filled hybrid UHMWPE composites in water. Tribol Int

Werner P, Altstädt V, Jaskulka R, Jacobs O, Sandler JKW, Shaffer MSP, Windle A (2004) Tribological behaviour of carbon-nanofiber-reinforced poly(ether ether ketone). Wear 257:1006–1014

Xin X, Xu CG, Qing LF (2007) Friction properties of sisal fibre reinforced resin brake composites. Wear 262:736–741

Yousif BF (2009) Frictional and wear performance of polyester composites based on coir fibres. Proc Inst Mech Eng, Part J: J Eng Tribol 223:51–59

Yousif B, El-Tayeb N (2007) Tribological evaluations of polyester composites considering three orientations of CSM glass fibres using BOR machine. Appl Compos Mater 14:105–116

Yousif BF, Lau STW, McWilliam S (2010) Polyester composite based on betelnut fibre for tribological applications. Tribol Int 43:503–511

Zhang H, Friedrich K (2007) Science and effect of fiber length on the wear resistance of short carbon fiber reinforced epoxy composites. Compos Sci Technol 67:222–230

Zhang LC, Zarudi I, Xiao KQ (2006) Novel behaviour of friction and wear of epoxy composites reinforced by carbon nanotubes. Wear 261:806–811

Mechanical and Tribological Properties of Utilized Natural-$CaCO_3$ and Potassium-Rich Polymeric Fillers by VATRM and IM Techniques

Ramdziah Md. Nasir, Syahrain Sadali, and Aslina Anjang Ab Rahman

Abstract In the fast-developing world, the environmental pollution, preservation of renewable and biodegradable resources has motivated researchers to develop novel eco-friendly materials and products based on sustainability principles. In this work, natural sources such as inner banana trunk polymer matrix composites (IBTPMC) and eggshell polymer matrix composites (ESPMC) are prepared by injection molding (IM) and vacuum assisted resin transfer molding (VARTM). The objective is to evaluate feasibility of the materials using IM and VARTM techniques. The specimens were subjected to the mechanical testing for its strength. The tensile modulus for ESPMC composites is 897 MPa compared to IBTPMC only 561 MPa. In izod impact test, IBTPMC shows the highest impact strength of 0.362 J, which is superior then in ESPMC or pure polypropylene (PP). The flexural strength of the pure PP is 58 MPa while for IBTPMC and ESTPMC, both are 42 MPa and 53 MPa respectively. Wear rate for ESPCM has reduced by 64–66% by increasing the load. Wear rate for IBTPCM shows increased by 13 times the initial wear as loading increases and reduced from 1–1.3 times as sliding distance increases. It shows that the IBTPMC fibre ruptured easily leading to higher wear in comparison to ESPCM structure due to fibre wetting properties and distribution of fibre in the matric interface. Coefficient of friction (COF) for IBTPMC shows increment from 0.38 to 0.55. While the COF for ESPCM increased from 0.28 to 0.35. Finally, Scanning Electron Microscope (SEM) to analyze the structure of the fractured surfaces was carried out.

Keywords VATRM · POD · Natural-$CaCO_3$ composites · Potassium-rich composites · Wear · COF

R. Md. Nasir (✉) · S. Sadali
School of Mechanical Engineering, Engineering Campus, Universiti Sains Malaysia, Nibong Tebal, 14300 Pulau Pinang, Malaysia
e-mail: ramdziah@usm.my

S. Sadali
e-mail: mesyahrain@gmail.com

A. A. A. Rahman
School of Aerospace Engineering, Engineering Campus, Universiti Sains Malaysia, Nibong Tebal, 14300 Pulau Pinang, Malaysia
e-mail: aeaslina@usm.my

M. T. Hameed Sultan et al. (eds.), *Tribological Applications of Composite Materials*, Composites Science and Technology, https://doi.org/10.1007/978-981-15-9635-3_5

1 Introduction

Kaizen in present material and the growth in classifying structural material has been associated with technological advancement. In modern minimalist design, optimizing composite materials with minimum deficiencies and lean wastage has been an inspirational. Polymer matrix composites (PMC) are considered as near natural materials because of a prolonged evolution process and ability to mimic the nature characters. Waste products such as eggshells are found easily in food industries, houses, restaurant and bakery while banana trunk at the plantation are cut down after the banana has been harvested. Eggshells has high compressive strength but it can create pollution to the environment due to its high calcium content. Eggshell powder is considered as an alternative and the eggshell wastage can be reduced. In fact, eggshell is a waste product for the regular use and a cheapest material for innovative and renewable conducive flame-retardant materials. Flora and fauna fibers have proven to be biodegradable and recyclable. They have generally great strength, stiffness, and one of renewable crude materials. Plant fibers, for example inner banana trunk, are usually consisting of cellulosic patterns usually used in various applications.

In modern living, renewable PMC is preferred because it has high strength, low cost and simple manufacturing method. Constituting reinforcements in composites to upgrade general mechanical properties of matrix and offer quality to composites. In this project, the composites were fabricated by using injection molding (IM) and vacuum assisted resin transfer molding (VARTM) techniques. The mechanical, wear and friction properties of polymer matrix composite produced by these two methods are studied, result were reported and comparative studied are made based on existing references. Focused on smart material sources for better invention, environmental assessable and naturally decomposed has become an impetus for using normal disposal. Eggshells and banana trunk which is cost-effective, natural and have no health hazard are easily found at food industries, houses, restaurant and banana plantations. Mechanical properties would be very important concerning recycling whereby the strength is the focal concern. Updated, there are only few references on specific wear, friction and mechanical properties of both inner banana trunk polymer matrix composite (IBTPMC) and eggshell polymer matrix composite (ESPMC).

The main objective is to evaluate the wear, friction and mechanical properties of IBTPMC and ESPMC using IM and VARTM techniques. In order to carry out this research, composite polymer matrices are produced by injection molding (IM) and vacuum assisted resin transfer molding (VARTM) methods are used. Before that, the composites polymer matrices are cut into desired shape per testing being perform. Four tests are conducted which is impact, tensile, flexure, wear and friction test. Analysis is done by capturing fracture structure of the composites after test by using Scanning Electron Microscopy (SEM).

One of the oldest cultivated plants in the world is banana. 'Banana' comes from the Arabic language means 'finger'. There are approximately 300 species and it belongs to the Musaceae family, only 20 varieties are used for consumption. The nutritional facts of banana (100 g pulp) are as follows: carbohydrates—18.8 g; protein—1.15 g;

fat—0.18 g; water—73.9 g; vitamins C1 B1 B2 B6 E, other minerals—0.83 g and 81 kcal (Bilba et al. 2007). At present banana fiber are used in producing ropes, mats, wall hangings, handbags, cloth making and paper industry (Venkateshwaran and Elayaperumal 2010). Tables 1 and 2 show the physical and mechanical properties of banana natural fibers (Satyanarayana et al. 1984, 1990).

Justiz-Smith et al. conducted standardized characterization tests on fiber sample, such as ash and carbon content, water absorption, moisture content, tensile strength, elemental analysis, and chemical analysis. The micrographs of longitudinal section of banana fiber strands and cross-sectional area of fiber strands is analyzed and the presence of metal element as ions in the natural fibers are tabulated in Table 3 (Jústiz-Smith et al. 2008).

Investigation on the cellular structure of banana fiber is conducted using optical microscopy by Kulkarni et al. (1982) and it shows banana fiber consist of four kinds of cells namely xylem, phloem, schlerenchyma, and parenchyma. Further, which

Table 1 Common physical and mechanical properties of some natural fibers (Venkateshwaran and Elayaperumal 2010)

Fibers	Width or diameter (um)	Density (kg/m^3)	Initial modulus (GPa)	Ultimate tensile strength (MPa)	Elongation (%)
Banana	80–250	1350	7.7–20.0	54–754	10.35
Coir	100–450	1150	4–6	106–175	17–47
Sisal	50–200	1450	9.4–15.8	568–640	3–7
Pineapple leaf	20–80	1440	34.5–82.5	413–1627	0.8–1
Palmyra	70–1300	1090	4.4–6.1	180–215	7–15

Table 2 Physical properties of different cellulosic fibers (Jústiz-Smith et al. 2008)

Fibers	Moisture content (wt%)	Ash content (wt%)	Carbon content (wt%)	Water absorption (wt%)	Tensile strength (MPa)
Banana	85.6	8.3	50.9	40	142.9
Coconut	27.1	5.1	51.5	169	138.7
Bagasse	52.2	4.5	53	235	29.6

Table 3 Metal elements present as ions in cellulosic fibers (Jústiz-Smith et al. 2008)

Fibers	AI^{3+}	Ca^{+}	Mg^{+}	Na^{+}	Si^{+4}
Banana	85.6	8.3	50.9	40	142.9
Coconut	27.1	5.1	51.5	169	138.7
Bagasse	52.2	4.5	53	235	29.6

Table 4 Mechanical properties of banana fibers of different diameters

Sample number	Diameter of fiber (um)	Initial Young's modulus (GN/m^2)	SD initial Young's modulus (GN/m^2)	Breaking strength (MN/m^2)	SD breaking strength (MN/m^2)	% Strain	SD % strain
1	50	32.703	8.190	779.078	209.300	2.750	0.957
2	100	30.463	4.689	711.661	239.614	2.469	0.798
3	150	29.748	8.561	773.002	297.104	3.583	1.114
4	200	27.698	7.083	789.289	128.558	3.340	0.688
5	250	29.904	4.059	766.605	165.515	3.244	1.284

is shown in Table 4, they observed how the mechanical property varied about the diameter of the fiber.

Prediction of various mechanical properties like tensile strength, flexure strength of banana fiber and banana fiber reinforced with polymer have been made by number of investigation. From all the results shown, banana fibers exhibited excellent mechanical properties. Murali et al. carried out a comparative study of stress and strain on various natural fibres especially vakka, sisal, bamboo and banana (Murali Mohan Rao et al. 2010). According to ASTM-D 3379–75, the tensile test was conducted. It is found that banana have a stress value 560 MPa when the percentage of strain is 3.5% implying that banana fibers have strength and stiffness compared to sisal as shown in Table 5.

Based on experiment conducted by Geethama et al. (1998) estimated the tenacity and elastic modulus of banana fiber are in the range of 529–759 MPa, and 8–20 GPa, respectively. Banana fiber has the percentage elongation at break varies between 1.0 and 3.5. Igwe and Onuegbu studied the mechanical and end-use properties of eggshell powder filled polypropylene have been determined at filler contents, 0 to 40 wt. %, and particle sizes, 0.150, 0.30, and 0.420 μm. Adding eggshell and fish bone powder into polypropylene resulted in improvement mechanical properties and water absorption. However, the elongation at break of the composites was decreased with increase in filler contents, and particle sizes (Igwe and Onuegbu 2012). Bootklad and Kaewtatip investigated the effect of the eggshell powder (EP) on thermoplastic

Table 5 Tensile properties of natural fibers. The specimen diameter of banana is 0.06–0.08 mm (Kulkarni et al. 1982)

Weight of filler	%Tensile strain	Average tensile strength (MPa)	Average tensile modulus (GPa)	Specific tensile strength (MPa/(kgm^{-3}))	Specific tensile modulus (MPa/(kgm^{-3}))
10	3.36	600	17.85	0.4444	13.22
20	5.45	567	10.40	0.3910	7.17
Coconut	20.00	500	2.50	0.4348	2.17

starch (TPS) prepared using compression molding and compared with the effect of commercial calcium carbonate (CC). Resultant is a strong organic adhesion between the eggshell powder and the TPS matrix, therefore delayed the biodegradation of TPS (Bootklad and Kaewtatip 2013). Two types of eggshell biofiller: unmodified, and chemically modified with isophthalic acid were studied by Kumar et al. Modified eggshell (PP-MES) composite possessed slightly higher tensile strength than the unmodified eggshell (PP-ES) composite. They proposed that better bonding gives rise to a higher flexural modulus, i.e., higher stiffness, and that this behavior allows composites to withstand greater loads (Kumar et al. 2014). Hassan et al. studied 10 to 50 wt% uncarbonized and carbonized eggshell particles as reinforcement in polyester matrix. Carbonized eggshell shows higher value of mechanical properties compared to uncarbonized eggshell for most of the mechanical properties such as tensile strength, compressive strength, and flexural strength (Hassan et al. 2012). The investigation of tensile and thermal properties of chicken eggshell powder (ESP) filled polypropylene composite at different filler loading, which is 0 to 40 part per hundred resin (phr) conducted by Long (2016) shows improvement in its performance but both decreases at higher than 40 phr. ESP loading. Toro et al. investigate tensile value on different proportions of chicken eggshell(ES) in polypropylene (PP) composite were compared with different particle sizes and proportions of commercial talc and calcium carbonate fillers. Again bio-filler outnumbered carbonate fillers with different particle sizes used as high as 75% ES could maintain a similar stiffness and E values compared to the talc composites due to ES/matrix interface geometric ratio (Toro et al. 2007).

The fabrication and mechanical testing of eggshell particles(ESPs) and graphite-reinforced Al composites were conducted by Almomani et al. Mechanical properties shows improvement after adding hybrid ES confirming that the incorporation of waste ESPs in the Al matrix serve as reinforcements (Almomani et al. 2020). Boopathi et al. reported that upon usage of natural banana-hemp glass fiber shows superior properties in term of interfacial characteristics, internal structures, fiber failure mode and fractured surfaces (Bhoopathi et al. 2014). Alavudeen et al. works dealt with the effect of weaving patterns and random orientation on the mechanical properties of banana, kenaf and banana/kenaf fiber-reinforced hybrid polyester composites. Composites were prepared using hand lay-up method with two different weaving patterns, namely, plain and twill type. Tensile test, flexural tests and impact tests were performed according to ASTM:D-790–10 (Alavudeen 2015). Gowtham et al. conducted a wear and friction test on banana-borassus fruit fiber reinforced composite to explore the influence of fiber and matrix properties in wear and friction performance of banana and borassus fruit fibers reinforced polymeric composite Composite is prepared with 20% banana fiber, 20% borassus fiber and 60% polyester matrix. The friction and wear tests are conducted as per the ASTM D2734 standard on a pin on disc apparatus for various sliding distance, disc speed and axial load. The result showed that when the load is increased the wear rate also increased. The wear rate was calculated by the following Eq. (1) (Gowtham 2014).

$$W = \frac{\text{Weight loss}}{\text{Applied load}} \tag{1}$$

The frictional force was determined the following Eq. (2).

$$\mu = \frac{\text{Measured frictional force}}{\text{Normal applied load}} \tag{2}$$

All mechanical, physical, chemical, geometrical surface contact and conditioning atmosphere could affect surface interaction and tribological system. The first order classification of wear distinguishes adhesive wear, abrasive wear, wear caused by surface fatigue, and wear due to tribochemical reactions. The purpose of tribotesting is to increase the fundamental and general understanding of how a material behaves in tribological applications. Niklas and Ha designed measurement of friction and wear to stimulate abrasive wear, erosive wear, wear in sliding and rolling contacts, and mild wear (Niklas and Ha 2001). Nirmal et al. investigated the adhesive wear and frictional performance of Bamboo Fibres Reinforced Epoxy (BMBFRE) composite. Wear and frictional test of the BMBFRE composite was conducted on Pin on Disc (POD) tribo test machine. Specific wear rate (Ws) was calculated using Eq. (3):

$$Ws = \frac{\Delta V}{FnD} \tag{3}$$

The performance of BMBFRE composite was found to be superior for AP-O. The friction performance of BMBFRE composite was improved by about 44% at low sliding velocity for AP as compared to the higher sliding velocity (Nirmal et al. 2012). Nirmal and Ahmad has also reported tribological performance of palm fibres. One of the test which to investigate adhesive wear characteristics is using a pin on disc (POD) to stimulate the wear of the composite at dry contact condition using randomly distributed palm fibres as reinforcement in polyester (Nirmal et al. 2015). El_Tayeb has explored the possibility of using sugarcane fibre to reinforce polyester and thus opens a new way to implement locally available inexpensive fibres and produce a new candidate tribo-material for bearing applications. Sugarcane fibre/polyester (SCRP) and glass fibre/polyester (GRP) were prepared using compression mould and hand-lay-up techniques. Adhesive wear tests were performed using a pin-on-disc tribo-test machine. Results of friction and wear proved that SCRP composite is a promising composite which can be a competitive to GRP composite. Specifically, untreated sugarcane fibre (SCF) was used in two form to reinforce polyester (SCRP). Mechanical tests were performed to investigate tensile and compression strength of the Chopped-SCF reinforced polyester composites (C-SCRP) and Undirectional-SCF reinforced polyester composites (U-SCRP). Wear of SCRP composite was sensitive to variations of load, fibre length and fibre orientation and less sensitive to sliding velocity (El-Tayeb 2008). Babu et al. has investigated the tensile and wear characterization of polymer composites made by reinforcing Calotropis Gigantea fruit fiber as a new natural fiber into a polyester resin. To determine wear properties, wear test is performed using Tribometer (DUCOM;TL-20) by measuring the wear groove with a profilometer and measured the amount of material removal. Since pins were fabricated from a wide range of materials virtually any combination of metal, glass, plastic, composite, or ceramic substrates can be tested. Thus, the variation of mean tensile strength and tensile modules with varying fiber was clearly with increasing the fiber content in the polyester matrix, the tensile strength also increasing (Babu et al. 2014). Recent review by Rajiv et al. shows that within 5 consecutive years, tribological study on natural fibres has been exceeding the synthetic fibres for more than 50% and the numbers has been projected to increase by year 2050 as advancement in technology and know-how (Rajiv and Ankush 2019).

2 Methodology

The preparation process adopted is injection molding (IM) and vacuum assisted resin transfer molding (VARTM).The composite materials are gained from local resources. The banana inner trunk was extracted roughly in the width of 1–2 cm before dried under the sun to remove the moisture for more than 24 h. Meanwhile for eggshell, it is washed by using tap water to clean all it contents and then dried under the hot sun for more than 1 days until its completely dried. All composites for IM process was blended to 150 micron mesh size. The composites was dried again in the oven for 1 day before mixing. Injection molding specimens were prepared by ratio of 70:30 of polypropylene and composites material. For VARTM specimen, the polymer matrix composites consist of 1 layer. The materials were impregnated with unsaturated polypropylene. The banana fibers are mounted on the base plate which is placed on a table, and then filled with the polypropylene resin. Before the resin dried up, the second layer was mounted accordingly to ensure adhesiveness between interfaces is intact. The process was repeated until the last layers accumulated to be approximately 2 cm thickness. The polypropylene resin applied was distributed to the entire surface by means of a roller and the air gaps formed between layers during fabrication are removed by squeezing gently. The process was ideally done at a temperature of 32 °C/50%RH conditioning. It is advisable to maintain the conditioning temperature and RH to ensure the curing process could be carried out properly and no additional curing agent needed. After all the process are done, the specimen was cut based on the test been performed. Specimen for wear test produced by injection molding process was taken from the part of the injection molding called runner as shown in Fig. 1.

Tensile strength used in structural applications i.e., the ability of a material to resist breaking under tensile load. The test specimen was prepared according to the ASTM D638 standard and performed on the Universal Testing Machine (UTM). During the application of tension, the elongation of the gauge section is recorded against the applied load. The dumbbell shape tensile specimens are produced by injection molding machine and the size of the specimens were prepared as shown as Fig. 2 and gauge length was 50 mm. The tests were conducted in 27 °C of room temperature. The tensile strength of the composites was measured with a computerized Instron universal testing machine in accordance with the ASTM: D-638 (Materials 1995) procedure at a crosshead speed of 5 mm/min. Figure 3 shows the UTM used at the

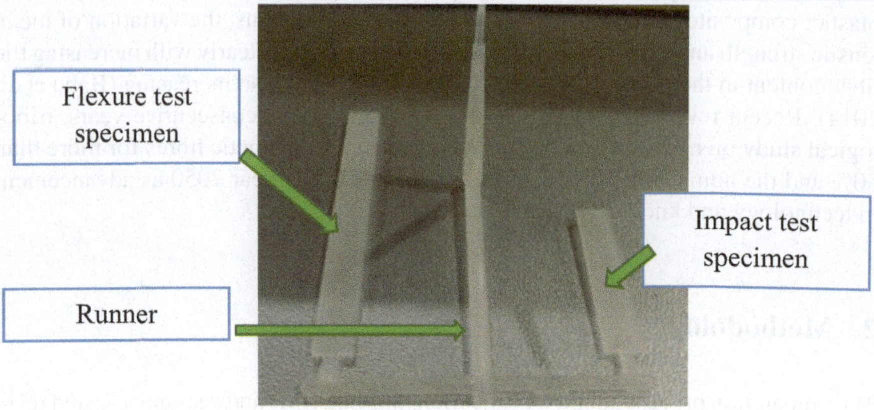

Fig. 1 Specimens produced by injection molding process. First session of injection molding process producing specimens for impact and flexure test

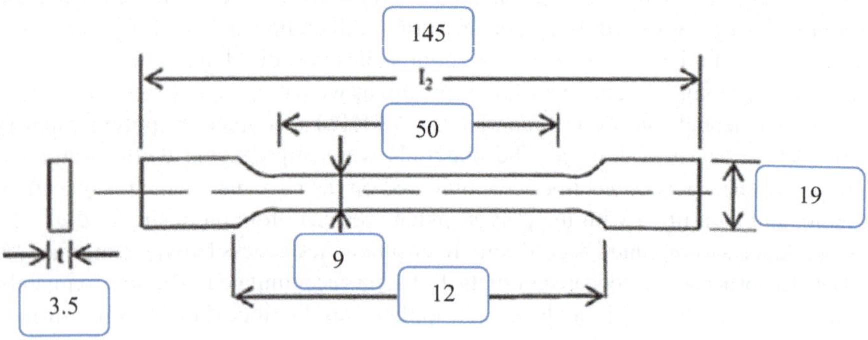

Fig. 2 The size of tensile test specimens. All dimensions are all in mm

vicinity of Universiti Sains Malaysia while Fig. 4 shows the equipment used for Izod impact test. Figure 5 and 6 shows the Pin-on-disk test machine.

Izod impact test machine was used to determining the impact resistance of composites specimen. The specimen was prepared accordance to ASTM D-256 (Materials 1995) by using an injection molding process. All impact test specimens were notched (2 mm depth) and the test specimen was supported as a vertical cantilever beam and broken by a single swing of a pendulum. The specimen was loaded in the Zwick impact machine and allows the pendulum until it fractures or breaks. An arm held at a specific height and released during the testing occurred. The weight of the hammers used were 7.5 and 15.0 J. Its impact energy is obtained from the energy absorbed by the composite or sample. The specimen dimension for impact test was 61.0 mm × 12.5 mm in rectangular shaped. The test was conducted in air room temperature condition of 27 °C. For each case, a total of five samples were tested and data is recorded and tabulated.

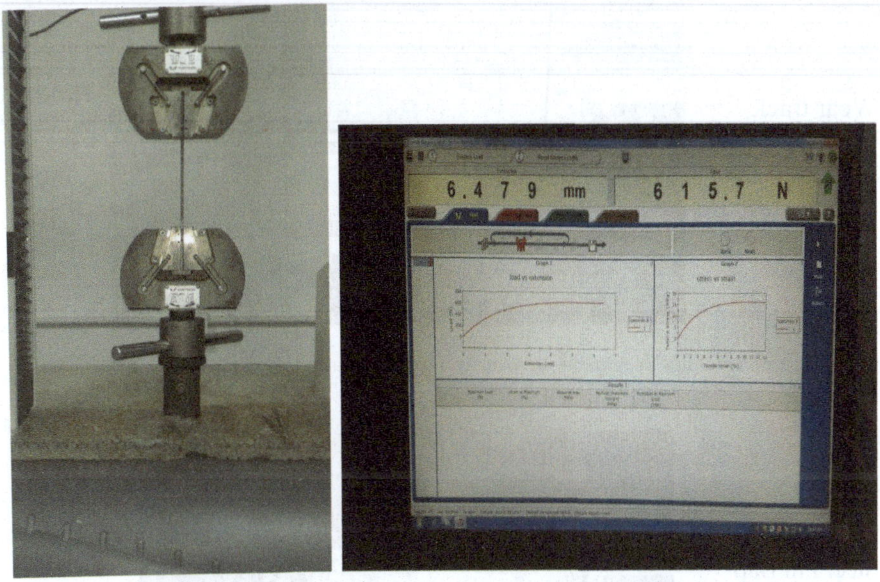

(a) Specimen gripped on (b) Computerized Instron UTM

Fig. 3 Tensile test methodology: **a** The specimen was placed and gripped onto the jaws on UTM before the test are performed and **b** computerized Instron UTM were used to recorded the data

Fig. 4 Impact Testing: **a** Zwick impact test machine at Mineral Resources School are used and **c** Specimen is loaded onto the impact machine. The notched area is aligned with the benchmark on the machine as shown on the arrow

Fig. 5 Injection molding specimens were tested on pin on disk wear test (POD) (rotary motion)

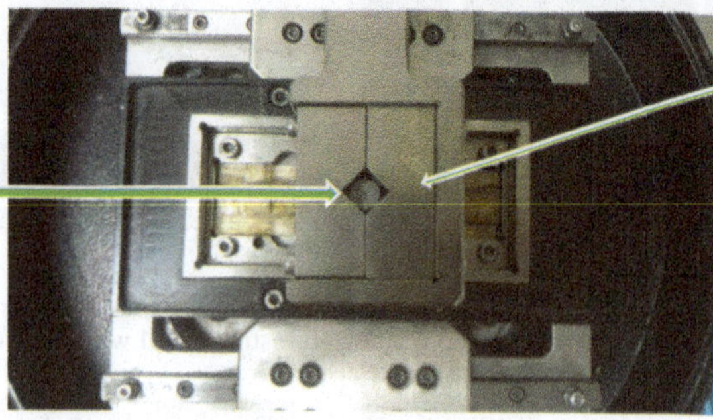

Fig. 6 VARTM specimens were tested on reciprocatory wear test

Instron universal testing machine was used to performed the flexural tests using the three-point bending fixture per ASTM D-790 (Materials 1995) and with a crosshead speed of 5 mm/min. The specimen was prepared using an injection molding process with dimensions 124.7 mm × 13.0 mm × 3.0 mm in a rectangular shaped three-point bending specimens. As per standard, with a ratio of 16:1, the distance between supports span was kept at 48 mm. The loading nose diameter used was 19 mm with speed. The experiments were performed at 27 °C of room temperature. For each test, five specimens were tested and the average values reported.

Wear test of polymer composite matrices produced by injection molding was done by rotary wear test on a pin-on-disc (POD) test machine. Pin on disc consist of load cell, speed disc etc. counter face was polished using acetone to get rid any possible contaminant or chips on it. POD wear test were done based on ASTM G99 (West 2005). Meanwhile, for specimens produced by VARTM, the wear test was done based on reciprocatory test. This due to the specimens could not be altered to pin shape and the size of the specimens were cut into dimension 100.0 mm × 25.0 mm × 2.0 mm in rectangular shaped. Mild steel ball pin was used to contact with the specimens in this case with a diameter of 10 mm. Wear test was carried for different load, disc speed, and constant sliding time 1800s. The weight loss for the specimen for the specimen was recorded.

The adhesive wear resistance of mild steel and its composites were studied by conducting dry sliding wear test. The wear testing specimen is prepared with diameter of 6 mm. The specimen is cut on upper surface by diamond cutter to give an evenly surfaces to contact with counter face. A POD wear testing configuration as shown in Fig. 5 (rotary) and Fig. 6 (reciprocatory). The specimens were rubbed against rotating hard steel disc without any lubricant. Tests were conducted for different variables such as speed ranging from 40 to 100 rpm, normal load form 10 to 50 N, and track diameter of 40 to 100 mm. The fractography studies of the fractured specimen were carried out using Scanning Electron Microscope (SEM).

3 Result and Discussion

According to Muller and Krobjilowski, tensile strength, flexural strength, and impact strength are an excellent measure of the degree of reinforcement provided by the fiber to the composite (Mueller and Krobjilowski 2003). The results indicated that the banana fibers exhibited excellent impact strength for both impact load. Meanwhile, PP has the highest tensile strength and flexural strength following by ESPMC and IBTPMC. Table 6 indicated the mechanical properties of polymer matrix composites and PP. The average value for each type of polymer matrix composites (PMC) and PP as tabulated in Table 7.

Tensile strength of polymer matrix composites and polypropylene as shown in Fig. 7. The tensile strength is highest for polypropylene compared to others composite due to its ductile properties. The reasons for the lower tensile properties are possibly

Table 6 Experimental findings of the composite specimens

Specimens	Tensile strength (MPa)	Impact strength (Joules)		Flexural strength (MPa)
Impact load		7.5	15	
IBTPMC	19.7469	0.362	0.472	38.26
ESPMC	24.6945	0.342	0.408	53.46
PP	31.5449	0.3	0.34	58.40

Table 7 The mechanical properties of PMC and PP

Samples	Maximum Load (N)	Extension Maximum at Load (mm)	Young's modulus (MPa)	Tensile strain (%)	Tensile strength (MPa)
IBTPMC	702.0899	5.4066	560.5131	10.8333	19.7469
ESPMC	834.9202	4.1034	897.4746	8.2068	24.6945
PP	1066.5322	7.3916	735.1214	14.78328	31.5449

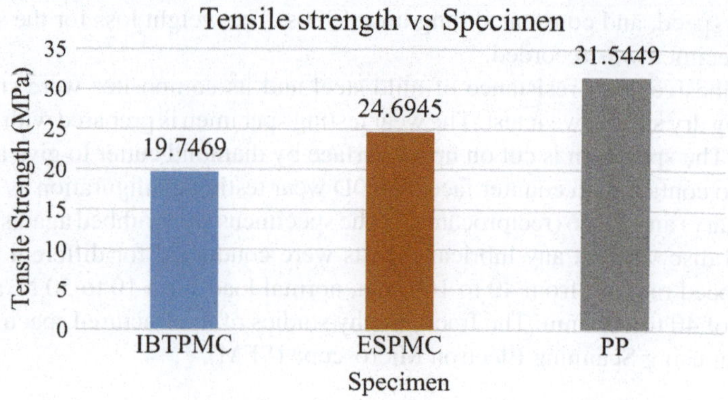

Fig. 7 The comparison tensile properties of PMC and PP based on average value

due to poor composite to matrix bonding, blister, void and composite or fiber pull out as shown in Figs. 8 and 9. The tensile strength of the pure polypropylene is 31.5 MPa. Meanwhile, the tensile strength of the IBTPMC and ESPMC composite are 19.7 MPa and 24.7 MPa respectively.

Furthermore, reduced in fiber dispersion partly caused by strong interfiber bonding incorporating fibers/thermoplastics layers. Inadequate bonding altered a significant increase in tensile strength (Kumar et al. 2013). Thus, fiber bonding increases with loading, mainly due to non-entrance of wetting problem as according to Bozzelli (2011).

Tensile strength of the PP is the highest in comparison to eggshell composites and banana trunk composites. This is due to the slippage and bad wetting agent for the composites resulting the breakage of the banana fibre and crushing happen to the eggshell composites. The breakage of the banana fibre is due to the maturing (aging property of the cellulosic) while the crushing property is due to the brittleness of the eggshell. Figure 8 shows the variation in tensile modulus with respect to types of specimen. It is observed that the tensile modulus or load bearing capacity increases with brittleness. The deformation resistance increases with filler content therefore the composite become stiffer (Raghavendra et al. 2012). The tensile modulus of the pure polypropylene is 735.1 MPa. The tensile modulus for ESPMC composites is 897.5 MPa which highest tensile modulus observed compared to IBTPMC only 560.5 MPa. Figure 9 shows the composites before and after testing.

Figure 10 shows the fractography using SEM to investigate the fractured surfaces after undergoes tensile loading. It is shown that there was a small hole in the polymer matrix which known as blister probably caused where air bubble trapped during filling up the molten plastic into the mold (Kumar et al. 2013). Air pocket is a root cause of bubbles and blisters. Water vapor, volatiles from the resin, or decomposing by-products could create sink and gas entrapment causing blister (Bozzelli 2011). Blister and crack propagation probably caused by incomplete wettability or bonding between matrix resin and fiber (Bozzelli 2011) as shown in Figs. 11 and 12. Figures 13

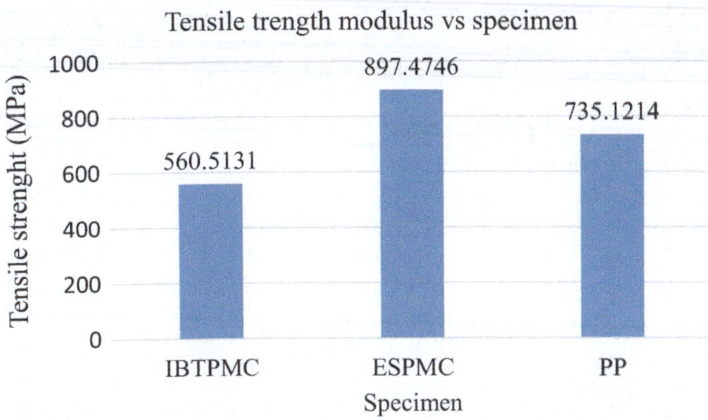

Fig. 8 The comparison tensile properties of PMC and PP based on average value

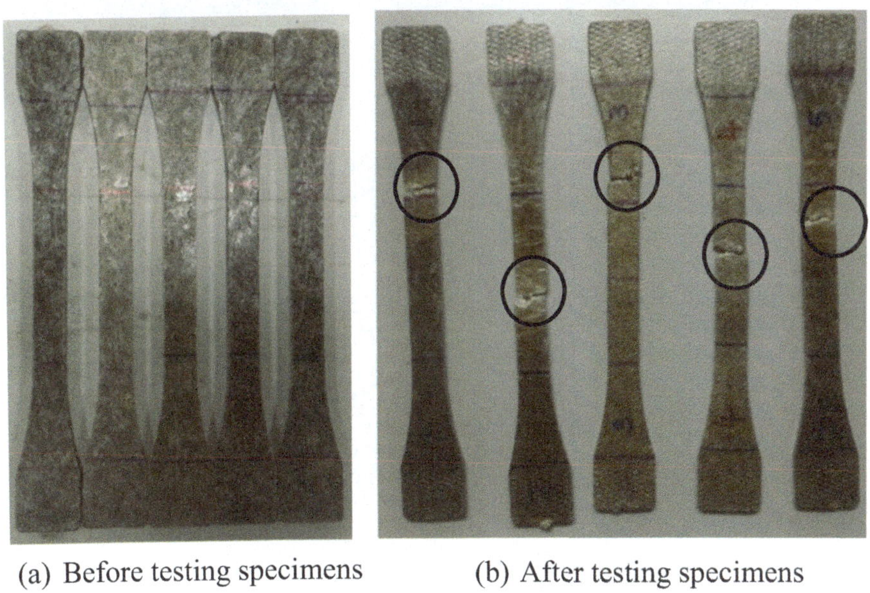

Fig. 9 Testing of tensile strength: **a** before testing specimens and **b** after testing specimen. The circles indicate the fracture area of the specimens

and 14 are the fractography using SEM to investigate the fractured surfaces after undergoes tensile loading.

From the izod impact testing machine, the energy loss could be obtained. The comparison of impact strength is presented in Figs. 15 and 16. It isobserved the IBTPMC performing better than ESPMC and PP which can hold the impact load from 7.5 to 15.0 J. In the impact resistance of the composite, the fibers play a vital

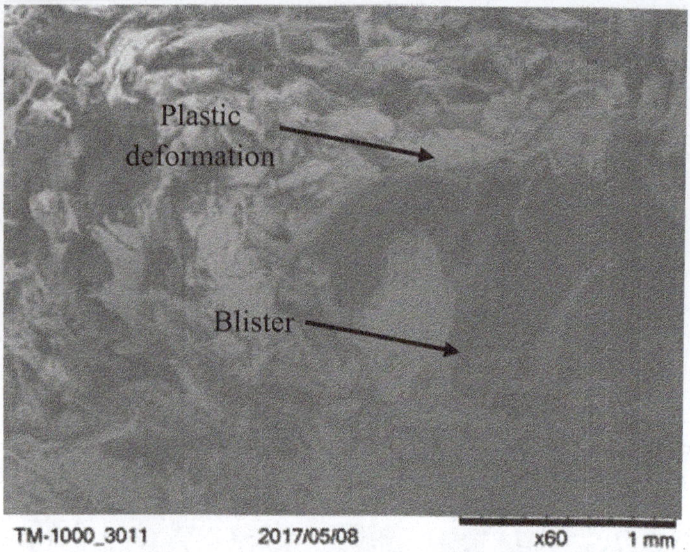

Fig. 10 The fractography of IBTPMC

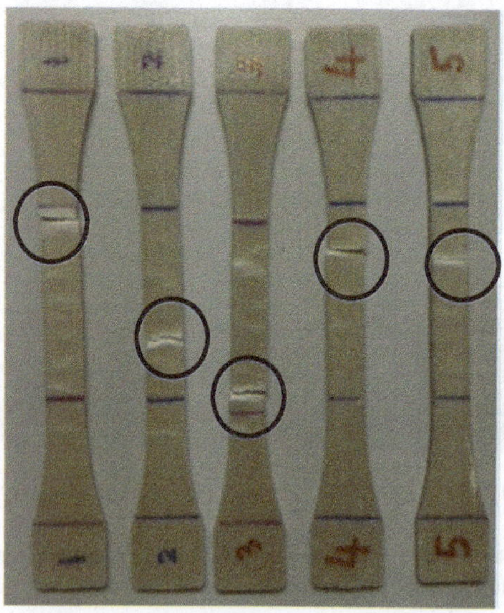

Fig. 11 After testing specimens. The circles indicate the fracture area of the specimens

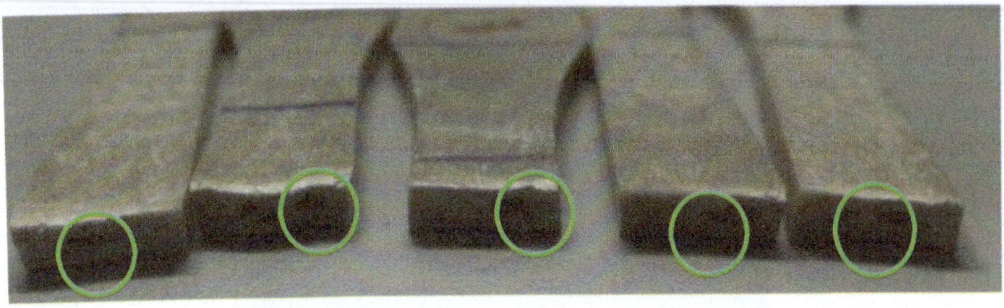

Fig. 12 Defects inside the specimens are shown by the circles

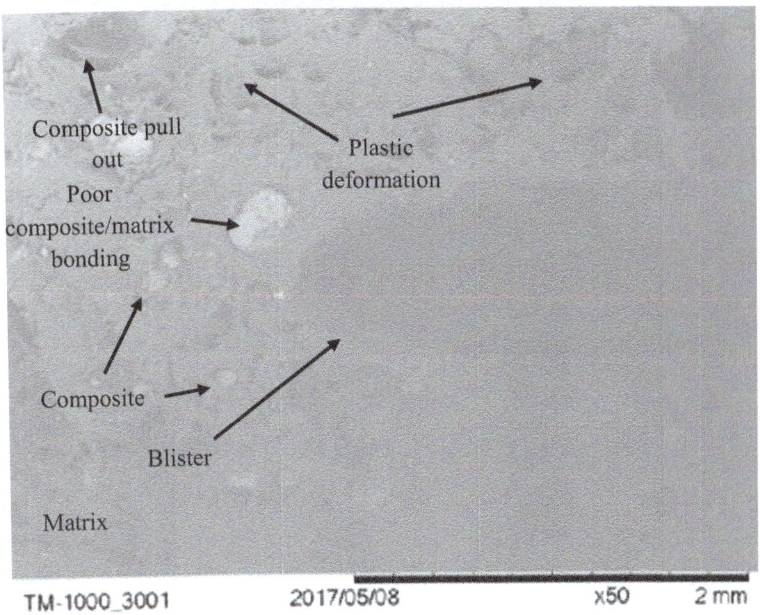

Fig. 13 The fractography of ESPMC

role as interaction with the crack formation in the matrix act as stress transferring medium. It is also observed that the impact strength increases with increasing fiber content (Mueller and Krobjilowski 2003).

For both experiments, IBTPMC shows the highest impact strength which for hammer 7.5 J, the impact strength was 0.362 J, meanwhile for hammer 15 J, the impact strength was 0.472 J. Meanwhile, polypropylene shows the lowest value of impact strength for both hammer. While Figs. 17, 18, 19 and 20 are the fractography using SEM to investigate the fractured surfaces after undergoes tensile loading. Flexural strength of composites and PP is shown in Fig. 20. The reasons for the lower flexural are possibly due to the lower fiber to fiber interaction, void and poor dispersion of fiber in the matrix (Mueller and Krobjilowski 2003). The flexural strength of the

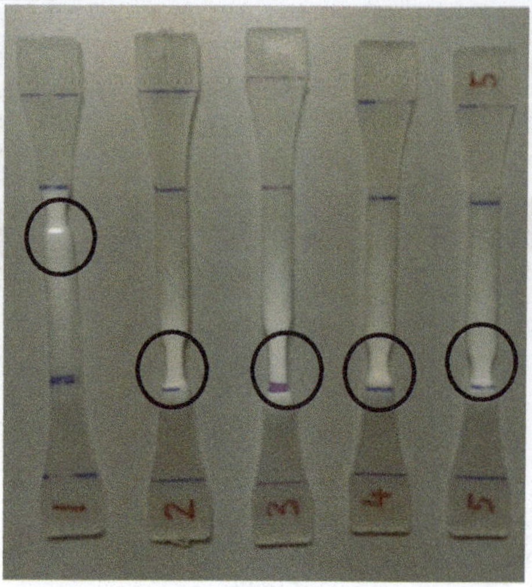

Fig. 14 Specimen after testing. The circles indicate the fracture/necking area of the specimens

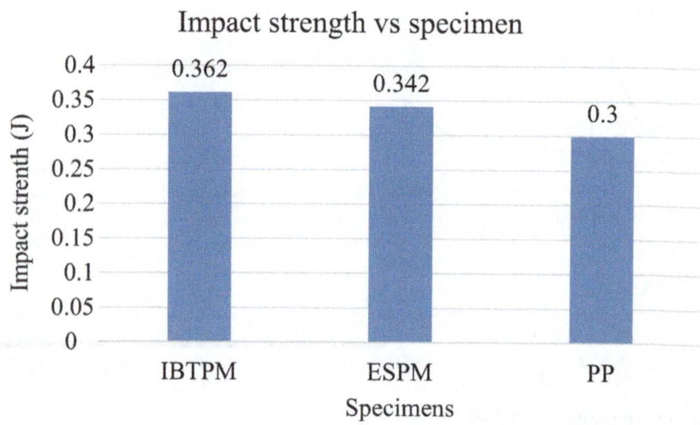

Fig. 15 Impact strength for 7.5 J load

pure polypropylene is 58.40 MPa. Meanwhile for IBTPMC and ESTPMC, both are 42.15 MPa and 53.46 MPa as shown in Table 8.

The wear performance of PMC and PP was observed in terms of applied load, and disc speed as shown in Figs. 21 and 22. Figure 23 shows the correlation of wear rate variation as the applied load and distance increases. It was observed during the test that there was a high material removal process from either matrix or fiber regions. It may be due to the increase in the applied load which influences the wear rate such a behavior was also observed (Jústiz-Smith et al. 2008) as shown in Table 9.

Wear rate for ESPCM has reduced by 64–66% by increasing the load and increased by 87–92% by increasing the sliding distance and time. Wear rate for BITPCM

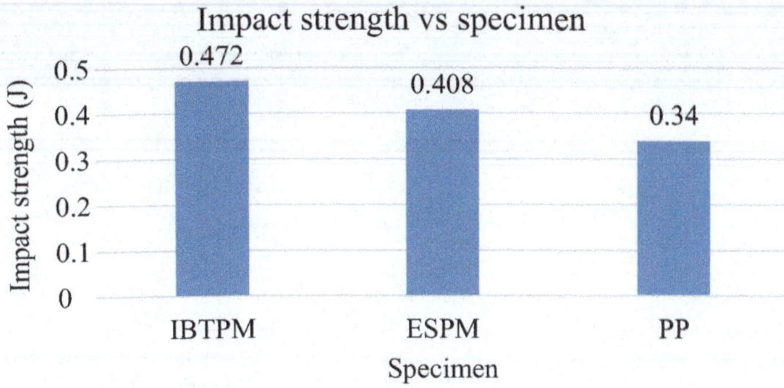

Fig. 16 Impact strength for 15 J load

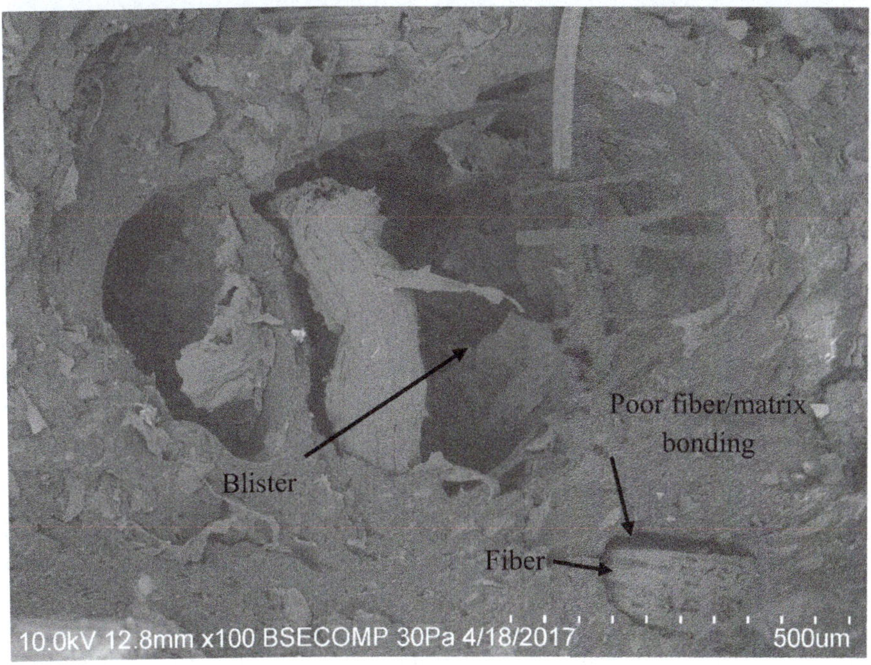

Fig. 17 The fractography of IBTPM for load impact 7.5 J

shows increased by 13 times the initial wear as loading increases and reduced from 1 to 1.3 times as sliding distance increases. From the wear rate Figs. 23 and 24, it shows that the banana fibre ruptured easily leading to higher wear in comparison to egg structure whereby it has lower wear rate compared to banana trunk fibre and polypropylene. On the other hand, eggs shell shows superior wear resistance due to its homogeneous powdered structures as shown in Table 10 and observed in Figs. 27, 28 and 29. Similar observation was found by Raghavendra et al. (2012) on banana

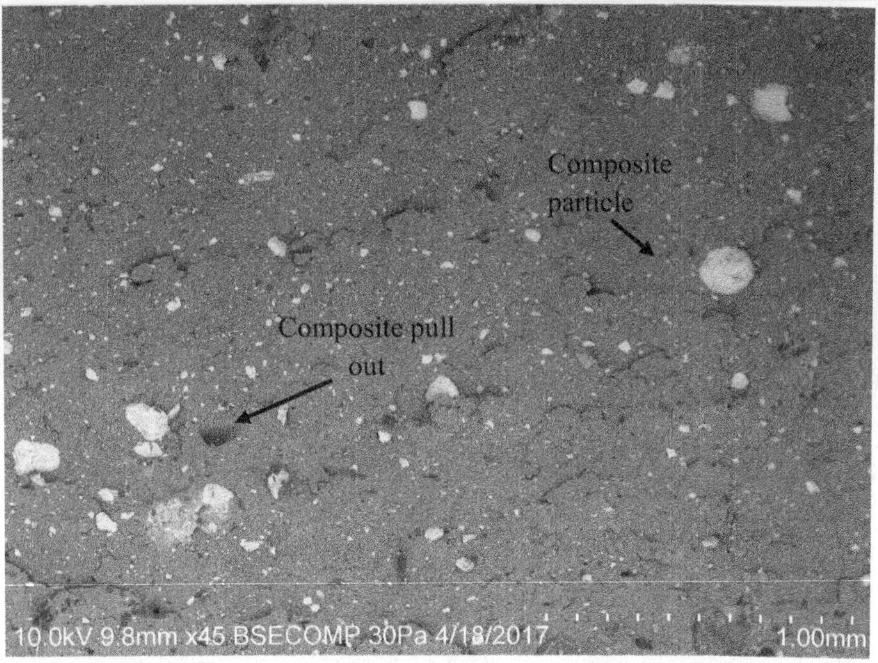

Fig. 18 The fractography of ESPMC for load impact 7.5 J at 1.00 mm magnification

length affecting tensile, (Nirmal et al. 2012) on bamboo fibres and also (El-Tayeb 2008) on sugarcane fibres. Hence, it is obvious that the fibre wettability plays a vital role to abrasion test character whereby eggshells powder distributed evenly to the PP easily to overcome the abrasion wear.

Coefficient of friction (COF) for IBTPMC shows increment from 0.38 to 0.45 with increasing the sliding distance and 0.39–0.55 with increasing loads. While COF of ESPCM shows reduction value from 0.28 to 0.18 with sliding distance and increased from 0.29 to 0.35 with increasing loads. Similar phenomena was reported by (Alavudeen 2015; Gowtham 2014) whereby after cyclic abrasion with increasing sliding distance, the fibre was abraded to form a thin film between the counter surfaces and after successfully removed first layer, it forms a smoother surface, therefore COF reduced after many cycles. Upon increasing the load, the pressure on the contact surface enhance the interaction between the counter surfaces and resulting the increasing on COF. From the frictional coefficient values as shown in Figs. 25 and 26, it shows that the banana trunk composites have higher COF compared to eggs shell and polypropylene composites due to the pultruding and cellulosic properties that resist the motion of the composites upon abrading on the pin on disk plates. In comparison to eggs shell and polypropylene whereby the structures are made of powdered and melted beads formed as pictorial shown in Figs. 27, 28 and 29 as discussed by (Almomani et al. 2020) for eggshells performance and again by (Bhoopathi et al. 2014) for banana-hemp glass fibre. From the wear rate in Fig. 30

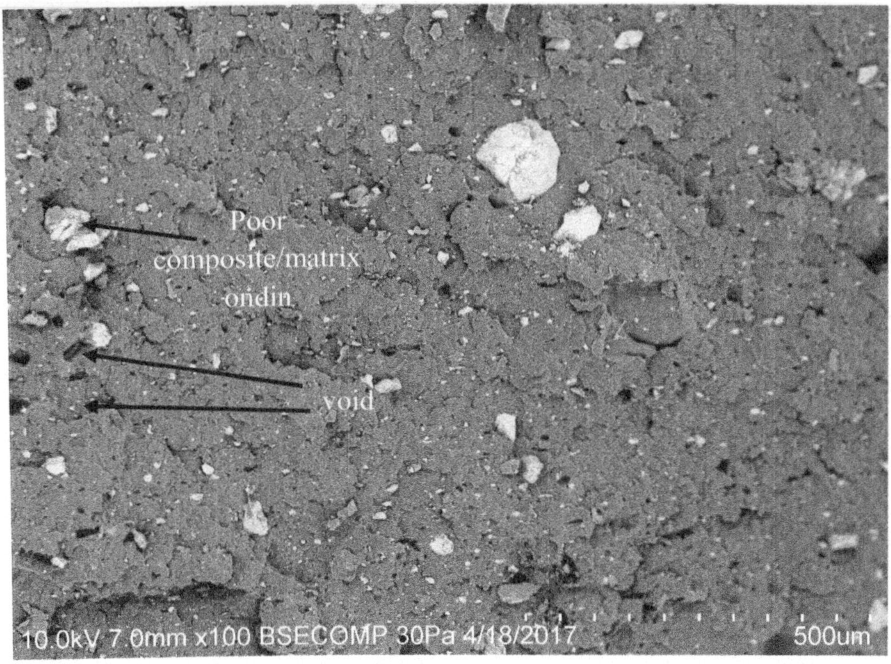

Fig. 19 The fractography of ESPMC for load impact 7.5 J at 500 um magnification

and summarized in Table 11, it shows that the banana fiber ruptured easily leading to higher wear rate with increases of load and speed. This due to cellulosic properties that resist the motion of the composites as proven by (Toro et al. 2007; Gowtham 2014).

The structure of the fractured surfaces due to the mechanical loading is observed through SEM analysis. The SEM micrographs are used to observe the internal cracks, fractured surfaces and internal structure of the tested samples of the composite materials. The SEM micrograph of the sample subjected to tensile loading, impact test, wear and friction is presented as shown in Fig. 31.

4 Conclusion

The material properties of fabricated natural fibre reinforced composites were successfully fabricated using IM and VATRM techniques. The results indicated that the banana fibers exhibited excellent mechanical properties on absorbed the energy during impact test. Meanwhile, eggshell composite has the highest tensile modulus as it is brittle compared to banana fiber and polypropylene. Tensile strength of the PP is the highest in comparison to eggshell composites and banana trunk composites

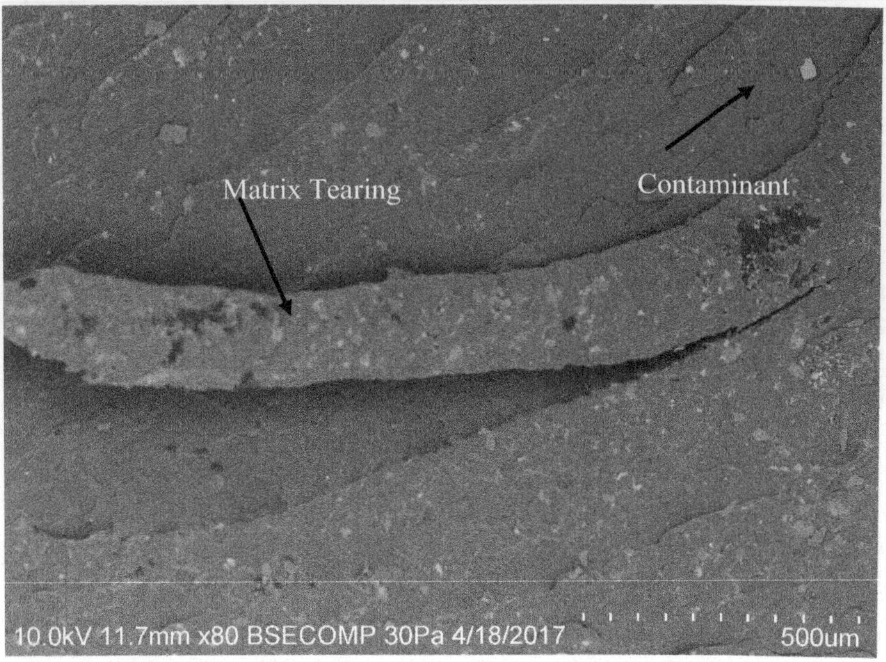

Fig. 20 The fractography of PP for load impact 7.5 J at 500 um magnification

Table 8 Flexural properties of PMC and PP

Sample Number	Flexure modulus (MPa)	Flexure Strain (mm/mm)	Flexure Strength (MPa)
IBTPMC	1314.07	0.06425	42.15
ESTPMC	1886.59010	0.07113	53.46
PP	1496.01520	0.0850	58.40

due to slippage and poor wettability properties of the banana fibre and cracks propagation in eggshell composites. Impact strength increases with the increase in the fiber content. IBTPMC shows the highest impact strength. Specific wear rate and the coefficient of friction of Injection Molding (IM) composites depends on the structure of polymer composite. Banana composite has the highest wear rate due to cellulosic properties. Eggs shell and polypropylene lower wear rate due to the structures are made of powdered and melted beads formed. Specific wear rate and the coefficient of friction of VARTM composite increased with increasing load and speed. The tensile modulus for ESPMC composites is 897 MPa compared to IBTPMC only 561 MPa. In izod impact test, IBTPMC shows the highest impact strength of 0.362 J, which is superior then in ESPMC or pure polypropylene (PP). The flexural strength of the pure PP is 58 MPa while for IBTPMC and ESTPMC, both are 42 MPa and 53 MPa respectively. Wear rate for ESPCM has reduced by 64–66% by increasing

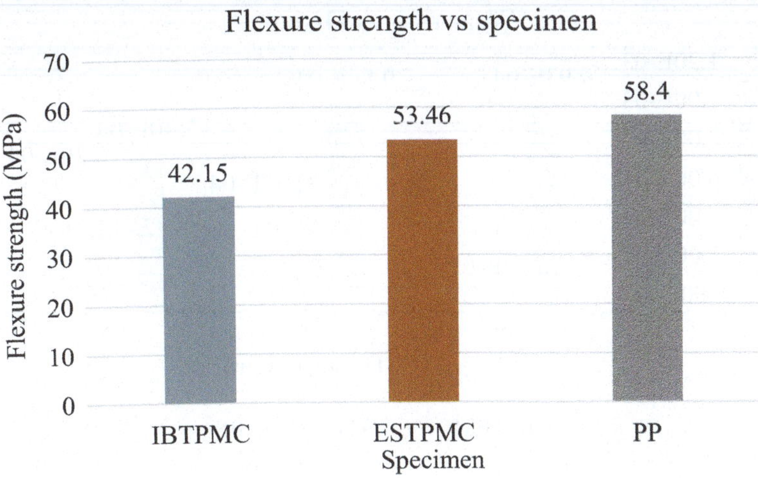

Fig. 21 Flexural strength versus specimen

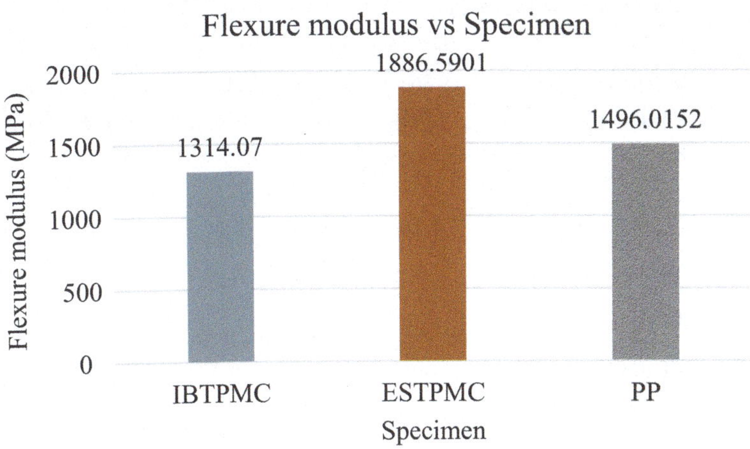

Fig. 22 Tensile modulus versus specimen

the load. Wear rate for IBTPCM shows increased by 13 times the initial wear as loading increases and reduced from 1 to 1.3 times as sliding distance increases. It shows that the IBTPMC fibre ruptured easily leading to higher wear in comparison to ESPCM structure due to fibre wetting properties and distribution of fibre in the matric interface. Coefficient of friction (COF) for IBTPMC shows increment from 0.38 to 0.55. While the COF for ESPCM increased from 0.28 to 0.35.

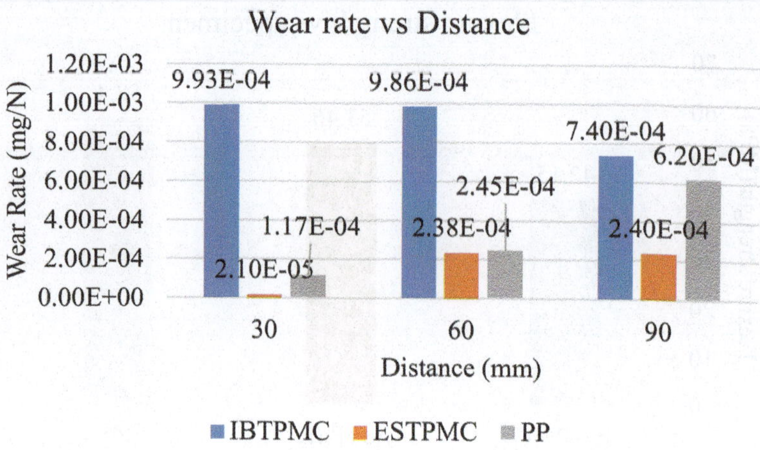

Fig. 23 Wear rate versus distance

Table 9 Table below shows the wear rate of PMC and PP

	Specimens	Distance (mm)			Load (N)		
		30	60	90	10	30	50
Wear Rate (mg/N)	IBTPM	9.93e-4	9.86e-4	7.4e-4	0.011	1.21e-4	9.05e-4
	ESTPM	2.1e-5	2.38e-4	2.4e-4	2.92e-4	0.63e-4	0.34e-4
	PP	1.17e-4	2.45e-4	6.2e-4	0.037e-4	0.057e-4	0.056e-4

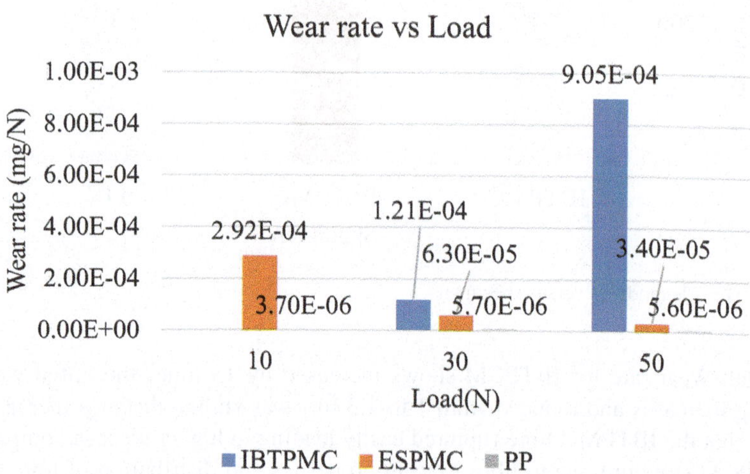

Fig. 24 Wear rate versus Load

Table 10 Table below shows the coefficient of friction (COF) PMC and PP

	Specimens	Distance (mm)			Load (N)		
		30	60	90	10	30	50
Wear Rate (mg/N)	IBTPM	0.38	0.45	0.41	0.39	0.45	0.55
	ESTPM	0.28	0.21	0.18	0.29	0.3	0.3
	PP	0.31	0.36	0.21	0.35	0.37	0.39

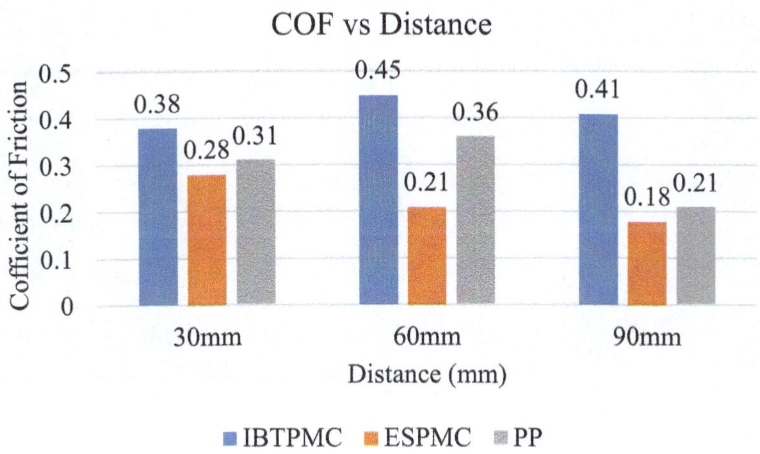

Fig. 25 COF versus Distance

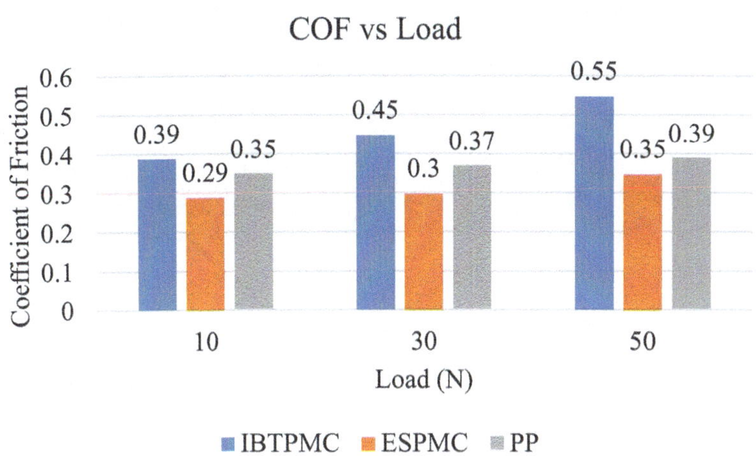

Fig. 26 COF versus Load

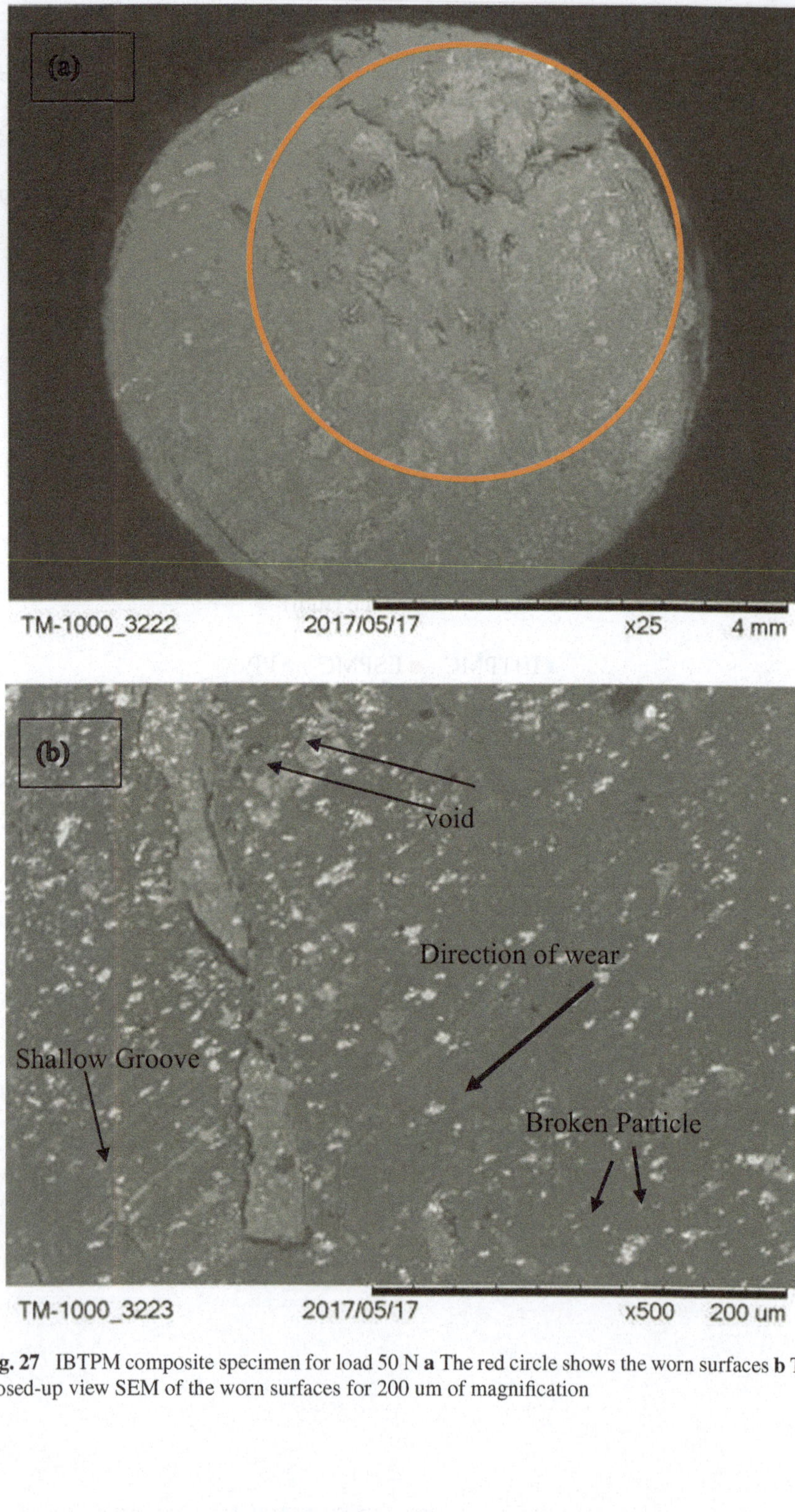

Fig. 27 IBTPM composite specimen for load 50 N **a** The red circle shows the worn surfaces **b** The closed-up view SEM of the worn surfaces for 200 um of magnification

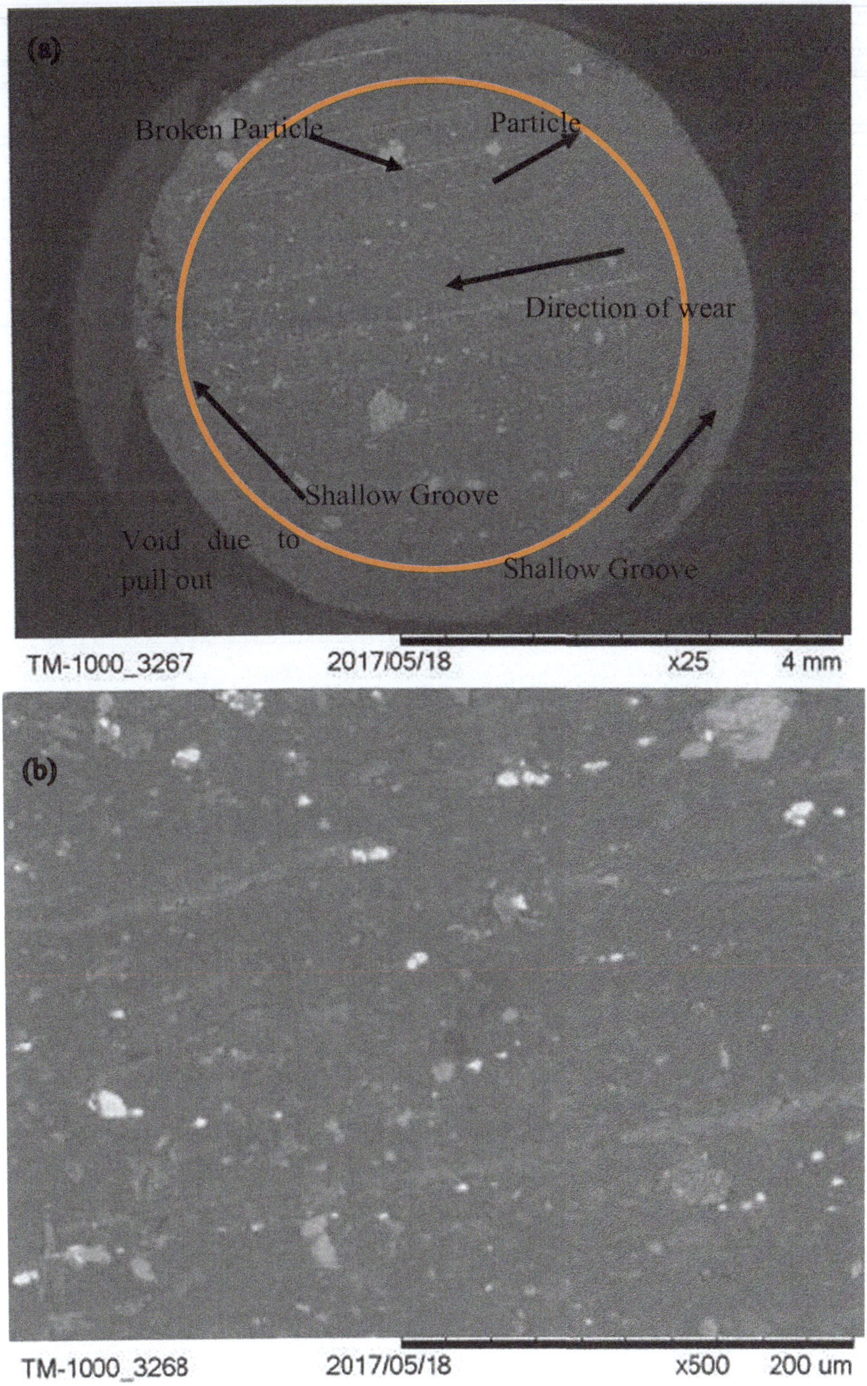

Fig. 28 ESPM composite specimen for load 50 N **a** The red circle shows the worn surfaces **b** The closed-up view SEM of the worn surfaces for 200 um of magnification

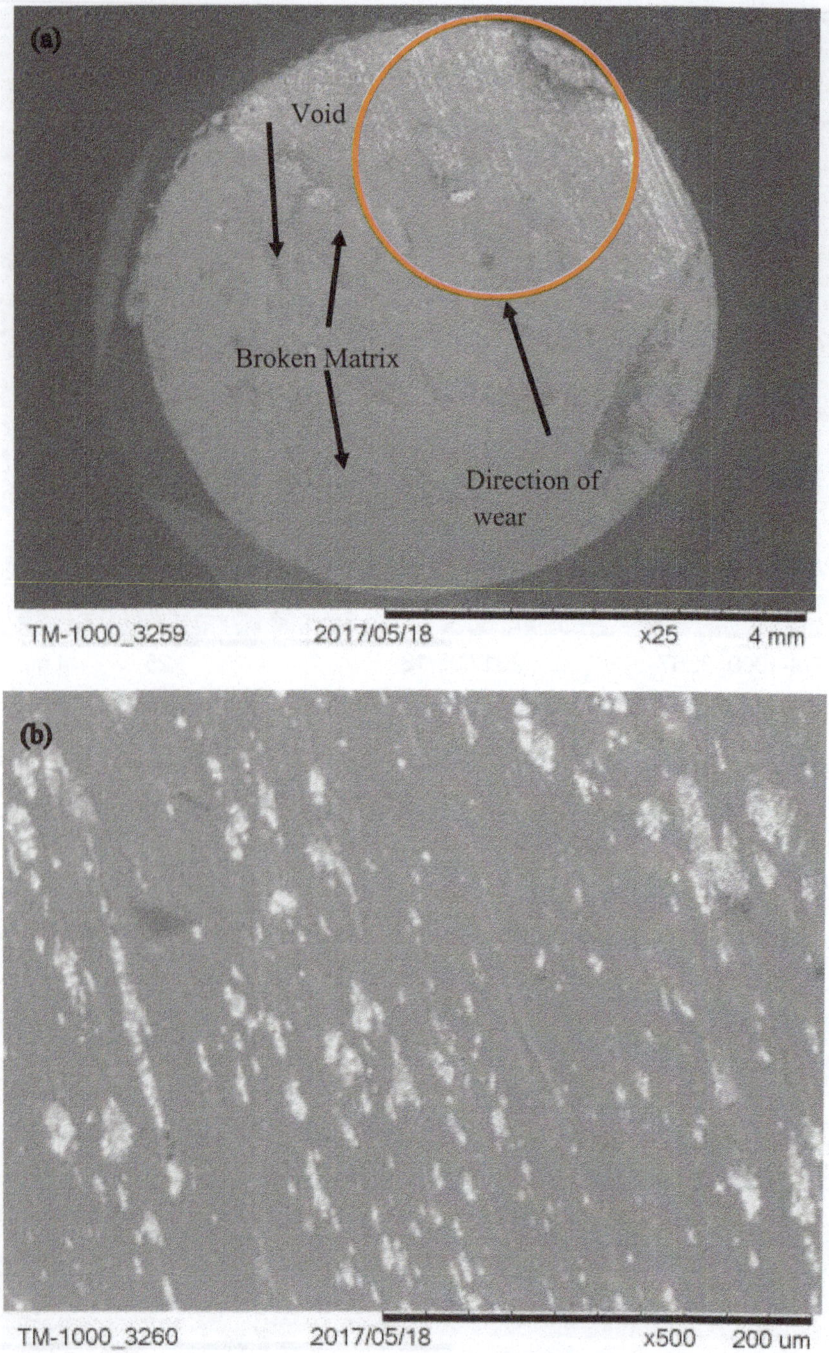

Fig. 29 PP composite specimen under SEM for load 50 N **a** The red circle shows the worn surfaces **b** The closed-up view SEM of the worn surfaces for 200 um of magnification

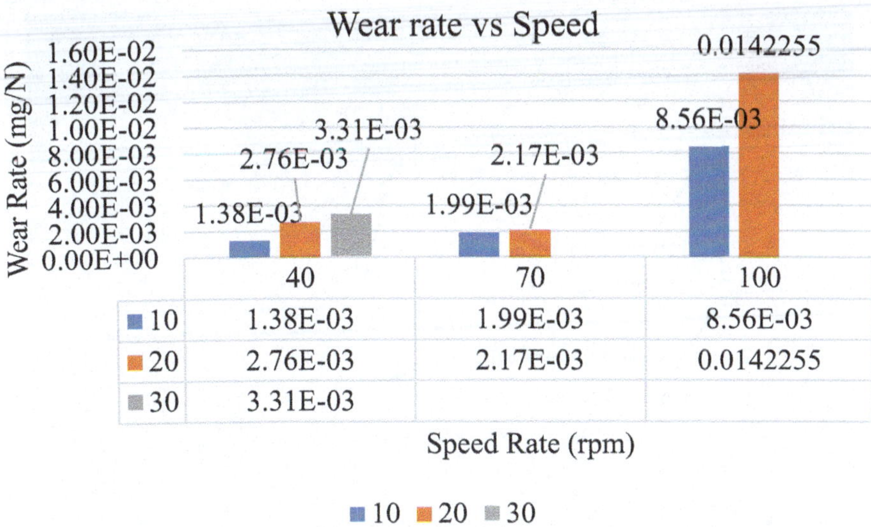

Fig. 30 Wear rate versus speed rate

Table 11 Wear rate of PMC and PP specimens

	Load (N)	Speed (rpm)		
		40	70	100
Wear Rate (mg/N)	10	1.3815e-3	1.990e-3	8.558e-3
	20	2.761e-3	2.1715e-3	0.0142255
	30	3.311e-3	0.00001	0.00001

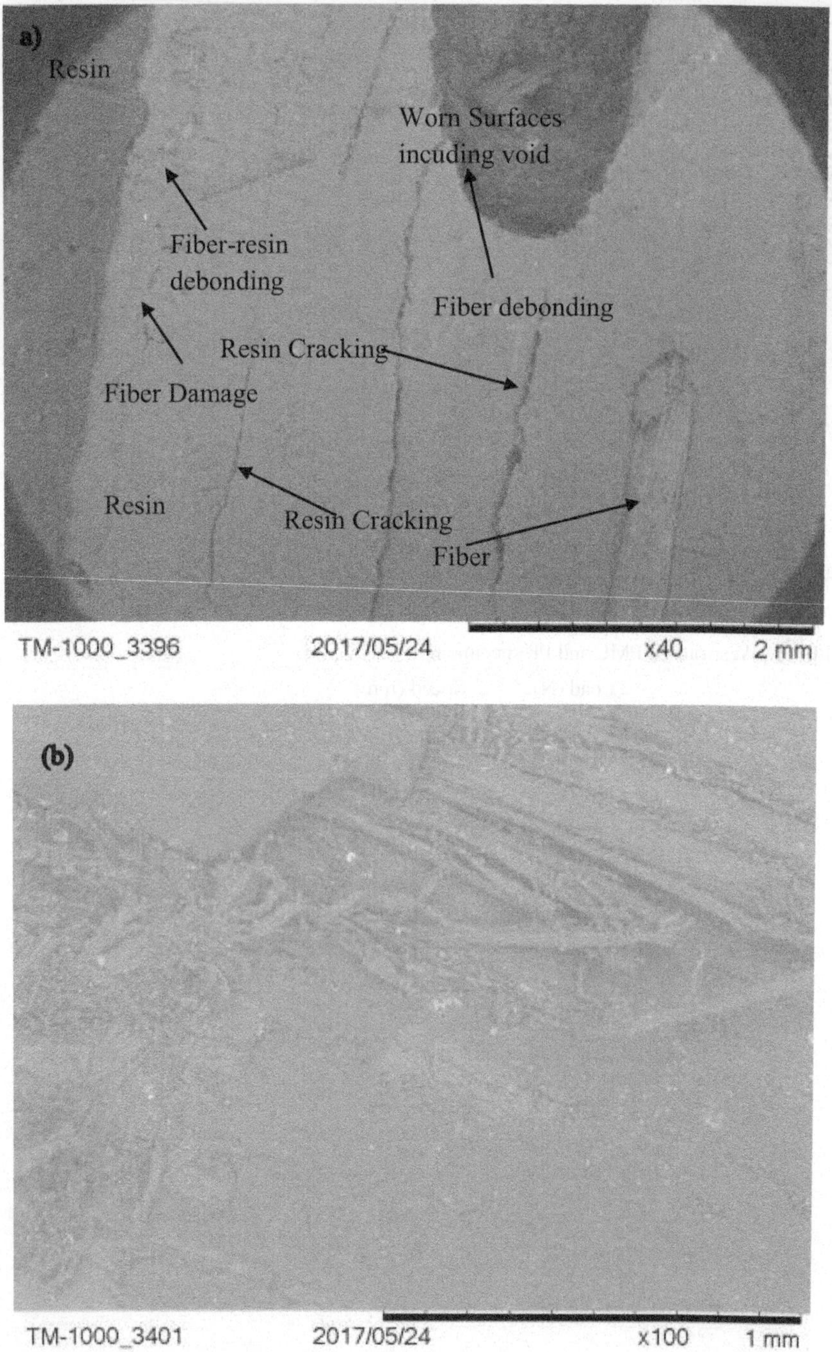

Fig. 31 **a** SEM on VARTM specimen for 40 rpm and 10 N. **b** SEM on VARTM specimen for 40 rpm and 30 N. Figure above shown some of the damage occurred on the worn surfaces

References

Alavudeen A (2015) Mechanical properties of banana/kenaf fiber-reinforced hybrid polyester composites: effect of woven fabric and random orientation. Mat Des 66:246–257

Almomani MA, Hayajneh MT, Al-Shrida MM (2020) Investigation of mechanical and tribological properties of hybrid green eggshells and graphite-reinforced aluminum composites. J Brazilian Soc Mech Sci Eng 42:45

Babu GD, Babu KS, Kishore PN (2014) Tensile and wear behavior of calotropis gigentea fruit fiber reinforced polyester composites. Proc Eng 97:531–535

Bhoopathi R, Ramesh M, Deepa C (2014) Fabrication and property evaluation of banana-hemp-glass fiber reinforced composites. Proc Eng 97:2032–2041

Bilba K, Arsene M-A, Quesanga A (2007) Study of banana and coconut fibers: botanical composition, thermal degradation and textural observation. J Bioresource Tech 98:58–68

Bootklad M, Kaewtatip K (2013) Biodegradation of thermoplastic starch/eggshell powder composites. Carbohydr Poly 97(2):315–320

Bozzelli J (2011) Eliminate bubbles, voids, sinks & blisters. Prospector Knowledge Center

El-Tayeb NSM (2008) A study on the potential of sugarcane fibers/polyester composite for tribological applications. Wear 265(1–2):223–235

Geethama VG, Mathew KT, Lakshminarayana R, Thomas S (1998) Composite of short coir fibers and natural rubber: effect of chemical modification, loading and orientation of fiber. J Poly Sci Pert B Polym Phys 65:1483–1491

Gowtham SMA et al. (2014) Wear and friction peformance of banana-borassus fruit fibre reinforced composites. Tech Lett 1(6):29–32

Hassan SB, Aigbodion VS, Ptrick SN (2012) Development of polyester/eggshell particulate composites. Tribol Industry 34(4):217–225

Igwe OI, Onuegbu CG (2012) Studies on properties of eggshell and fish bone powder filled polypropylene. Am J Poly Sci 2(4):56–61

Jústiz-Smith NG, Virgo GJ, Buchanan VE (2008) Potential of Jamaican banana, coir, bagasse fiber as composite materials. J Mat Char 59:1273–1278

Kulkarni AG, Satyanarayana KG, Rohatgi PK, Vijayan K (1982) Mechanical properties of banana fibers. J Mat Sci 18:2290–2296

Kumar NR, Prasad GR, Rao BR (2013) Investigation on mechanical properties of banana fiber glass reinforced hybrid thermoplastic composites. Int J Eng Res Tech (IJERT) 2(11):3701–3706

Kumar R, Dhaliwal JS, Kapur GS, Shashikant (2014) Mechanical properties of modified bio-filler polypropylene composite. Poly Comp 35(4)

Long CC (2016) Development popypropylene-modified chicken eggshell composites. https://eprints.utar.edu.my/2037/1/PE-2016-1105263-1.pdf. Accessed 18 Feb 2020

Materials, ASTM, Standard test method impact resistance of plastics and electrical insulating materials. D256-93a. . Philadelphia, Pa ASTM, 1995

Materials, ASTM, Standard test method for tensile properties of plastics. D-638. Philadelphia, Pa: ASTM, 1995

Materials, ASTM, Standard test method for Flexural properties of unreinforced and reinforced plastics and electrical insulating materials. D790-92. Philadelphia, Pa, ASTM, 1995

Mueller DH, Krobjilowski A (2003) New discovery in the properties of composites reinforced with natural fibers. J Ind Text 33(2):111–130

Murali Mohan Rao K, Mohana Rao K, Ratna Prasad AV (2010) Fabrication and testing of natural fibre composites: Vakka, sisal, bamboo and banana. Mat Des 31(1):508–513

Niklas A, Ha SSJ (2001) Modern tribology handbook. In: Bhushan B (ed) Friction and wear measurement techniques. CRC Press LLC

Nirmal U, Hashim J, Low KO (2012) Adhesive wear and frictional performance of bamboo fibres reinforced epoxy composite. Trib Int 47:122–133

Nirmal U, Hashim J, Megat AMMH (2015) A review on tribological performance of natural fibre polymeric composites. Trib Int 83:77–104

Raghavendra S, Balachanderashetty P, Mukunda PG, Sathyanarayana KG (2012) The effect of fiber length on tensile properties of epoxy resin composite reinforced by the fibers of banana. Int J Eng Res Technol (IJERT) (16)

Rajiv K, Ankush A (2019) Tribological behavior of natural fiber reinforced epoxy based composites: a review. Mat Today Proc 18(2019):3247–3251

Satyanarayana KG, Sukumaran K, Kulkarni AG, Pillai SGK, Rohatgi PK (1984) Performance of banana fabric polyester composites. In: Marshall IH (ed) Proceedings of second international conference on Comp Str, pp 535–537

Satyanarayana KG, Sukumaran K, Mukherjee PS, Pavithran C, Pillai SGK (1990) Natural fiber-polymer composites. J Cement Concrete Comp 12:117–136

Toro P, Quijada R, Yazdani-Pedram M, Arias JL (2007) Eggshell, a new bio-filler for polypropylene composites. Mat Lett 61(22):4347–4350

Venkateshwaran N, Elayaperumal A (2010) banana fiber reinforced polymer composites—a review. J Reinf Plas Comp 29(15):2387–2396

West CP (2005) Standard test method for wear testing with a pin-on-disk apparatus ASTM G99-05. ASTM Int

Friction and Wear Performance of Nano Hydroxy Apatite (nHAp) Polyoxymethylene Composites on 316L Steel

Shubrajit Bhaumik, Rajeswar Bandyopadhyay, Tanveer Ahamed Rohit, Anik Banerjee, Helen Annal Therese, and Rajan Pathak

Abstract Biomedical applications are important aspects in mankind. The word "Biomedical" is recently been used in the researches where the materials are made or synthesized from biological sources or these materials are used for various treatment of living beings. Recently in biomedical applications, polymers have paved their way in various applications such as implants, grafts, connective tissues etc. and hence, understanding the tribological properties of these polymers becomes an important factor to estimate their performance index. Polymethyl methacrylate (PMMA), poly ether ether ketone (PEEK) and Ultra-high-molecular-weight polyethylene (UHMWPE) has been extensively used for biomedical applications but polyoxymethylene (POM) has not been explored in depth till date. This chapter focusses on investigating the tribological properties of nano hydroxy apatite (nHAp) added polyoxymethylene (POM) composites using a pin on disc tribometer. nHAp prepared from egg shells by wet-precipitation method, were added in various weight percentages (1%, 2%, 3%, 4%, and 5%) to POM. Several pins made of POM reinforced with nHAPwere used to investigate the tribological properties using a pin on

S. Bhaumik (✉) · R. Bandyopadhyay · T. A. Rohit · A. Banerjee
Tribology and Surface Interaction Research Laboratory, Department of Mechanical Engineering, SRM Institute of Science and Technology, Kattankulathur, India
e-mail: shubrajb@srmist.edu.in

R. Bandyopadhyay
e-mail: rajeswar.bandyopadhyay51@gmail.com

T. A. Rohit
e-mail: tanveerbunny@gmail.com

A. Banerjee
e-mail: banerjeeanik04@gmail.com

H. A. Therese
Department of Chemistry, Faculty of Engineering and Technology, SRM Institute of Science and Technology, Kattankulathur Campus, Kattankulathur, India
e-mail: helena@srmist.edu.in

R. Pathak
Research Institute, Central Research Facility, SRM Institute of Science and Technology, Kattankulathur, India
e-mail: rajang@srmist.edu.in

M. T. Hameed Sultan et al. (eds.), *Tribological Applications of Composite Materials*, Composites Science and Technology, https://doi.org/10.1007/978-981-15-9635-3_6

disc tribometer. A 316L stainless steel disc was used as counter surface to these pins during the tribo test. The shore hardness of the POM composites were measured which exhibited an increase in hardness with the increase in concentration of nHAp in POM. The tribo tests results exhibited a decrease in the coefficient of friction (CoF) up to a certain concentration of nHAp (4% this case), beyond which the coefficient of friction increased. In case of wear rate, the lower concentrations of nHAp showed less wear rate but the wear rate increased with the increase in concentration of nHAp. 1–4% nHAp included POM pins exhibited as low as 77% wear rate compared to POM compsite which did not contain any nHAp particles. The presented results indicates that POM can be a good choice for several biotribo pairs, however the biocompatibility of POM is to be explored. The present work will be helpful in making various composites for implants once the biocompatibility of POM is established.

Keywords Nano hydroxy apatite · Polyoxymethylene · Tribology

1 Introduction

This chapter aims to find a cost effective, easily available bio-material, for biomedical application in bones and teeth. The main constituent of human bones and teeth is calcium (Abou et al. 2016).With increasing harmful radiations and different kinds of pollution the human body is prone to various diseases among which, osteoporosis has become a widely progressive bone disease, which decreases bone density and mass (Letarouilly et al. 2018), resulting in bone fracture (Nishizwa et al. 2018) but with the advancement of medical field, the solution comes in the form of various implants. Titanium remained the most popular implant due to its ability to form excellent oxide layer (Williams 1981), followed by Tetragonal zirconia polycrystals with low porosity and high bending (Adatia et al. 2009) and Titanium zirconium alloys with better mechanical attributes (Chiapasco et al. 2011). Nowadays bioactive ceramics have gained much attention due to their osteo-conductive properties and ability to promote the formation of continuous bone–ceramic interface, therefore allowing the implant fixation mechanism (Pereira et al. 2003). Considerable research studies have been devoted to the synthesis of various bio-ceramics (Ayatollahi et al. 2015). The mechanical properties of bio-ceramics are usually quite different from those of natural tissue, in particular their high elastic modulus and low toughness and thus, restricts the use of these materials in a wide range of applications (Pielichowska 2012). Amongst different classes of bio-ceramics, hydroxy apatite (HAp) [$Ca_{10}(PO_4)_6(OH)_2$] is the most emerging bio-ceramic, which is extensively used in various clinical applications such asorthopedics and dentistry (Navarro et al. 2012). HAp is a crystalline form of calcium phosphate (Kay et al. 1964). The intense use of HAp has been due to its special chemical composition, and also its biological and crystallographic similarity with the mineral portion of hard tissues. There are two main methods for producing HAp; the first is inorganic synthesis such as wet-chemical method, sol–gel method and hydrothermal method (Fathi and Hanifi 2007; Cengiz et al. 2008; Monmaturapoj

2008). Natural raw materials such as eggshells, cuttlefish shells, and bovine bonesare commonly used to synthesize Hap (Ooi et al. 2007; Kamalanathan et al. 2014). Hydroxyapatite is an easily producible cost-effective material which has already been identified as very good substitute material for human bones and teeth due to its similarity in structure with bones and teeth (Vecchio et al. 2007).

HAp supports osteogenesis, a physiology term for bone formation and it is also osteoconductive which is a type of bone healing process in which the bones grow through the micro pores available in the implant and binds with the base material as its own (Ripamonti et al. 2009). It has also shown the property of forming osteoblast which is a cell that secretes the material of bone tissue (Liet al. 2018). A good bone replacement material is the one that will allow the new cells to easily bond with bone tissues and also facilitate the growth and penetration of new cells. And for this to happen the material must be osteophilic and porous. This will help the new tissues to grow easily and not let the implant loosen or move. Hydroxyapatite is observed to possess minerals found in bone and also have outstanding osteophilic properties. A huge amount of work had been done to develop porous scaffold of hydroxy apatite. The grain growth and sintering of hydroxy apatite were studied to find an optimum temperature for the preparation of porous scaffolds of hydroxyapatite (Saiz et al. 2007). These porosities should be interconnected to facilitate the ingrowth of cells, diffusion of the nutrients and also vascularization. Also, it had been observed that the porous interconnection which are less than 10 μm will not permit cell migration (Simske et al. 1997; Itälä et al. 2001; Tamai et al. 2001). Hence, an approximate size of around 100 μm is needed for the mineralized in growth as established by Klawitter et al. (1971). Biocompatibility is one of the most wanted properties in the bone replacements. Vecchio et al. (2007) synthesized biocompatible hydroxyapatite from seashells for bone implants. Shell due to their dense and tailored structure were selected as a natural source. The seashells were converted to hydroxyapatite by means of hydrothermal methods. The produced hydroxyapatite was mechanically tested and the results were found to be close to that of bones. Then it was tested on rat femoral for 6 weeks. At the end of 6 weeks the images of microtomography revealed that there were no displacements of the implants. Formation and growth of bone tissues around the implants were found in the histological studies. Fibrosis ring tissues were not found around the implants which indicated that there was no loosening of the implants. Thus, it indicates towards the excellent biocompatible and bioactive nature of hydroxyapatite and hence it's uses as bone implants. Hydroxyapatite from egg shells using phosphoric acid were prepared by (Lee and Oh 2003). The synthesized hydroxyapatite was highly sinterable and pure. The microstructure of the produced powder was studied using XRD(X-Ray Diffraction) and SEM(Scanning Electron Microscope) images, which concluded that egg shell is a potential source for the synthesis of hydroxy apatite.

The advantage of natural raw materials for synthesizing HAp (Shashvatt et al. 2017) and better bonding of HAp with bone tissues (Swetha et al. 2010) makes HAp favorable but the intrinsic brittleness and poor strength of sintered HAp restricts its clinical applications under loadbearing conditions (Curtin and Sheldon 2004; White et al. 2007). However, polymer matrices reinforced with a bioactive phase such as

HAp combine the typical bioactive behavior of bio-ceramics with enhanced mechanical properties that make them comparable to human tissue (Shuai et al. 2016).In the cartilage tissue-engineering study, a recently adopted perfusion bioreactor system is in use to grow cell-scaffold composite constructs (Baskaran et al. 2003). The mechanical properties of polyoxymethylene, its tensile properties (76 MPa ultimate at 23 °C), machining characteristics, shear strength, toughness, stiffness, and thermal properties, and also its biocompatible nature were recorded through various characterizations. Focusing on the medical adaptability of implants, scaffold or the broken bones and cartilage has witnessed a wide density of mesenchymal cell with uniform spatial density.

In 2006, researches have obtained a method of vacuum-aided seeding technique to produce Hyaff-sponges (Hyaff is a hyaluronic acid-based biomaterial) and also tested its properties and claimed that it is safe and most importantly reproducible and the process is rapid as required, with controlled cell loading and uniform distribution it can be installed in human body in places of scaffold (Solchaga et al. 2006). Studies were conduction based on the two-phase composite grafting made up of calcium phosphate and Hyaff sponges in the osteochondral defects and experimentation has claimed that it actually provides a mechanical base to the superficial scaffold or broken bone, cartilage to repair(Garg and Hales 2004).Constructs were made based on mesenchymal stem cell which are used to repair cartilage and bones by use of a chondrogenic medium (Solchaga et al. 2002).Polymer POM, acetal homopolymer, polyacetal is unstable, and reverts to the monomer upon heating at 120 °C (Solchagaet al. 2006). Laluppa et al. (1997) also claimed through his research that POM is not biocompatible in human body but Penick et al. (2005) repeated the same experiment and found completely opposite results as previously mentioned. He confirmed it after testing his models with mesenchymal cells derived from human bone that with the POM and secondly with human articular chondrocytes and POM and clearly explained that the base, he used in his experiment is least harmful for human body even after repeated machining and autoclaving (Penick et al. 2005). POM was converted into prototypes of various body part and samples were drilled and holes were made, threads were tapped to coin down results of various experiments conducted, which results showed up excellent impact strength, low friction, and good electrical insulator when constrained in various directions (Agnihotri et al. 1999). It is dimensionally stable, wear-resistant to an extent and chemically inert to solvents. Its melting point is (as result obtained in Differential Scanning Calorimeter (DSC)) 167 °C (Fig. 1), which is well above the normal steam autoclave temperature and thus, has the ability to retain its shape and withstand high heat (Fister et al. 1985).

POM also has some excellent biocompatibility and the characteristics which is proved and has already been studied (Agnihotri et al. 1999). POM has a negligible porosity and low moisture. The US food and drug administration under Code of federal regulations under title 21 has clearly justified the use of POM in Food Packaging under various condition and concluded its safety as food packaging. POM also has long history of uses in animal studies and a long-term implant material in variety of medical application. These includes cardiac valve prostheses (Björk 1972),

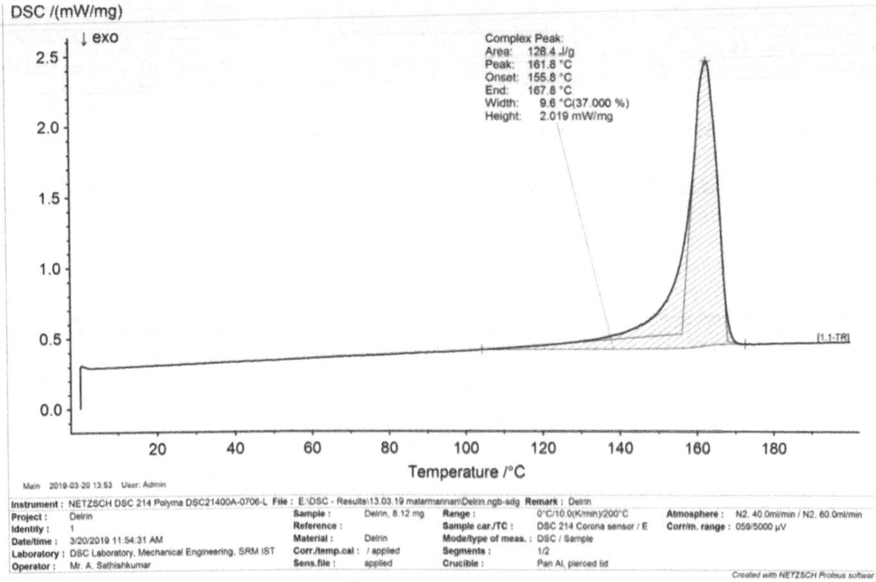

Fig. 1 Differential Scanning Calorimeter reading of POM sample used in this work

temporomandibular joint reconstruction (MacAfee and Quinn 1992) and also tilting disc valve prosthesis (Björk 1969).

2 Fabrication of Nano Hydroxy Apatite (nHAp)POM Composites

nHAp was prepared from egg shells (Rivera et al. 1999) by wet precipitation method (Cahyaningrumet al. 2018). X-Ray Diffraction was used to determine the phases of the synthesized. Figure 2 shows the steps that was followed to produce nHAp in this work. Egg shells were collected and washed using distilled water to remove any dirt or impurities present and then were dried at room temperature. Then the egg shells were crushed finely and were kept in alumina crucible which were thermally decomposed to calcium oxide by heat treatment in muffle furnace at 900 °C for 8 h starting with first heating the sample to 450 °C in 2 h then isothermal heating at 450 °C for 2 h and then again heating the sample from 450 to 900 °C in 2 h and then again isothermally heating the sample for 2 h at 900 °C. It was then cooled inside the furnace overnight and the resultant product formed was calcium oxide.

A solution of calcium hydroxide was prepared using the produced calcium oxide and distilled water stoichiometrically as per requirement. Phosphoric acid was added dropwise to the mixture at 60 °C with a constant stirring using glass rod till the pH of the mixture was 7. The pH was measured using table pH meter. Then heat was

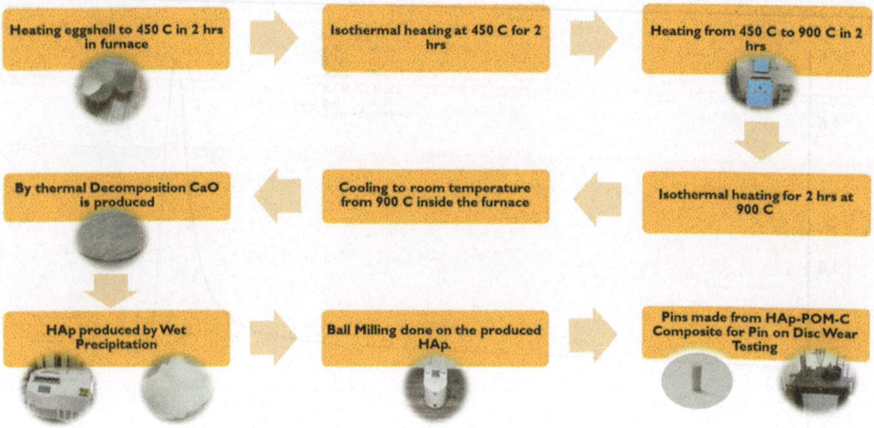

Fig. 2 Flowchart of the nHAp produced in this work

removed and then sodium hydroxide solution was added to the mixture till the pH was 11. After then the mixture was left overnight for 24 h to form proper precipitate. After proper precipitate formation the solution was decanted and washed using distilled water until the pH was bought down to 7. At every wash the water was filtered out using filter paper. After the required pH i.e. 7 was obtained, the precipitate was allowed to dry normally in a petri dish. The dry precipitate was then kept in alumina crucible and was sintered isothermally inside muffle furnace at 900 °C for 2 h. It was allowed to cool overnight inside the furnace. At the end of the heat treatment a fine white powder was obtained which is nothing but HAp. Then the sintered HAp was made finer by ball milling in ball mill apparatus for 3 min in the presence of toluene. Figure 3 shows the scanning electron microscope images and XRD analysis of nHAp. The SEM images shows that the size of the nHAp varies from 79 to 500 nm

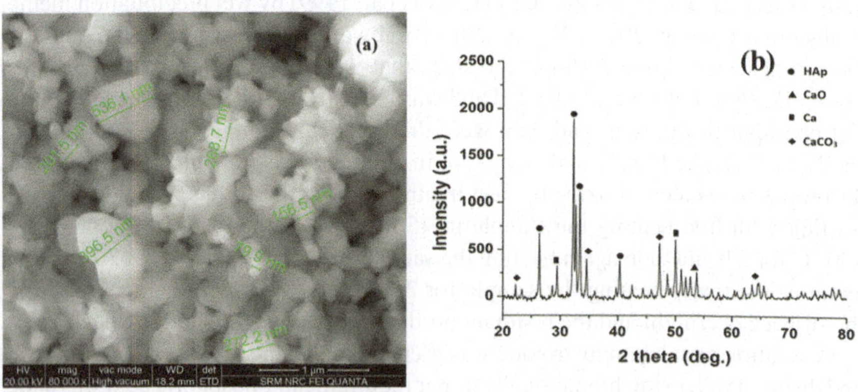

Fig. 3 **a** Scanning electron microscope images **b** XRD of the nano nHAp particles

Table 1 Shore hardness of the pins

Composite	Shore D hardness
0% nHApPOM	70.8
1% nHApPOM	72.6
2% nHApPOM	73.6
3% nHApPOM	74.8
4% nHApPOM	74.2
5% nHApPOM	74.6

(Fig. 3a). Sharp peaks of CaO, $CaCO_3$, and nHAp were detected using XRD (Fig. 3b) indicating acceptable purity of the synthesized nHAp powders.

The synthesized nHAp was then mixed with POM by weight percentages (0%, 1%, 2%, 3%, 4%, and 5%) (Pielichowska 2012). The mixture was melted and stirred thoroughly for homogeneous composition and was then made into pins as per the required dimensions (10 mm diameter and 25 mm length) using injection molding method (500 bar and 170 °C). POM is a high-performance acetyl resin with several desirable physical, mechanical and biomedical properties. Its approach and compatibility with the human body is already well established (Pielichowska 2012). The shore hardness of the pins was found out using as per ASTM (American Society for Testing and Materials) D2240 hardness test method and tabulated in Table 1. The tribological tests were carried out in pin-on-disc tribometer under dry conditions with low load 80 N and 156 rpm (Vinoth and Datta 2019). A disc of SS 316L was used as it is widely used in biomedical applications (Chakraborty et al. 2016; Amanov et al. 2017).

3 Investigating the Tribological Properties of nHAp-POM Using a Pin on Disc Tribometer

The anti-wear properties of the nHAp-POM composites were investigated using a pin on disc tribometer (Fig. 4). The pins were made using injection molding as discussed in Sect. 2. The disc made of 316L was rotated at a speed of 156 rpm under 8 N. The tests were performed in dry conditions. Each surface of the pin was tested for 3600 s. It is to be noted that the polymers have a tendency to align themselves along the direction of sliding (Luengo et al. 2000) which affects the wear behavior hence, the pin was rotated 90° after every 900 seconds in order to avoid the unidirectional alignment of the polymer structures so that the polymer does not get oriented in a single direction due to the constant force being applied in a particular direction.

From Fig. 4a, it can be seen that nHAp did not influence the coefficient of friction till the concentration of nHAp reached 4%. The polymer composite containing 4% nHAp showed a decrease of 21% CoF as compared to pure POM pin. The specific wear rate of the pins was calculated as per Eq. 1.

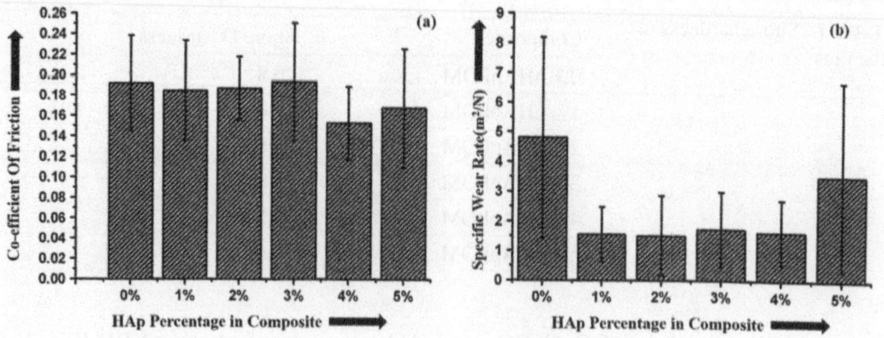

Fig. 4 **a** Coefficient of friction, **b** Specififc wear rate of nHApPOM composites

$$\text{Specific wear rate} = \frac{\Delta w}{\rho L D} \tag{1}$$

where, Δw is the difference in weight of the pin before and after the test, ρ is the density of the pin, L is the applied load and D is the wear track diameter.

The specific wear rate of the pins is shown in Fig. 4b. It can be seen that the nHAp composites have exhibited less wear rate than pure POM. The wear rate was less in lower concentrations of nHAp but increase with the increase in concentration of nHAp. 1–4% nHAp included POM pins exhibited as low as 77% wear rate compared to POM pin which did not contain any nHAp particles. The wear rate of the pins increased as the nHAp concentration in the pin reached 5%. Even though the wear rate was less in all nHAp-POM pins by 22–77% as compared to the 0% nHApp in containing but an increase in the wear rate can easily be seen, indicating failure at higher concentrations of nHAp in the polymer (5% in this case). Looking into the formation of the surfaces after the test (Fig. 4b), it can be seen that 2% and 4% nHAp-POM pin exhibited better anti wear behavior as compared to others. More damaged surfaces can be seen in 5% nHAp-POM where a large amount of material accumulated at the end of the contact surfaces. Chunk of material also seem to have been removed from the surface in case of 5% nHAp-POM pin. Therefore, it can be seen that nHAp nano particles can enhance the tribological properties of POMup to a certain concentration.

4 Understanding the Tribo Mechanism Involved During the Test

The behavior of polymers during the tribo test depends on many factors, but the foremost task is to analyze the structure of the polymer (Pielichowska et al. 2015). Polyoxymethylene is a (–CH_2O) polymeric chain classified under thermoplastic polymer, with wide range of application in field of mechanical, electrical, biomedical and automobile (Shi et al. 2009). Kern et al. (1961) reported that copolymer shows a better thermal and mechanical stability over its homopolymer. POM used in this work is a copolymer in nature. Hu et al. (2007) reported that POM is a crystalline polymer having excellent friction and wear property. POM has a hexagonal structure with folded crystal chain (FCC) and extended chain crystal (ECC), (Iguchi 1973) which provides POM a stable crystalline structure and high dimensional stability (Li et al. 2005).The mechanical properties of a homopolymer is very different from a copolymer and thus, the polymer stability depends on the processing of the polymer. According to Brew and Ward (1978), processes like extrusion, cold and hot drawing enhances the mechanical properties of POM polymer along with high resistance to wear and indentations and hence, the process of extrusion has been adopted in this work to prepare the nHAp-POM pins. nHAp played a major role due to its biocompatibility and according to Pielichowska et al. (2012) it's resistance to change the crystallinity of the polymer.

In the present work, there has been a gradual decrease in the wear rate from 0 to 4%. The decreasing trend can be explained due to crystallinity and transfer mechanism in between polymer and metal hard surface (Vande and Bahadur 1995). Addition of nHAp enhanced the mechanical properties of the polymer and the hexagonal structure became more stable due to the effective nucleating site in the polymerization process (Pielichowska et al. 2011). Addition of nHAp also increases the Young's Modulus of the pin thus, decreasing the chances of failure and hence has low wear rate (Pielichowska 2012). On investigating the hardness of the samples (Table 1), it was observed that there was a slight increase in the hardness with the addition of nHAp in POM. The hardness is the resistance of deformation of the materials, thus the normal hardness measured due to deformation of the surface due to the penetration of the indenter seems to have also played an important role in controlling the wear rate. The sliding of the pin on the disc caused a dynamic deformation of the surface specifically the ploughing mechanism due to the interfacial friction between the pin and disc (Ayatollahi et al. 2015). As observed from Fig. 4a, it can be seen that deep grooves were formed on the surfaces of the pins which did not contain nHAp, while the surface of the pin containing 4% nHAp did not show any damage, thus indicating a better wear resistivity of the nHAp containing samples up to a certain nHAp concentration.

The formation of transfer layer also depends on the type of the tribo surfaces. In the present work the counter bodies are polymer and stainless steel 316L, which is a low carbon steel specifically used for biomedical applications (Chandrasekaran 2019). Due to low carbon content 316L is less brittle and has low hardness (austenized steel), (Song et al. 2011). This makes SS 316L suitable for high machining operations which helps to gain highly polished surface with minimal irregularities or defects, The presence of 16.30% chromium makes the fatigue failure stress high compared to many other steel which ultimately makes the wear rate low when POM is used as pin (Li and Bell 2004). In case of polymer interacting with hard surface, there are high chances of transfer layer of polymer on the hard surface. This transfer mechanism in polymer over a hard smooth surface takes place as follows:

1. With the sliding the polymer particle get transferred to the disc surface.
2. Due to continuous sliding the transferred polymer is oriented in the sliding direction.
3. Due to continuous sliding, also the later layer of transferred material gets agglomerated and create grooves along the sliding direction (Sethuramiah 2003).

POM has a continuous structure and hence, a smooth layer is transferred on the disc (Kern et al. 1961). It is a known fact that adhesion occurs more in similar type of materials, hence initially the interaction between metal and the polymer will lead to a less of adhesive wear but as the layers of POM gets transferred on steel the adhesive wear increases (Hirst and Hollander 1974). In polymer-metal surface interaction adhesion do not have a very strong role and so with the transfer of any further layer on the previous layer the microgranules agglomerate and cause deep grooves on the pin. The agglomerated polymer molecules have an increased strength than the pin due to strain hardening and thus creates the plastic deformation in pure POM. Figure 5 shows the Raman spectra of the wear track where a pin made of POM without nHAp moved over the 316L disc. Peaks at 508, 1295 and 1443 cm^{-1}

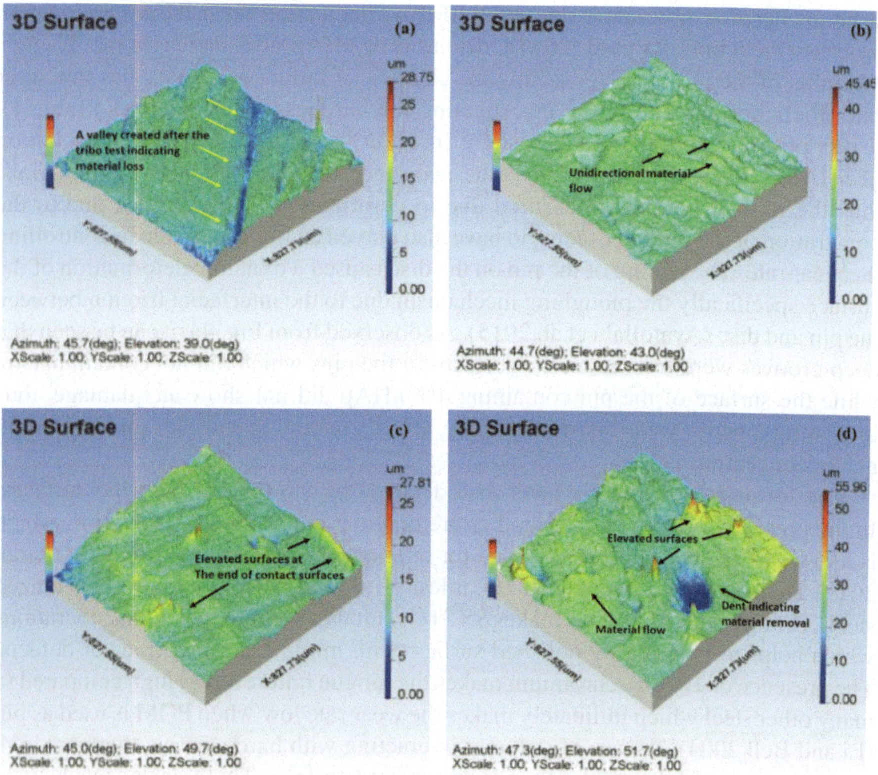

Fig. 5 Surface profiles **a** 0% nHAp **b** 2% nHAp **c** 4% nHAp **d** 5% nHAp

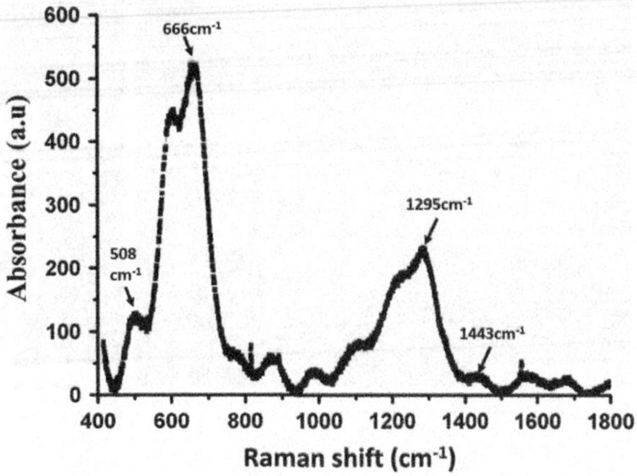

Fig. 6 Raman spectra of the wear track on 316L with POM pin indicating transferred POM on the wear track

are close to those reported by Koenig (1971) and Sugeta et al. (1969), indicating deposition of POM on the wear tracks (Fig. 6).

The nHAp added in POM behaved as nucleating agent and was an effective nucleating site for the polymer. The nucleating agent has an ability to control and regulate the crystallinity of a material and hence regulate its mechanical properties (Niaounakis 2015). HAp also facilitated formation of an apatite layer on the polymer pin surface which facilitates any kind of bioactivity on the surface (Pielichowska and Blazewicz 2010). It has been reported that nHAp inhibits the formation of transfer layer (Blanchet and Kennedy 1992) and the subsurface crack growth in the composite pin material resulting a hindrance in material transfer (Marcus and Allen 1994). The sudden increase of the specific wear in 5% is due to the heterogeneous nucleation. Though crystallization enhances with inclusion of nHAp but it is heterogeneous in nature and higher nHAp concentrations will hinder the crystallization process resulting in more defect in the crystal (Czichos 1983) leading to an increase in residual stress. Further to that higher percentage of nHAp have caused localized vibration and chattering of polymer-metal interface (Buciumeanu et al. 2018).

Further to the transfer layer by POM, pure POM also undergoes continuous plastic deformation due to the applied load (Friedrich 2018). Melick et al. (1999) reported that yield point of the polymer tends to decrease with increasing plastic deformation, this is intrinsic strain softening which occurs due to progressive re-hardening at large strain. In strain softening the surface in contact gets removed due to declination of uniaxial stress resulting in increasing strain and ultimately material failure resulting in voids but the micro-void size of pure POM is very less (Myshkin and Kovalev 2018). The nucleating agent nHAp which regulates the crystallinity of the polymer, increases the brittleness and hence inhibits the material transfer but the subsurface deformation in the polymer continues (Pielichowska 2012). These dislocations pile up to a finite distance from the surface and ultimately lead to the formation of micro-voids which generally occur in between polymer layers.

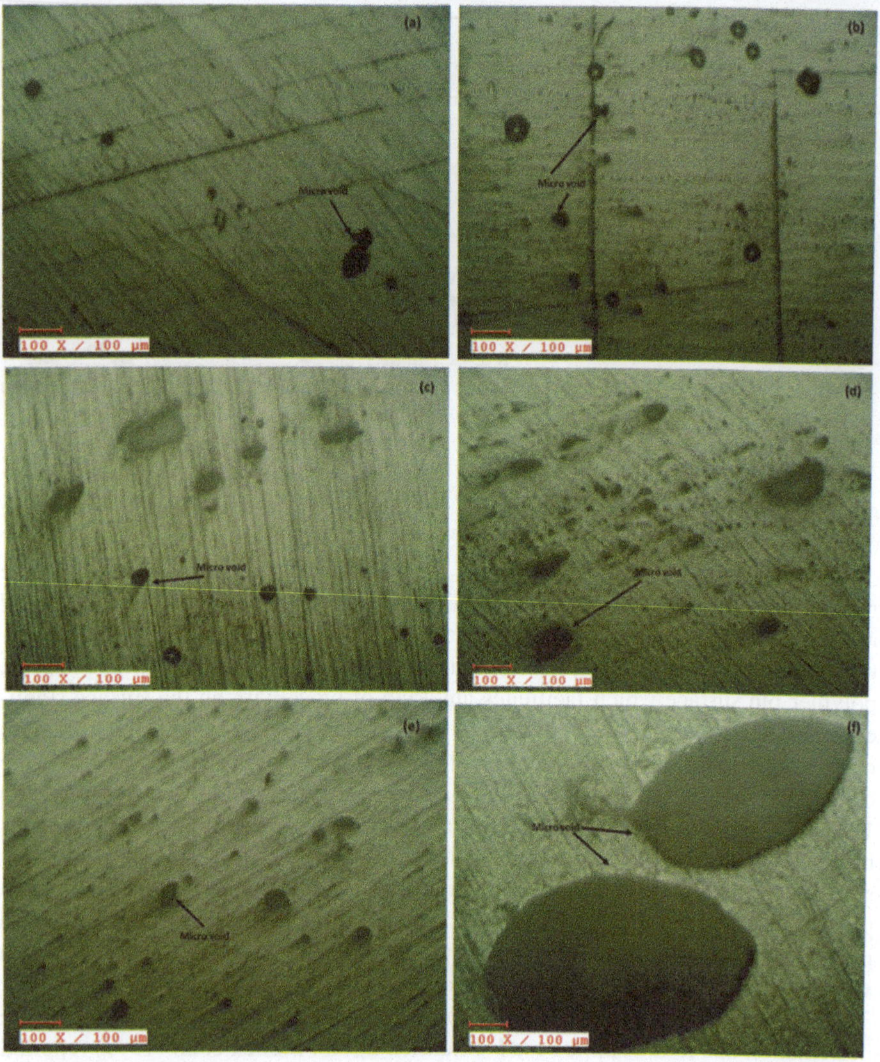

Fig. 7 Optical microscope images indicating the presence of voids in **a** POM without nHAp **b** 1% nHAp **c** 2% nHAp **d** 3% nHAp **e** 4%nHAp **f** 5% nHAp

Figure 7 shows the presence of micro voids in the POM pins. The size of the micro-void in 1% nHApPOM to 4% nHApPOM was observed to be in between the range of 18–30 μm. The micro-voids in 5% nHApPOM sample was large in the range 40–45 μm. Czichos (1983) reported that crystallization process gets hindered in the presence nHAp due to heterogeneous nucleation of POM while it was cooled down to make the pin resulting in residual stresses within the layers. The high energy facture mechanism responsible for this surface failure can be micro void coalescence, which is found evident in many engineering polymers (Williams 1981). Another reason can be transverse crack growth which takes place in the lateral plain of thc polymer and micro void is generated relatively bigger (Rowe 2014).

5 Conclusions

The following can be concluded from the present work reported:

i. nHAp did not influence the coefficient of friction till the concentration of nHAp reached 4%.
ii. The polymer composite containing 4% nHAp showed a decrease of 21% CoF as compared to POM pin without nHAp.
iii. The wear rate was less in lower concentrations of nHAp but increase with the increase in concentration of nHAp. 1–4% nHAp included POM pins exhibited as low as 77% wear rate compared to POM pin which did not contain any nHAp particles.
iv. Transfer layer of POM formed on the surface of 316L, nucleation of nHAp particles in the POM pin and concentration of pin plays a major role in controlling the friction and wear of the POM pins.

Acknowledgements The authors acknowledge the support of Central Research Facility, SRM Institute of Science and Technology, Kattankulathur, for providing the Micro Raman facility.

References

Abou NE, Aljabo A, Strange A, Ibrahim S, Coathup M, Young A, Mudera V (2016) Demineralization–remineralization dynamics in teeth and bone. Int J Nanomed 11:4743–4763

Adatia ND, Bayne SC, Cooper LF, Thompson JY (2009) Fracture Resistance of Yttria-stabilized zirconia dental implant abutments. J Prosthodontics 18(1):17–22

Agnihotri N, Kisaalita WS, Keith CH (1999) Micro-perfusion flow cell for imaging cultured cells. Biotechniques 27(722–726):728

Amanov A, Lee S, Pyun Y (2017) Low friction and high strength of 316L stainless steel tubing for biomedical applications. Mater Sci Eng C 71:176–185

Ayatollahi MR, Yahya MY, Asgharzadeh SH, Hassan SA (2015) Mechanical and tribological properties of hydroxyapatite nanoparticles extracted from natural bovine bone and the bone cement developed by nano-sized bovine hydroxyapatite filler. Ceram Int 41(9):10818–10827

Baskaran H, Solchaga L, Penick K, Berilla J, Welter J (2003) Substrate mass transport in tissue engineered cartilage. In: Hoffner L (ed) Proceedings of the Arthritis Research Conference; Arthritis Foundation, Atlanta, GA, P 2.32

Björk VO (1969) A new tilting disc valve prosthesis. Scand J Thorac Cardiovasc Surg 3(1):1–10

Björk VO (1972) Delrin as implant material for valve occluders. Scand J Thorac Cardiovasc Surg 6(2):103–107

Blanchet TA, Kennedy FE (1992) Sliding wear mechanism of polytetrafluoroethylene (PTFE) and PTFE composites. Wear 153(1):229–243

Brew B, Ward IM (1978) Study of the production of ultra-high modulus polyoxymethylene by tensile drawing at high temperatures. Polymer 19(11):1338–1344
Buciumeanu M, Almeida S, Bartolomeu F, Costa MM, Alves N, Silva FS, Miranda G (2018) Ti6Al4V cellular structures impregnated with biomedical PEEK—new material design for improved tribological behavior. Tribol Int 119:157–164
Cahyaningrum SE, Herdyastuty N, Devina B, Supangat D (2018) Synthesis and characterization of hydroxyapatite powder by wet precipitation method. IOP Conf Ser Mater Sci Eng 299:012039
Cengiz B, Gokce Y, Yildiz N, Aktas Z, Calimli A (2008) Synthesis and characterization of hydroxyapatite nanoparticles. Colloids Surf A 322(1–3):29–33
Chakraborty R, Sengupta S, Saha P, Das K, Das S (2016) Synthesis of calcium hydrogen phosphate and hydroxyapatite coating on SS316 substrate through pulsed electrodeposition. Mater Sci Eng C 69:875–883
Chandrasekaran M (2019) Forging of metals and alloys for biomedical applications. Metals Biomed Devices 293–310
Chiapasco M, Casentini P, Zaniboni M, Corsi E, Anello T (2011) Titanium-zirconium alloy narrow-diameter implants (Straumann Roxolid®) for the rehabilitation of horizontally deficient edentulous ridges: prospective study on 18 consecutive patients. Clin Oral Implant Res 23(10):1136–1141
Curtin WA, Sheldon BW (2004) CNT-reinforced ceramics and metals. Mater Today 7(11):44–49
Czichos H (1983) Influence of adhesive and abrasive mechanisms on the tribological behaviour of thermoplastic polymers. Wear 88(1):27–43
Fathi MH, Hanifi A (2007) Evaluation and characterization of nanostructure hydroxyapatite powder prepared by simple sol–gel method. Mater Lett 61(18):3978–3983
Fister JS, Memoli VA, Galante JO, Rostoker W, Urban RM (1985) Biocompatibility of Delrin 150: a creep-resistant polymer for total joint prostheses. J Biomed Mater Res 19(5):519–533
Friedrich K (2018) Polymer composites for tribological applications. In: Advanced industrial and engineering polymer research
Garg HG, Hales CA (2004) Chemistry and biology of hyaluronan. Elsevier Inc
Hirst W, Hollander AE (1974) Surface finish and damage in sliding. Proc R Soc A Math Phys Eng Sci 337(1610):379–394
Hu P, Shang T, Jiang M, Chen M (2007) Investigation of the thermal decomposition properties of polyoxymethylene. J Wuhan Univ Technol Mater Sci Ed 22(1):171–173
Iguchi M (1973) Growth of needle-like crystals of polyoxymethylene during polymerisation. Br Polym J 5(3):195–198
Itälä AI, Ylänen HO, Ekholm C, Karlsson KH, Aro HT (2001) Pore diameter of more than 100 μm is not requisite for bone ingrowth in rabbits. J Biomed Mater Res 58(6):679–683
Kamalanathan P, Ramesh S, Bang LT, Niakan A, Tan CY, Purbolaksono J, Teng WD (2014) Synthesis and sintering of hydroxyapatite derived from eggshells as a calcium precursor. Ceram Int 40(10):16349–16359
Kay M, Young RA, Posner AS (1964) Crystal structure of hydroxyapatite. Nature 204(4963):1050–1052
Kern W, Cherdron H, Jaacks V (1961) Polyoxymethylene. AngewandteChemie 73(6):177–186
Koenig JL (1971) Raman scattering of synthetic polymers—a review. Appl Spectrosc Rev 4(2):233–305
Klawitter JJ, Hulbert SF (1971) Application of porous ceramics for the attachment of load bearing internal orthopedic applications. J Biomed Mater Res 5(6):161–229
Laluppa JA, McAdams TA, Papoutsakis ET, Miller WM (1997) Culture materials affect ex vivo expansion of hematopoietic progenitor cells. J Biomed Mater Res 36(3):347–359
Lee SJ, Oh SH (2003) Fabrication of calcium phosphate bioceramics by using eggshell and phosphoric acid. Mater Lett 57(29):4570–4574
Letarouilly JG, Broux O, Clabaut A (2018) New insights into the epigenetics of osteoporosis. Genomics

Li C, Bell T (2004) Corrosion properties of active screen plasma nitrided 316 austenitic stainless steel. Corros Sci 46(6):1527–1547

Li Q, Wen Z, Chen J, Huang H, Shi X, Zhang Q (2018) Preparation of controllable hydroxyapaptite nanoparticles with abalone shells. Mater Lett

Li Y, Xie T, Yang G (2005) Studies on novel composites of polyoxymethylene/polyamide 6. J Appl Polym Sci 99(1):335–339

Luengo G, Heuberger M, Israelachvili J (2000) Tribology of shearing polymer surfaces, 2. Polymer (PnBMA) sliding on Mica. J Phys Chem B 104(33):7944–7950

MacAfee KA, Quinn PD (1992) Total temporomandibular joint reconstruction with a Delrin titanium implant. J Craniofacial Surg 3(3):160–169

Marcus K, Allen C (1994) The sliding wear of ultrahigh molecular weight polyethylene in an aqueous environment. Wear 178(1–2):17–28

Melick HGHV, Casteren IAV, Govaert LE, Meijer HEH (1999) The influence of intrinsic strain softening on the macroscopic deformation behaviour of amorphous polymers. Eindhoven University of Technology, Materials Technology

Monmaturapoj N (2008) Nano-size hydroxyapatite powders preparation by wet- chemical precipitation route. J Metals Mater Minerals 18:15–20

Myshkin N, Kovalev A (2018) Adhesion and surface forces in polymer tribology—a review. Friction 6(2):143–155

Navarro CH, Moreno KJ, Chávez-Valdez A, Louvier-Hernández F, García-Miranda JS, Lesso R, Arizmendi-Morquecho A (2012) Friction and wear properties of polymethyl methacrylate–hydroxyapatite hybrid coating on UHMWPE substrates. Wear 282–283:76–80

Niaounakis M (2015) Compounding and additives. Biopolymers: processing and products, pp 215–262

Nishizawa Y, Miura M, Ichimura S, Inaba M, Imanishi Y, Shiraki M, Takada J, Chaki O, Hagino H, Fukunaga M, Fujiwara S, Miki T, Yoshimura N, Ohta H (2019) Executive summary of the Japan osteoporosis society guide for the use of bone turnover markers in the diagnosis and treatment of osteoporosis (2018 edn), ClinicaChimica Acta 498:01–107

Ooi CY, Hamdi M, Ramesh S (2007) Properties of hydroxyapatite produced by annealing of bovine bone. Ceram Int 33(7):1171–1177

Penick KJ, Solchaga LA, Berilla JA, Welter JF (2005) Performance of polyoxymethylene plastic (POM) as a component of a tissue engineering bioreactor. J Biomed Mater Res Part A 75A(1):168–174

de Pereira M, Oréfice RL, Mansur HS, Lopes MTP, Turchetti-Maia RMDM, Vasconcelos AC (2003) Preparation and biocompatibility of poly (methyl methacrylate) reinforced with bioactive particles. Mater Res 6(3):311–315

Pielichowska K (2012a) The influence of molecular weight on the properties of polyacetal/hydroxyapatite nanocomposites. Part 1. Microstructural analysis and phase transition studies. J Polym Res 19(2):9775

Pielichowska K (2012b) The influence of molecular weight on the properties of polyacetal/hydroxyapatite nanocomposites. Part 2. In vitro assessment. J Polym Res 19(2):9788

Pielichowska K (2015) Preparation and characterization of polyoxymethylene nanocomposites. In: Manufacturing of nanocomposites with engineering plastics, pp 103–125

Pielichowska K, Blazewicz S (2010) Bioactive polymer/hydroxyapatite (nano) composites for bone tissue regeneration. In: Advances in polymer science, pp 97–207

Pielichowska K, Dryzek E, Olejniczak Z, Pamuła E, Pagacz J (2012) A study on the melting and crystallization of polyoxymethylene-copolymer/hydroxyapatite nanocomposites. Polym Adv Technol 24(3):318–330

Pielichowska K, Szczygielska A, Spasówka E (2011) Preparation and characterization of polyoxymethylene-copolymer/hydroxyapatite nanocomposites for long-term bone implants. Polym Adv Technol 23(8):1141–1150

Ripamonti U, Crooks J, Khoali L, Roden L (2009) The induction of bone formation by coral-derived calcium carbonate/hydroxyapatite constructs. Biomaterials 30(7):1428–1439

Rivera EM, Araiza M, Brostow W, Castaño VM, Díaz-Estrada J, Hernández R, Rodríguez JR (1999) Synthesis of hydroxyapatite from eggshells. Mater Lett 41(3):128–134
Rowe WB (2014) Mechanics of abrasion and wear. In: Principles of modern grinding technology, pp 349–379
Saiz E, Gremillard L, Menendez G, Miranda P, Gryn K, Tomsia AP (2007) Preparation of porous hydroxyapatite scaffolds. Mater Sci Eng C 27(3):546–550
Sethuramiah A (2003) Wear of non-metallic materials. In: Lubricated wear-science and technology, pp 171–201
Shashvatt U, Aris H, Blaney L (2017) Evaluation of animal manure composition for protection of sensitive water supplies through nutrient recovery processes. In: Chemistry and water, pp 469–509
Shi J, Jing B, Zou X, Luo H, Dai W (2009) Investigation on thermo-stabilization effect and non-isothermal degradation kinetics of the new compound additives on polyoxymethylene. J Mater Sci 44(5):1251–1257
Shuai C, Shuai C, Wu P, Yuan F, Feng P, Yang Y, Gao C (2016) Characterization and bioactivity evaluation of (polyetheretherketone/polyglycolicacid)-hydroyapatite scaffolds for tissue regeneration. Materials 9(11):934
Simske SJ, Ayers RA, Bateman TA, Liu DM, Dixit V (1997) Porous materials for tissue engineering, vol 250, 151
Solchaga LA, Tognana E, Goldberg VM, Caplan AI, Welter JF (2002) Tissue engineered cartilage implants: in vitro chondrogenic pre-conditioning. In: Proceedings of the 4th symposium of the international cartilage repair society, Toronto/CA, 15–18 June 2002, p 271
Solchaga LA, Tognana E, Penick K, Baskaran H, Goldberg VM, Caplan AI, Welter JF (2006) A rapid seeding technique for the assembly of large cell/scaffold composite constructs. Tissue Eng 12(7):1851–1863
Song R, Xiang J, Hou D (2011) Characteristics of mechanical properties and microstructure for 316L austenitic stainless steel. J Iron Steel Res Int 18(11):53–59
Sugeta H, Miyazawa T, Kajiura T (1969) Laser Raman scattering of polyoxymethylene, vol 7, pp 251–253
Swetha M, Sahithi K, Moorthi A, Srinivasan N, Ramasamy K, Selvamurugan N (2010) Biocomposites containing natural polymers and hydroxyapatite for bone tissue engineering. Int J Biol Macromol 47(1):1–4
Tamai N, Myoui A, Tomita T, Nakase T, Tanaka J, Ochi T, Yoshikawa H (2001) Novel hydroxyapatite ceramics with an interconnective porous structure exhibit superior osteoconductionin vivo. J Biomed Mater Res 59(1):110–117
Vande VJ, Bahadur S (1995) The growth and bonding of transfer film and the role of CuS and PTFE in the tribological behaviour of PEEK. Wear 181–183:212–221
Vecchio KS, Zhang X, Massie JB, Wang M, Kim CW (2007) Conversion of bulk seashells to biocompatible hydroxyapatite for bone implants. Acta Biomater 3(6):910–918
Vinoth A, Datta S (2019) Design of the ultrahigh molecular weight polyethylene composites with multiple nanoparticles: an artificial intelligence approach. J Compos Mater 1–14
White AA, Best SM, Kinloch IA (2007) Hydroxyapatite-carbon nanotube composites for biomedical applications: a review. Int J Appl Ceram Technol 4(1):1–13
Williams DF (1981) Implants in dental and maxillofacial surgery. Biomaterials 2(3):133–146

Influence of Surface Modification on Tribological Properties of Elastomer Composites

Dariusz M. Bielinski, Mariusz Sicinski, and Jacek Jagielski

Abstract Friction of elastomer materials, apart their volume composition and structure, should be also considered from a point of view of their surface layer. Similarly, to the migration of low molecular weight components in rubber mixes, the surface segregation also takes place in polymer blends. The produced surface layer can either be of an amorphous or a crystalline character, lubricating the tribological contact or making it stiffer adequately, in the both cases acting to reduce friction. The surface hardness gradient of rubber vulcanizates is the result of the surface profile of their crosslink density and structure. It can be modified either by some changes to the composition of a curing system or by the application of an appropriate post-treatment, realized by various chemical and physical modifications (halogenation, plasma, laser or ion beam), very often accompanied by the changes to the surface microroughness and energy. Studies on the surface composition, structure and morphology and the related changes to hardness, microroughness and the surface energy of elastomer composites revealed a great potential to modify and/or to control their friction, what was discussed on some examples. The research methodology included an equilibrium swelling and a thiol-amine analysis, infra-red-, Raman-, X-rays photoelectron-, secondary ion mass- and Rutherford Back Scattering spectroscopies, atomic force-, scanning electron- and optical microscopies, a micro indentation, the surface profilometry and the contact angle determinations. The experimental results were correlated to the friction of various elastomer composites determined with a block-on-ring tribometer. The relations observed can be used for

D. M. Bielinski (✉) · M. Sicinski
Institute of Polymer and Dye Technology, Faculty of Chemistry, Lodz University of Technology, Stefanowskiego 12/16, 90-924 Lodz, Poland
e-mail: dariusz.bielinski@p.lodz.pl

M. Sicinski
e-mail: mariusz.sicinski@p.lodz.pl

J. Jagielski
Laboratory of Structural Research and Materials Characterization, Lukasiewicz Research Network—Institute of Electronic Materials Technology, Wolczynska 133, 01-919 Warsaw, Poland
e-mail: jacek.jagielski@itme.edu.pl

Material Physics Department, National Centre for Nuclear Research, Andrzeja Soltana 7, 05-400 Otwock, Poland

M. T. Hameed Sultan et al. (eds.), *Tribological Applications of Composite Materials*, Composites Science and Technology, https://doi.org/10.1007/978-981-15-9635-3_7

a knowledge-based designing of elastomer materials, their processing, modification and exploitation, enabling for tailoring friction of these materials.

Keywords Elastomer · Rubber · Surface · Modification · Friction

1 Introduction to the Friction of Elastomers: A Role of the Surface Layer

Despite a long history of elastomer tribology, started from the pioneering works of Schallamach (Schallamach 1957, 1971), most of the efforts have been focused on studying the effect of mechanical properties of the materials on friction, taking into account only their elasto-plastic bulk behaviour. In the seventies, Moore (1975) proposed a fundamental equation, dividing friction force into two components: adhesional and hysteretical, which in the version modified by Roberts (1979), considering the possibility of lubrication, it has been commonly used up to now. The proposed friction mechanism describes the phenomenon as the movement of material folds, associated with the dissipation of deformation energy, attributing the effect to the bulk properties of elastomers. According to Persson (1998), the energy is dissipated by internal friction, dependent on their stiffness, which in turn is related to the frequency of deformations. The relationship between frequency and elastic modulus varies between different elastomers, based on their morphology and structure, represented by macromolecular rubber structure, crosslink density and structure, filler loading, distribution and dispersion as well as rubber-filler interactions.

The situation changes when the stiffness of the friction contact increases or lubricating conditions appear, due to the effect of the surface hardness gradient (Bieliński 2001a, b) or surface migration/segregation of low molecular weight components of elastomers (Bieliński et al. 2005a, b) respectively.

1.1 Surface Gradient of Chemical Composition and Structure of Rubber Vulcanizates

The surface layer of polymer materials undergoes significant physical and chemical changes already during compounding, processing (the so-called "active" processing) or the exploitation of elastomer composites (blooming, aging). Unfortunately, even nowadays, the role of their surface layer is underestimated or even neglected in polymer technology and engineering.

Surface migration of low-molecular weight components of the curing system and products created during vulcanization clearly changes the character of the surface layer of rubber vulcanizates, what brings important consequences for their tribological properties. The observed changes concern not only surface polarity (Zaborski et al. 1991), but also modify surface microroughness, morphology and mechanical characteristics of the surface layer (Bieliński 2001a, b). Crosslinking substances are generally polar in nature and therefore manifest limited solubility in a non-polar rubber matrix. Used in excess (in the case of sulfur curing systems, this limit usually does not exceed 5 phr) are ineffective, migrating to the surface layer of material, making it enriched with curing agents compared to their bulk content. Simultaneously, the crystalline by product of sulfur vulcanization or low molecular weight components of rubber mixes, driven to the surface by the favourable crystallization phenomenon, change the surface character of the vulcanizates. The most often blooms of stearic acid, zinc stearate or microcrystalline isoparaffin waxes (added in order to physically protect the rubber from ozone aging) respectively, can be observed on the surface of vulcanizates—Fig. 1.

The limited thermal conductivity of elastomers makes the temperature of their surface layer, in contact with the surface of the metal mould during vulcanization higher, compared to the bulk of the materials. For large moulded goods, the difference

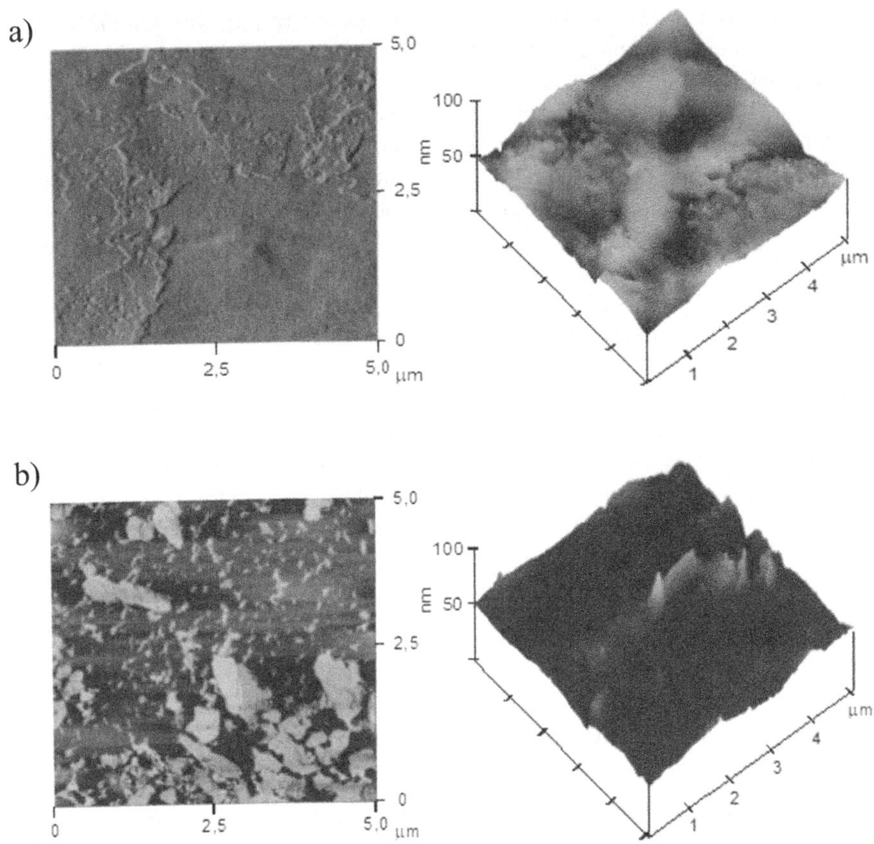

Fig. 1 Blooms of zinc stearate (**a**) and isoparaffin wax (**b**) on the surface of sulfur styrene-butadiene rubber (SBR) vulcanizates. ***Atomic force microscopy (AFM)—tapping mode***

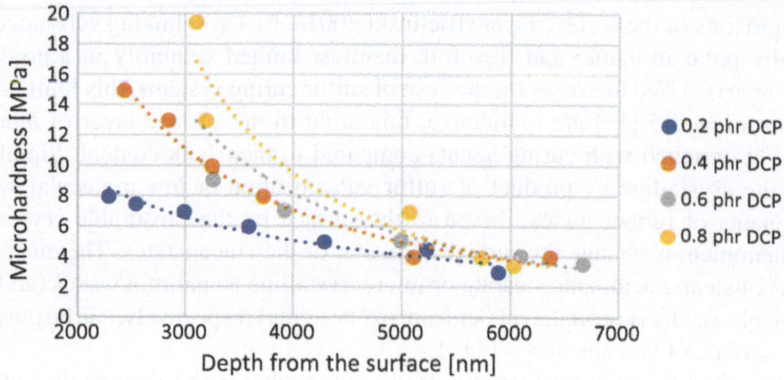

Fig. 2 Surface hardness profile of peroxide isoprene rubber (IR) vulcanizates. ***Spherical microindentation***

can achieve even up to 10°. Both of these effects are responsible for the surface hardness gradient, as demonstrated for isoprene rubber (IR) cured with dicumyl peroxide (DCP)—Fig. 2.

The hardness of cured rubber is proportional to its crosslink density (Boochathum and Prajudtake 2001). However, in the case of sulfur vulcanizates, despite the surface migration and temperature gradient, the phenomenon called maturing of networks, based on braking long polysulfidic crosslinks and creating in their place more shorter crosslinks, mainly monosulfidic ones, takes place in the surface layer of materials—Fig. 3.

It is reflected by the surface gradient of the hardness of sulfur vulcanizates, as demonstrated for isoprene rubber (IR) crosslinked with sulfur and diphenylguanidine (DPG) system—Fig. 4.

The second degree polynominal, representing the IR/S+DPG vulcanizate, is the result of the effect of the maturing of network, superimposed on the effect of the crosslink density gradient. The gradient character of vulcanization, affecting the dynamic and tribological properties of rubber vulcanizates to limited extent, has a considerable impact on their friction studied in a microscale (2001a, b). Despite the highest polarity and lability of crosslinks—reflected by the highest bulk hysteresis, polysulfide rubber vulcanizates characterize themselves by the lowest microfriction among the elastomers studied.

2 Influence of Chemical Treatment on Friction

Chemical modification, due to its versatility and low cost is the most popular surface treatment of finished elastomer products. Usually, their surface is subjected to chlorination (Perera 1987; Extrand and Gent 1988; Martin-Martinez et al. 1991a, b; Pastor-Blas et al. 2000), chloro-sulfonation (Roberts and Brackley 1990; Martin-Martinez

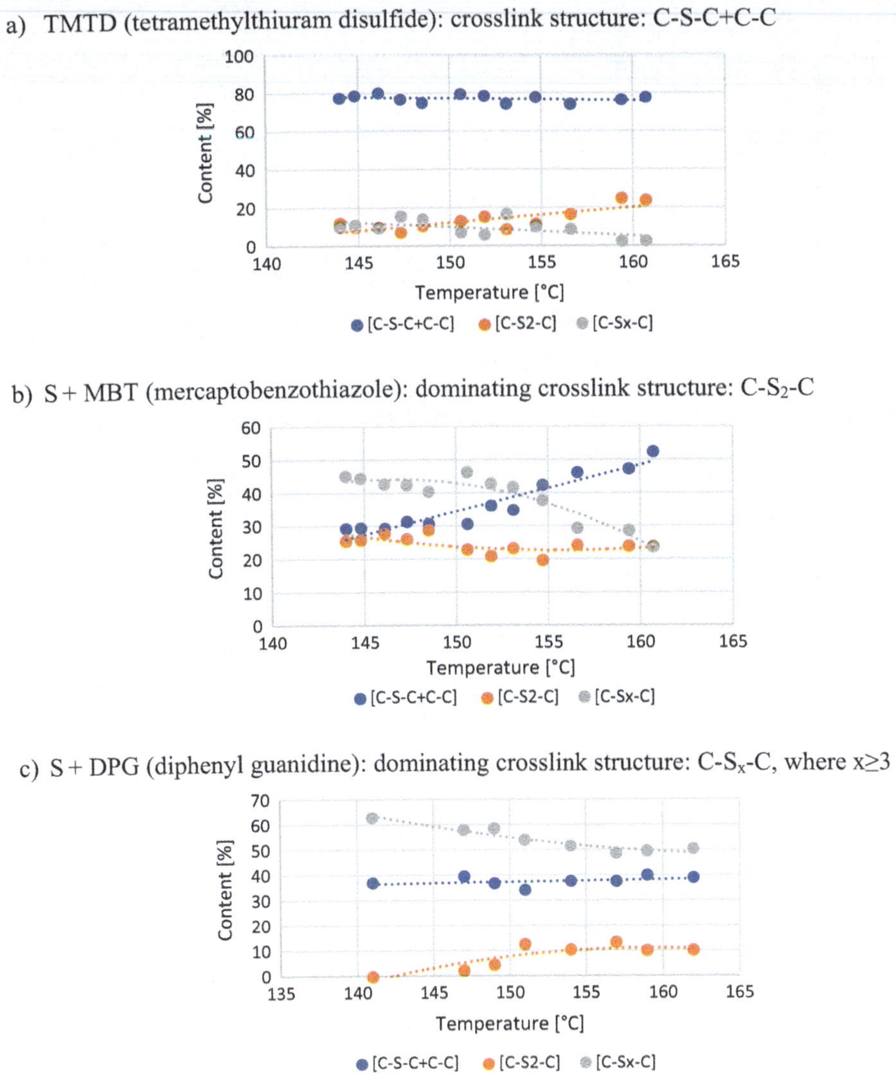

Fig. 3 Influence of temperature (vulcanization under a temperature gradient in a steel mould) on the structure of crosslinks of sulfur styrene butadiene rubber (SBR) vulcanizates

et al. 1991; Martin-Martinez et al. 1992) or sulfonation (Oldfield and Symes 1983; Hace et al. 1990), while rather rarely to fluorination (Vega-Cantú et al. 2003; Schlögl et al. 2011) or iodination (Bieliński et al. 1997a, b, 1998a, b). Halogenation is usually accompanied by strong surface oxidation (Bieliński et al. 1995), manifesting itself by a significant increase in material hardness, which can be tested with an elastomer made of natural rubber (NR) subjected to chlorination—Fig. 5.

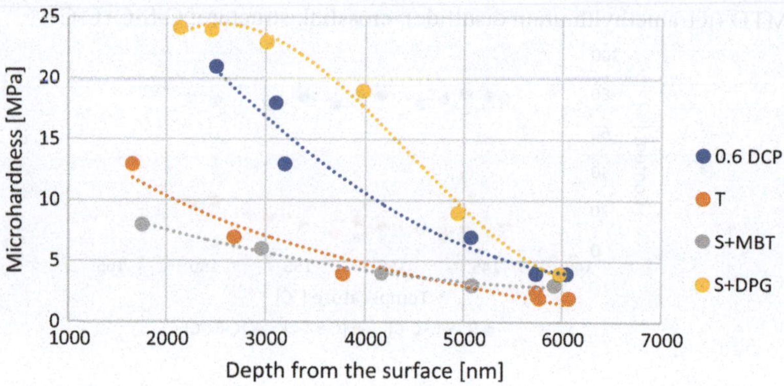

Fig. 4 Surface hardness profile of sulfur isoprene rubber (IR) vulcanizates. ***Spherical microindentation***

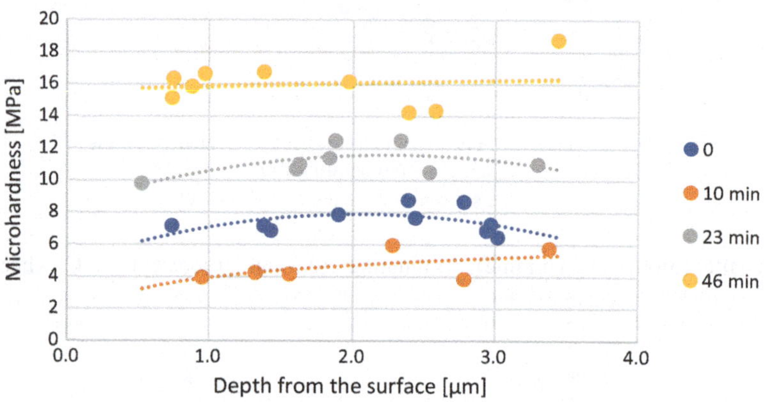

Fig. 5 The influence of the time of the so-called wet chlorination on the hardness of the windshield wipers made of natural rubber (NR) filled with carbon black. ***Spherical microindentation***

Excessively long residence time of NR-based elastomers in chlorine solution is not beneficial due to cracks appearing on their surface, which are the result of the formation of thick and hard "skin", badly cooperating with the elastic bulk of the material. Extending of the duration of the modification leads to degradation of rubber macromolecules and lowering of elastomer hardness. Gas fluorination of rubber is not as often used as chlorination. However, it is becoming more and more popular as a treatment that reduces the friction coefficient of finished elastomer products, e.g. seals, gaskets and hoses. Taking into account that the thermal stability and fuel resistance of elastomers (including those containing bio-components) can be further improved in this way, this type of modification is nteresting to the automotive industry (Bieliński et al. 2006a, b). Generally speaking, fluorination reduces material hardness—Fig. 6, develops surface roughness—Fig. 7 and modifies the chemical composition of the surface layer of elastomers.

Interesting results were obtained for acrylonitrile-butadiene rubber (NBR) elastomer or its hydrogenated analogue (HNBR), applying iodination (Bieliński et al. 1997a, b; Ślusarski et al. 1998). Immersion of the NBR sample in Lugol's solution (Polish Farmakopea 1970) makes its surface geometry developed, whereas the HNBR one in the other way round—its smoothing—Fig. 8.

In addition to the development of surface microroughness, fluoro-carbon compounds of very good lubrication properties are formed in the surface layer of the NBR elastomer, as confirmed by the FT-IR studies.

Attempts to apply iodination or bromination to modify the surface of elastomers based on different rubbers, using respectively a gaseous phase (iodine vapours) or a solution (bromine water), have failed (Bieliński 1993). The surface of the materials undergoes deep modification, reaching even several hundred micrometers. The

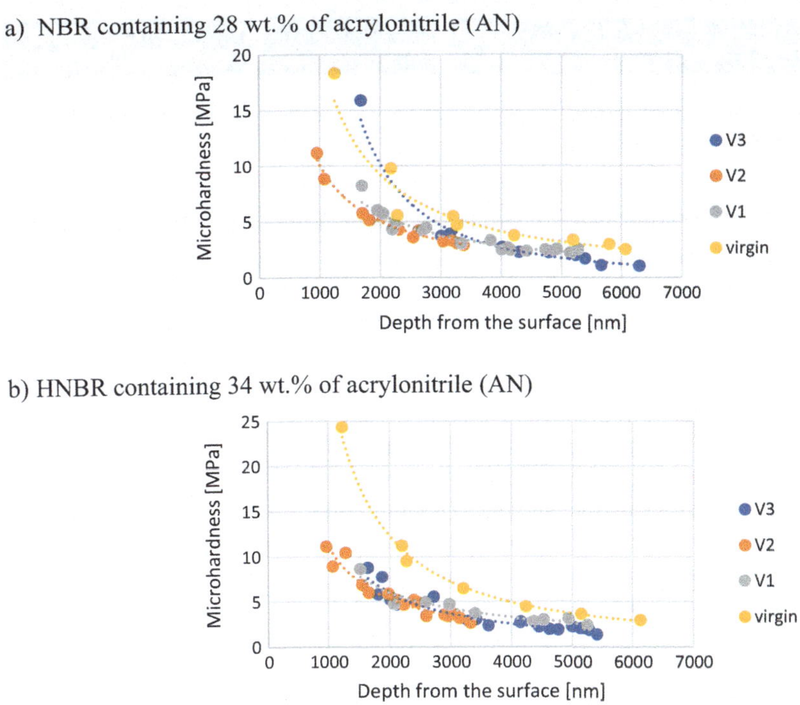

V1 – single-stage process; 5% concentration of fluorine in a chamber
V2 – process V1 carried out after pumping air out of a chamber
V3 – two-stage process; both 5% concentration of fluorine and oxygen in a chamber

Fig. 6 Influence of gas-phase fluorination on the hardness profile of selected butadiene-acrylonitrile rubber elastomers. *Spherical microindentation*

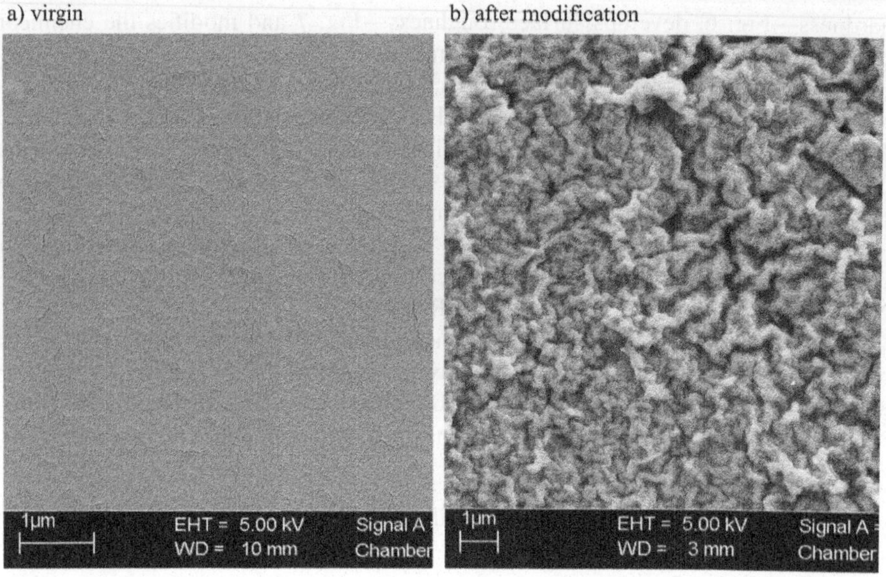

Fig. 7 Modification of the surface morphology of a gas-phase fluorinated acrylonitrile-butadiene elastomer (SEM)

accompanying oxidation causes a strong shrinkage of the surface layer, manifesting itself by its visible cracking—Fig. 9.

Contrary to halogenation, the sulfonation or chloro-sulfonation of elastomers generally leads to a smooth surface. The associated increase of hardness is greater compared to the effect that can be achieved by any halogenation, but thanks to the shallower modification range, the hard „skin" does not crack under load, working well with the elastic bulk of elastomers under dynamic conditions prevailing during friction. Effects of different chemical treatments on the friction of some synthetic elastomers is presented in Fig. 10.

In general, the chlorination, and the iodination—but applied only to acrylonitrile-butadiene rubber, proved to be most effective. In the latter case, it consists of the development of surface geometry, combined with its curing by means of specific interactions between iodine ions and nitrogen atoms from acrylonitrile groups. As the result, a paracrystalline phase is formed, which increases the mechanical properties of the surface layer of the modified material (Bieliński et al. 1998a, b). In the case of a chlorination, the best results are achieved by optimising the process time. Tests on natural rubber vulcanizates show, that if the wet chlorination process (solution of chlorine in water) is too short, strong oxidation and degradation of the elastomer prevails. If the treatment time increases, radical cross-linking occurs, which makes the elastomer hardness increased. However, too long process leads to a thick and rigid layer, that shrinks (strong crosslinking effect) and breaks eventually. The effect is more pronounced when bromine water (bromine is a stronger oxidant compared to chlorine) is used to modify the surface of elastomers. Very good tribological

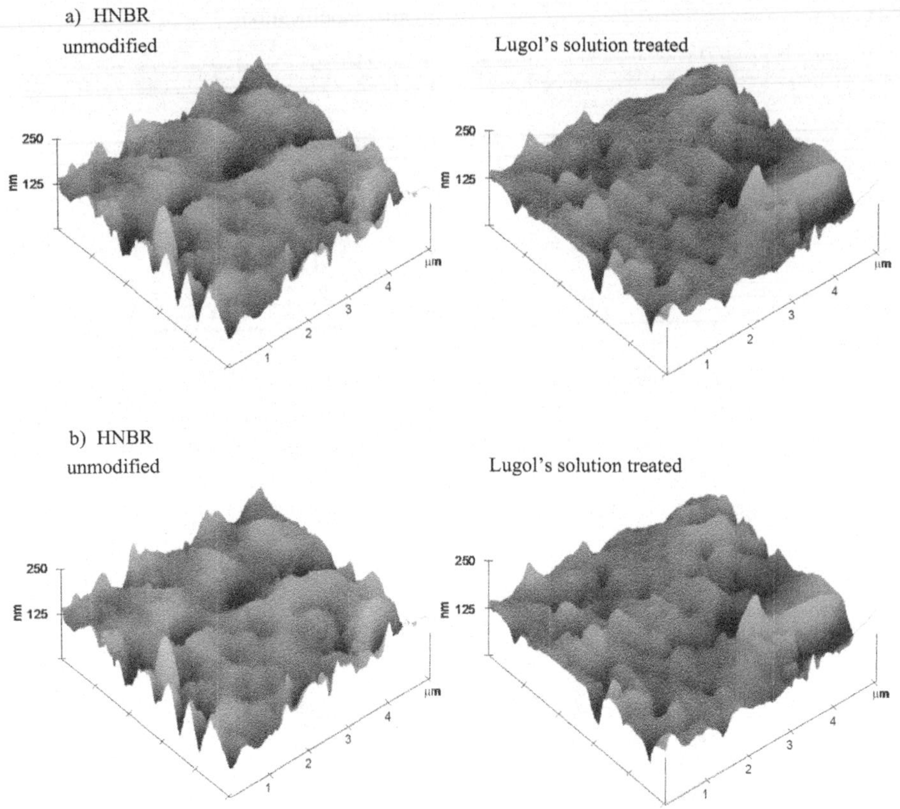

Fig. 8 Influence of various iodine treatments on the surface geometry of acrylonitrile-butadiene rubber elastomers. ***AFM—tapping mode***

effects, with the potential to reduce friction, can also be achieved by sulfonation (e.g. using the so-called 100% sulfuric acid, i.e. a mixture of concentrated and fuming sulfuric acid (Gibson and Bailey 1980) or chloro-sulfonation of the surface (applying chlorosulfonic acid). However, not only the adhesion component (F_A) but also the hysteretical component (F_H) of friction of elastomers sliding on hard substrates, depends on the properties of their surface layer, because energy dissipation occurs also in the surface layer of elastomers (Moore and Geyer 1974; Bieliński et al. 1998a, b). Due to the existence of surface gradient of mechanical properties, the hysteresis component plays a decisive role in the friction of elastomers analyzed on a macroscopic scale—Fig. 11.

The data presented show, that despite a similar surface energy value, chemically modified elastomers clearly indicate a reduction in the coefficient of friction after reaching a certain degree of the stiffness of the materials.

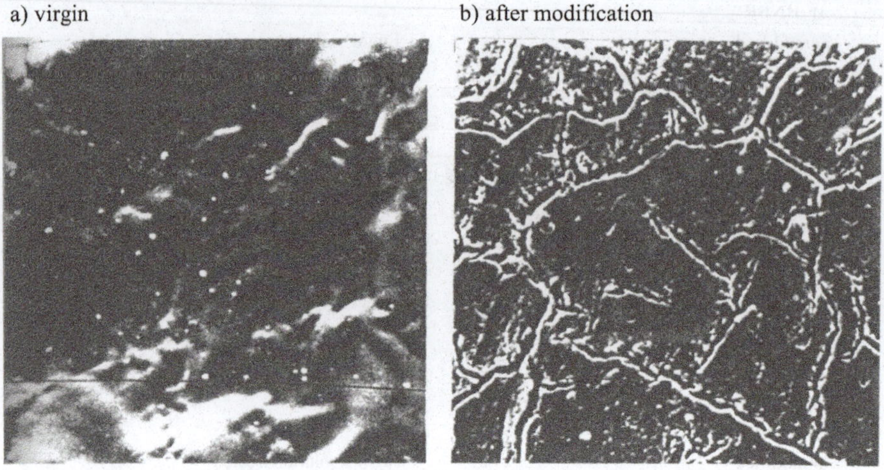

Fig. 9 Surface geometry of styrene-butadiene rubber (SBR) elastomer immersed in bromine water (SEM)

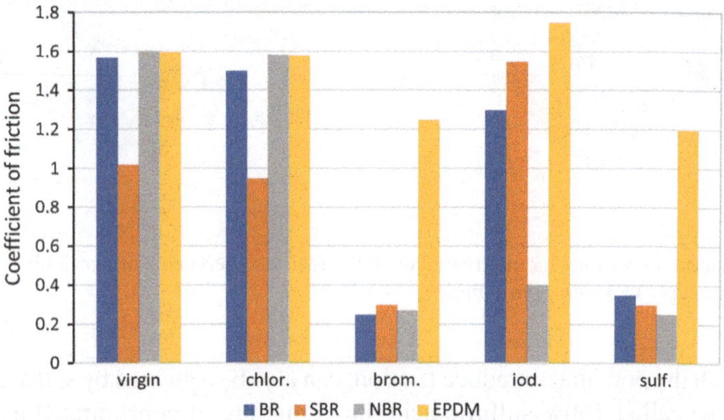

Fig. 10 Effects of different chemical treatments on the friction of selected synthetic elastomers

3 Influence of Physical Treatment on Friction

The effects of physical modification of a surface very often concern changes to its chemical composition, structure and morphology, therefore the division into chemical and physical modification of a material is a contractual matter.

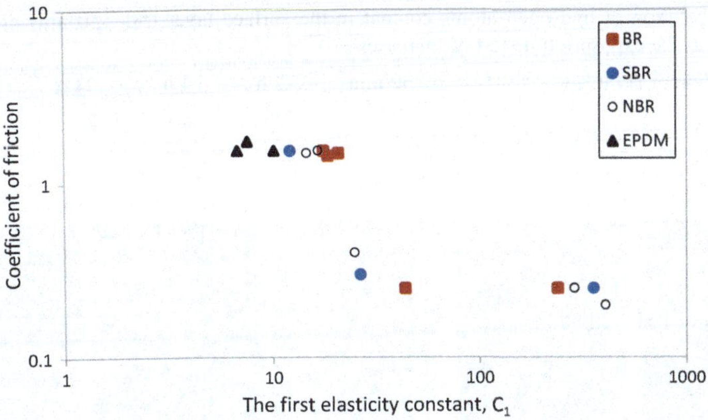

Fig. 11 Effect of surface layer stiffness of elastomers (C_1) subjected to various chemical modifications ($\gamma_S = 38 \div 41$ mJ/m^2) on their coefficient of friction against steel

3.1 UV Radiation

Ultraviolet radiation is commonly considered to be a negative factor, that causes the ageing of polymers. However, when used in a limited dose, it turns out to modify elastomers (Bieliński and Głąb 2005), which may be an advantage from a tribological point of view—Fig. 12.

As a result of exposure of rubber products to UV radiation, hydrogen atoms are released and the surface layer of material is gradually graphitized, which can be clearly seen from the data of nuclear reactions analysis (NRA) (Feldman and Picraux 1978)—Table 1. The range of modification, limited to a depth of a few micrometers

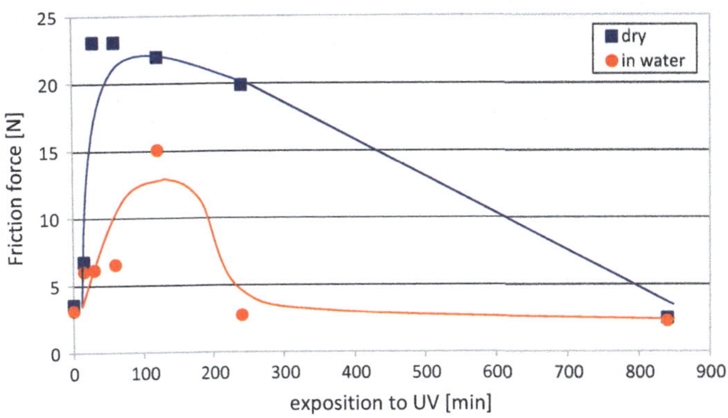

Fig. 12 Effect of exposition time to UV radiation on the coefficient of friction of carbon black filled natural rubber (NR) vulcanizates against glass plate

Table 1 Evolution of hydrogen atoms content in the surface layer (50–150 nm) of sulfur NR vulcaniztes as the exposure time to UV increases

Concentration of H (at.%)	0	30 min	2 h	4 h	24 h	72 h
	66	54	42	40	27	23

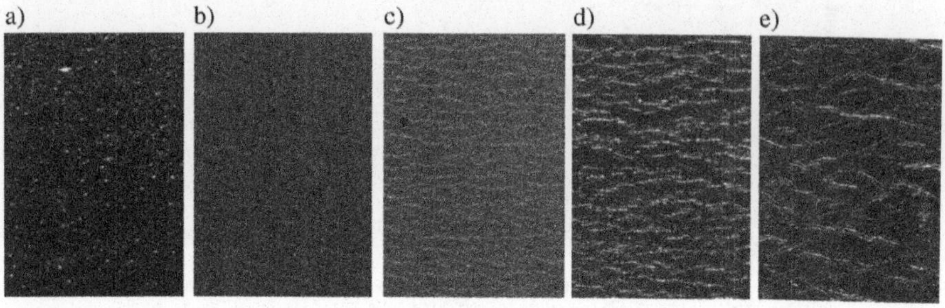

Fig. 13 Microcracks generated on the surface of NR windscreen wipers as a result of their ozone aging: virgin sample (**a**), 2 h exposure (**b**), 4 h exposure (**c**), 24 h exposure (**d**), or 72 h exposure (**e**)

suggests, that the top layer of graphite may not be chemically bonded to the substrate, forming a thin layer of grease on the surface of rubber vulcanizates.

Natural rubber (NR) is one of the polymers that softens (degradation of macromolecules) when aged. If only the effect of plasticization of its surface layer does not obscure the effect of lowering the friction coefficient following graphitization, the favourable tribological effect can be obtained, which is clearly visible from the analysis presented in Fig. 12 and Table 1. In the case of polymers which become harder because of aging (as a result of macromolecules being crosslinked), e.g. polybutadiene (BR), the effect obtained is problematic due to the accompanying excessive abrasive wear of the materials. The surface of the materials starts to be covered with a grid of ozone cracks, which give rise to their destruction. Initial microcracks are less visible to butadiene copolymers, like SBR and NBR, for which a low dose of UV radiation is beneficial from a tribological point of view. The appearance of the surface of car windscreen wiper blades after their exposure to UV is demonstrated on pictures from optical microscopy—Fig. 13.

3.2 *Plasma Treatment*

The application of low temperature radio frequency (RF) plasma for surface treatment of polymer materials has been tried since many years, mainly as an effective way to improve the adhesion of polymers to other materials, in order to control their surface energy and wettability or friction modification. The three main effects resulted from the modification can be achieved and namely:

- control of surface energy and wettability (Egitto et al. 1990),
- development of surface geometry, promoting mechanical adhesion (Morra et al. 1989), and
- plasma pre-treatment, promoting chemical adhesion (plasma induced modification) (Dierkes et al. 2019) or plasma polymerization (Yasuda 1985).

The first one is realized by oxidation of the surface layer—Fig. 14, which combined with modification of the surface microroughness of elastomers—Fig. 15, makes possible to control their wettability. An increase in the content of carbonyl, carboxyl and hydroxy groups of polar character, is clearly visible in the XPS spectra.

As the surface microroughness increases—its R_a parameter ca. doubles, aa a result of 6 min of plasma treatment, the energy of its surface changes, allowing to control the wettability of the surface. The polarity of the surface acts towards the hydrophilicity of the elastomers, whereas the microroughness facilitates their hydrophobicity.

However, the effect observed proved to be impermanent—water contact angle changed from 13° to 70° after 70 h, confirming the tendency of plasma treated elastomer surfaces to return to their original wettability with storage time (Liu et al. 2016).

Plasma etching is shown to be useful in exposing of coarse morphological details of semicrystalline polymers, such as spherulites, and in providing a convenient way to rapid stripping of polymer materials (Garton et al. 1978). The increased surface microroughness and accompanied chemical modification, together with decreased frictional contact between hard asperities protruding from the polymer surface and the counter-face, can facilitate decrease of a friction coefficient. Comprehensive summary of plasma technology applications for selective surface etching of polymer

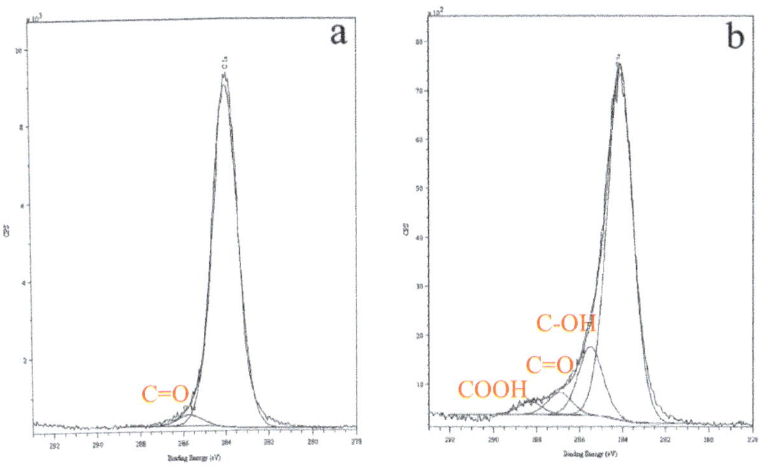

Fig. 14 XPS spectra of ethylene-propylene-diene (EPDM) elastomer before (**a**) and after treatment with nitrogen RF plasma (**b**)

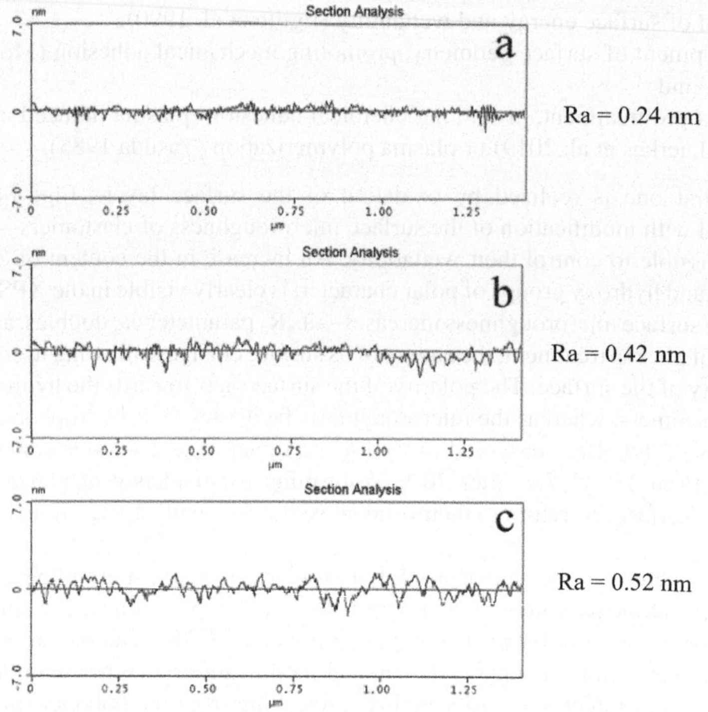

Fig. 15 Effect of RF nitrogen plasma treatment time: 2 min (**a**), 4 min (**b**) or 6 min (**c**), on surface microroughness of ethylene-propylene-diene (EPDM) elastomer

materials and plasma polymerization for advanced applications, has recently been presented by Puliyalil and Cvelbar (2016).

Frictional performance of several elastomers after plasma treatment. has been examined by Martin-Martinez et al. (2012). In all cases, the treated elastomers showed better performance than the corresponding untreated ones. Stronger treatments, in terms of a longer process time and/or a higher substrate bias voltage, however, led to a greater reduction in their coefficient of friction. This was interpreted as an enhancement of rubber crosslinking and a reduction in elastomer tackiness due to plasma exposure. The degree of improvement of a certain process depends on the maximum working temperature of the materials. Therefore, the treatment is more effective for more sensitive elastomers—Fig. 16.

3.3 Laser Texturing

Among the various surface treatment technologies used to obtain specific morphologies, laser ablation of material surface is particularly useful. Laser micromachining

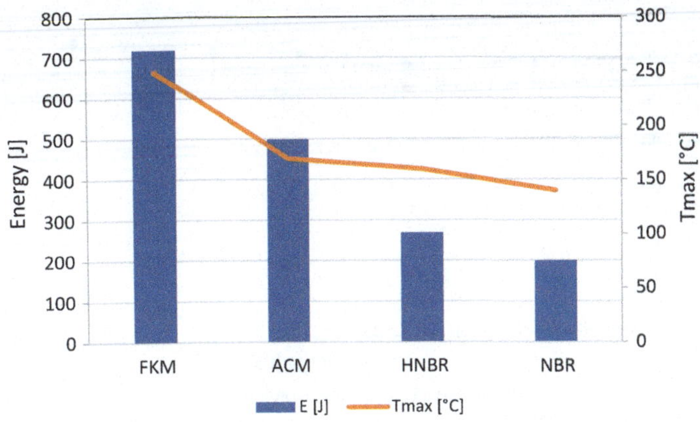

Fig. 16 Energy consumed during tribological tests for various elastomers after plasma treatment (bias voltage of 600 V/25 min of Ar, followed by 10 min of Ar/H_2)

can also be used to increase adhesion (Korzeniewska et al. 2014) or to shape structures on polymer substrates (Pawlak et al. 2017). Surface modification using laser ablation is similar to plasma treatment (Khorasani et al. 2006; Song-Hua et al. 2011; Sun et al. 2016) however, laser application can lead to the diverse effects, e.i. to reduce surface wettability. The proper selection of the laser wavelength (photon energy) and laser pulse duration, allows modifications of the surface chemistry and topography, in order to control the wettability of elastomers. The application of a nanosecond-pulsed IR laser beam towards elastomers filled with carbon fillers or pigments, capable of absorbing its energy (e.g. iron oxide black), may not cause any chemical changes to the materials, leaving a hydrophobic effect only to the geometric pattern on the surface. Subtle changes in the laser beam pulse energy, while preserving the same hatching distance did not significantly affect the contact angle value, however increasing the hatching distance, already by 0.02 mm, makes the contact angle increased significantly (even by more than 20°). For styrene-butadiene rubber or acrylonitrile-butadiene rubber filled with multiwalled carbon nanotubes and carbon black (SBR/MWCNT/CB and NBR/MWCNT/CB respectively), the relationship between the contact angle (representing the wettability of a surface) and pulse energy could be approximated by a linear function for the former and by a logarithmic one for the latter—Fig. 17.

Based on the results obtained, fibre-pulsed laser texturing of surfaces can be considered a very useful method for producing elastomers with a time-stable, superhydrophobic surface. The most likely the effect is the result of generation a hierarchical structure of MWCNT agglomerates, as demonstrated schematically in Fig. 18.

Elastomers filled only with CB, also exhibit improved thermal conductivity, but due to completely different morphology of the filler particles, the hierarchical structure was not produced. In the case of rubber filled with structural CB, the filler's

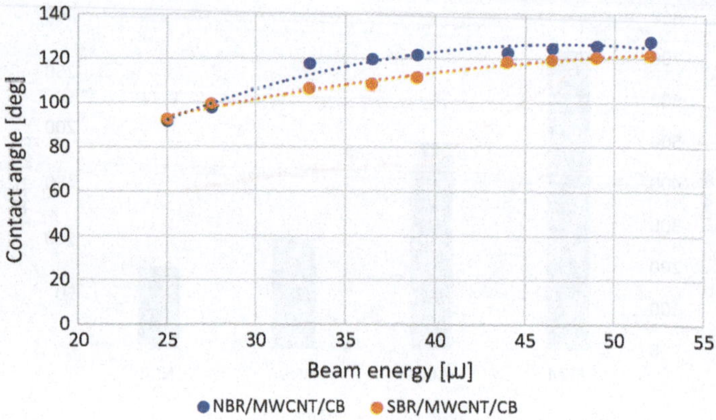

Fig. 17 Effect of beam pulse energy on the contact angle of the elastomer composites studied

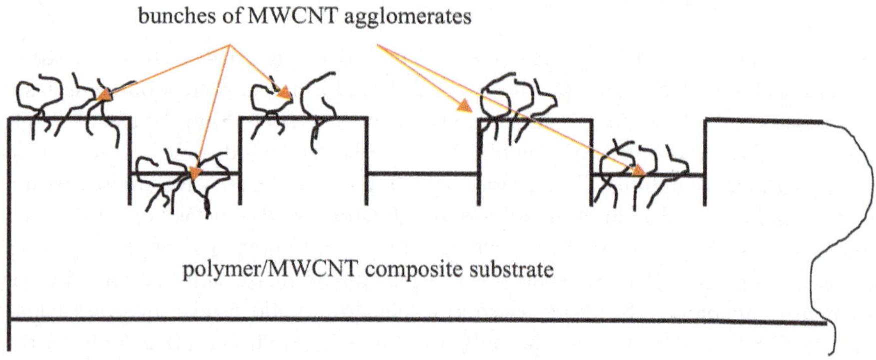

Fig. 18 Schematic representation of the hierarchical structure on the surface of an elastomer filled with MWCNT after laser ablation

tendency to form the internal structure prevailed over the agglomeration of its particles, therefore the hierarchical structure, contrary to MWCNT (as confirmed by SEM, Siciński et al. 2018) was not generated.

Surface wettability plays a very important role in the friction of elastomers, not only when friction takes place under lubricating conditions, determining the efficiency of the lubricant applied, but also because of the lubricating effect of water towards the polymers, as described by Borutto (Borutto et al. 1998).

3.4 *Ion Beam Bombardment*

The effect of surface layer graphitization of polymers was also observed when the surface was bombarded with high-energy particles, e.g. ion beams, which, due to their mass, unlike UV radiation or electrons (electron beam—EB), penetrate the material only to a limited depth, usually not exceeding some micrometers. In the Raman spectra obtained by a confocal microscope, the characteristic absorption bands D and G, derived from graphite, can be seen with wave numbers of 1350 i 1600 cm^{-1} respectively (Ferrari and Robertson 2000)—Fig. 19.

Graphitization of the external ends of the polymer macromolecules, while leaving their rest anchored in the subsurface layer, ensures the durability of the modification (Pieczyńska et al. 2012). The use of beams with an energy density <0.1 $\mu A/cm^2$, avoids excessive temperature rise of the material, leading to its destruction. As a result of the modification, hydrogen escapes from the polymer macromolecules, but the effect of the modification stops at an atomic composition C:H = 1:1, regardless of the type of ions and beam size, as can be seen for the example of high density polyethylene (HDPE) (Turos et al. 2003)—Fig. 20.

Graphitization of macromolecules is accompanied by degradation and crosslinking processes. The structure of the macromolecule determines which of them will prevail. In the case of chloroprene (CR) or natural rubber (NR), there is a reduction in hardness following the ion bombardment, while the surface layer of butadiene rubber (BR) and its styrene (SBR) and acrylonitrile (NBR) copolymers becomes additionally crosslinked. Crosslinking of macromolecules causes shrinkage of the surface layer of the material shrunk, what results in its cracking, ultimately increasing the surface microroughness. The effect is most visible for SBR vulcanizates, resembling broken ice pieces on the water surface, whereas for NR vulcanizates a more fine crack pattern is produced—Fig. 21.

The observed changes in surface microroughness, along with an increase in hardness, make the actual frictional contact decreased, which contributes to a decrease in the friction coefficient. The hardness of the elastomers studied changes—Fig. 22, influencing their tribological properties (Bieliński et al. 2014).

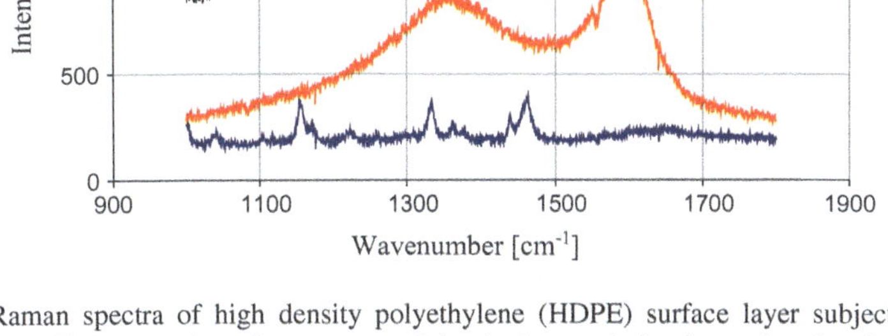

Fig. 19 Raman spectra of high density polyethylene (HDPE) surface layer subjected to ion bombardment. *For comparison the spectrum of the virgin sample has been placed at the bottom*

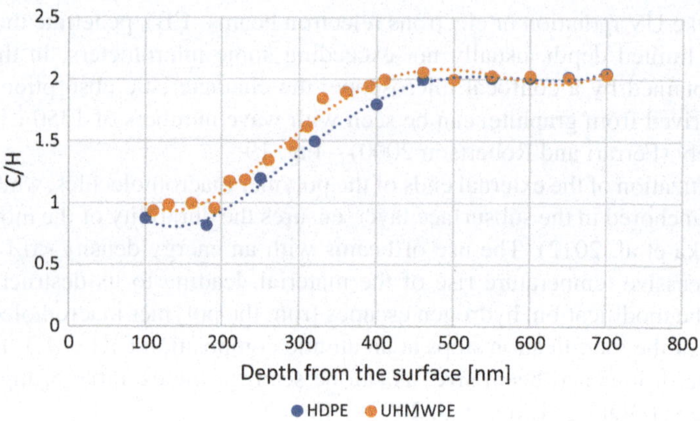

Fig. 20 Surface profile of the atomic hydrogen content of polyethylenes subjected to ion bombardment with a helium ion beam of energy 130 keV

However, the adhesive properties of elastomers also play a significant role in friction. By analogy to the polyolefins subjected to ion beam bombardment, graphitization makes progress, the degree of oxidation increases and the surface microroughness of elastomers develops. Oxidation, along with an increase in surface microroughness, are responsible for improving the wettability of materials, which occurs even for special elastomers—resistant to conventional chemical modifications—Fig. 23.

According to the hypothesis of Borutto (Borutto et al. 1998), the increase in wettabilily helps to reduce friction of elastomers—Fig. 24.

The ion bombardment has proven effective, e.g. reducing the friction of Simmer type seals made of fluorine rubber, working in power units by approx. 10–15%, or lowering the resistance of seals moving from static to kinetic friction. This is decisive from the point of view of choosing the size of hydraulic cylinders, that drive the mechanisms of panels and chassis. The seals used are made of special elastomers, which are difficult to modify in a conventional way, such as by chlorination, iodination or sulphonation, while in most cases the application of ion bombardment to modify their surface works—Fig. 25. In some cases, e.g. for thermoplastic polyurethane elastomer (TPU), the coefficient of friction could be reduced by over 2.5 times.

Modification by ion bombardment can be useful for special purpose elastomers, for which conventional, „wet" chemical treatment is ineffective, which becomes important for demanding aerospace and automotive applications. Friction modification can be obtained for any kinds of ions used for bombardment, but its effectiveness depends on the ion dose, what is demonstrated in Fig. 26.

The effect is of 0–1 character, becoming significant to the ion dose exceeding 10^{14} $1/cm^2$, regardless of the kind of irradiated elastomer or the kind of applied ions. A low coefficient of friction is likely to be associated with a graphite layer formed on the surface of an elastomer treated with a high dose ion beam. The reduction can be subscribed to changes in the mechanism of friction from the bulk—by Moore

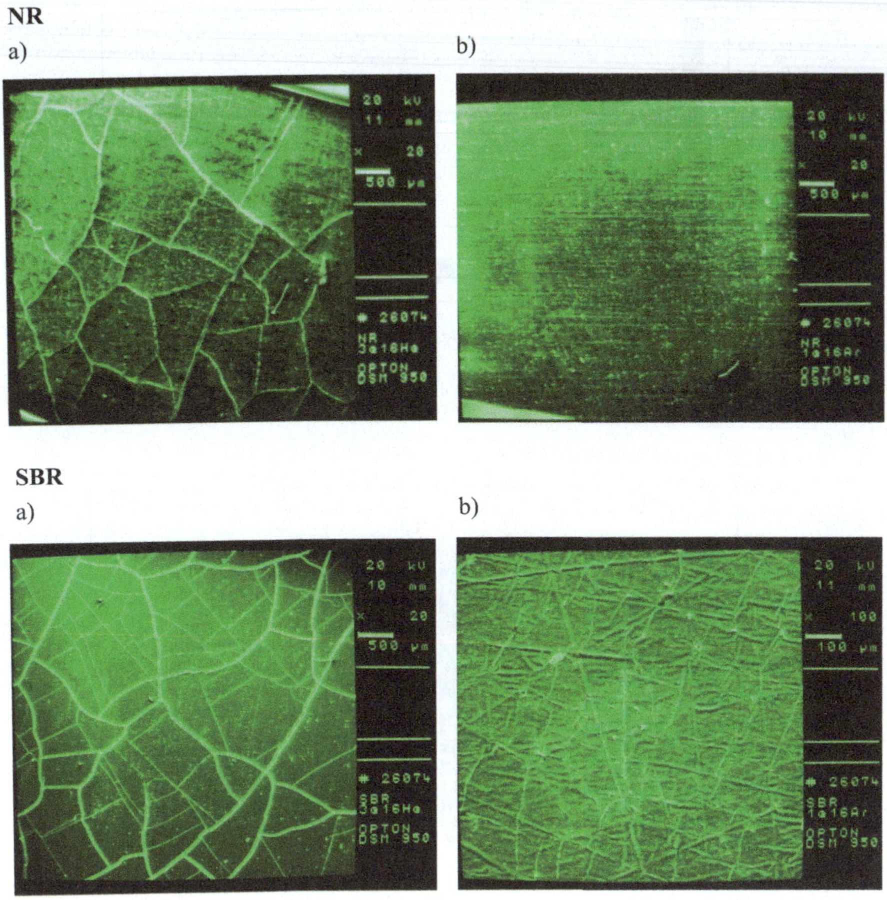

Fig. 21 Changes in surface microroughness of natural rubber (NR)-based and styrene-butadiene rubber (SBR)-based elastomers, subjected to ion beam treatment: **a** $He^{+}/3\times10^{16}$ cm^{-2}/130 keV or **b** $Ar^{+}/1\times10^{16}$ cm^{-2}/130 keV

(1975), Persson (1998) to the surface sensitive one—described by Bowden and Tabor (Persson 1998; Bowden and Tabor 1980; Rymuza 2004), shown in Fig. 27.

Dissipation of energy prevails over its accumulation in the case of elastomers with hard but very thin surface „skin", produced by ion bombardment, contrary to virgin samples, for which energy accumulation due to hysteresis of the materials is predominant (Moore 1975). The effect of the treatment is durable, despite the cracking of the elastomer surface presented above. The delamination wear, characteristic for conventional chemical modification of polymers can be avoided, due to a very shallow extent of changes (Jagielski et al. 2011).

Apart from the discussed above changes in the friction mechanism, lower friction of elastomers can also originate from a smaller actual tribological contact, as a result of increased surface microroughness and/or graphitization. Such possibility is

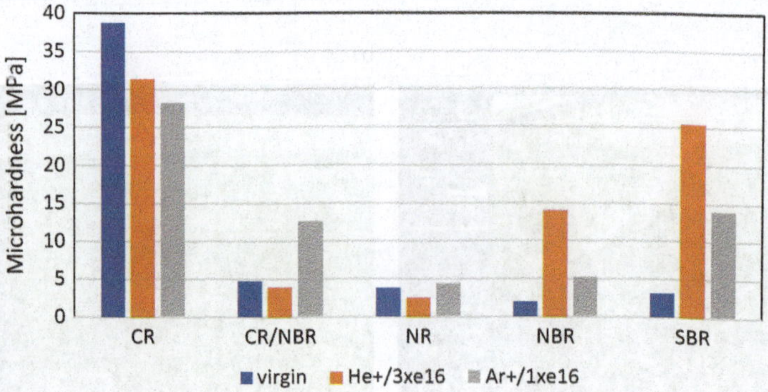

Fig. 22 Effect of ion bombardment on microhardness of the elastomers studied. *Spherical microindentation*

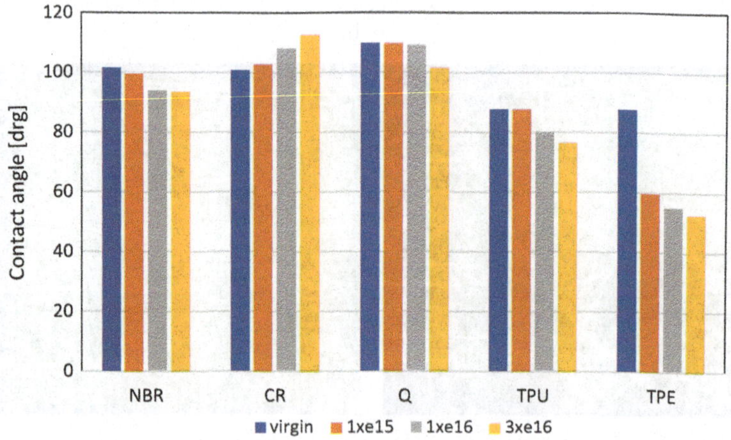

NBR - acrylonitrile-butadiene rubber vulcanizate, CR - chloroprene rubber vulcanizate, Q - silicone rubber, TPU - polyurethane elastomer, TPE - thermoplastic polyolefine elastomer

Fig. 23 Effect of ion bombardment (Ar^+ 3×10^{16} cm^{-2}/130 keV) on the water contact angle (wettability) of the special elastomers studied

demonstrated by the surface morphology shown in Fig. 28, and changes in the coefficient of friction presented in Fig. 29 for butadiene-acrylonitrile rubber/multiwalled carbon nanotubes composites (NBR/MWCNT), subjected to ion beam etching with He^+ or Ar^+ ions.

The surface irradiation of the composites with heavy Ar^+ ions exposes large carbon nanotube agglomerates, protruding above the elastomer substrate. Such „sliding islands" morphology of the surface tends to meet both of the above mentioned criteria of low friction: reduction of the actual frictional contact (islands) and lubrication (agglomerates of CNTs).

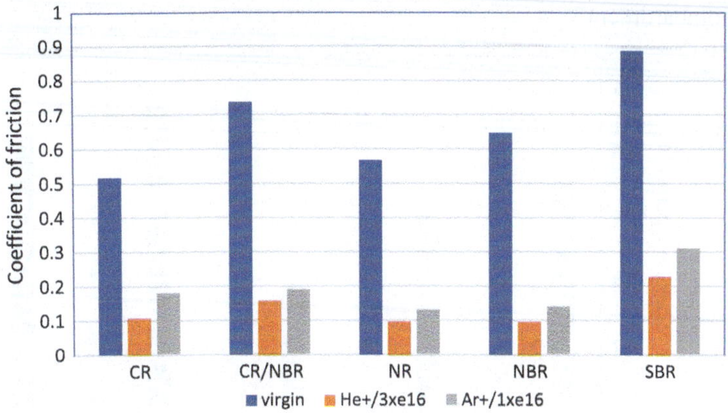

NBR - acrylonitrile-butadiene rubber vulcanizate, CR - chloroprene rubber vulcanizate, CR/NBR - 50:50 mixture of CR and NBR, NR - natural rubber vulcanizate elastomer, SBR - styrene-butadiene rubber vulcanizate

Fig. 24 The effect of ion bombardment on the coefficient of friction of various elastomers. *Steel block-on-elastomer sample ring*

4 Low Friction Coatings

Limited solubility of low molecular weight components of rubber mixes or elastomer composites, their crystallization or thermodynamical incompatibility with a nonpolar polymer matrix results in their surface migration. The surface layer of materials becomes enriched with chemicals, that change its mechanical properties, lubricate the surface or produce a kind of low-friction coatings.

4.1 Surface Migration of Low Molecular Weight Components

The blooming of low-molecular weight substances contained in the rubber mix concerns not only the components of the curing system and the products of their reaction. Nevertheless, the surface migration of softeners and protective waxes also plays an important role, as they simultaneously plasticize the rubber surface layer and modify its surface. It can be traced back to the surface images of SBR vulcanizates containing low-molecular-weight fatty acids, with increasing length of the backbone chain—Fig. 30 (Bieliński et al. 2005a, b).

Tribological properties of elastomers also depend on low molecular weight additives migrating to their surface, whose solubility limit in polymer has been exceeded. The driving force behind the migration is very often crystallization, which occurs more easily on the surface than inside a polymer matrix, or specific interactions such as hydrogen bonds. This can be traced back to paraffin waxes differing in the content

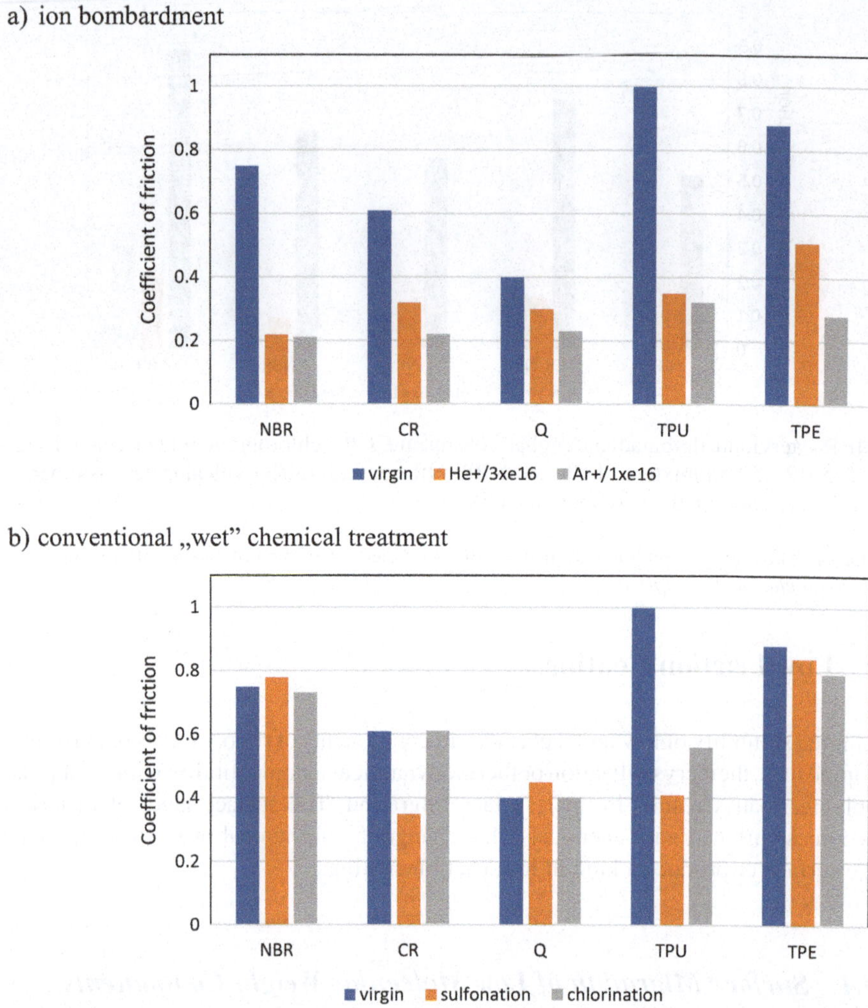

Fig. 25 Comparison of effectiveness of friction reduction by ion bombardment and conventional „wet", chemical treatment for the selected special elastomers. *Steel block-on-elastomer sample ring*

of linear phase (capable of crystallization), linear n-alcohols and n-carboxylic acids, with a backbone length of C11 to C18—Table 2.

The bloom thickness was calculated based on the spherical microindentation data, collected applying the so-called *load-partial unload technique* procedure (Briscoe et al. 1998). The tribological properties of the rubber are also influenced by the low molecular weight additives migrating on its surface, whose solubility limit in the rubber has been exceeded. The driving force behind the migration is very often crystallization, which occurs more easily on the surface than inside an elastomeric matrix, or specific interactions such as hydrogen bonds. This can be again traced back

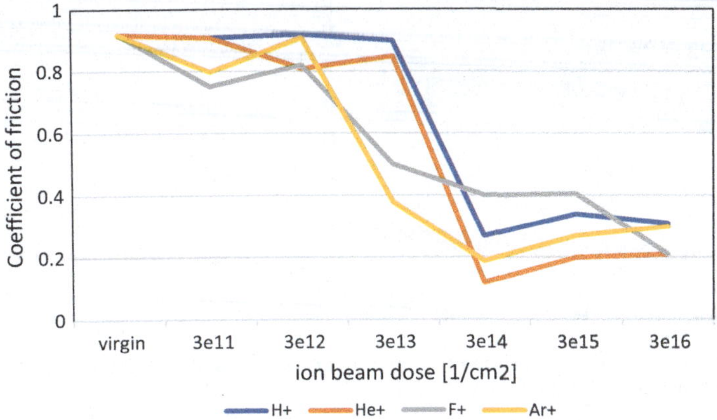

Fig. 26 Effect of various ion beam treatment on the coefficient of friction of styrene-butadiene rubber (SBR) vulcanizates. *Steel block-on-rubber sample ring*

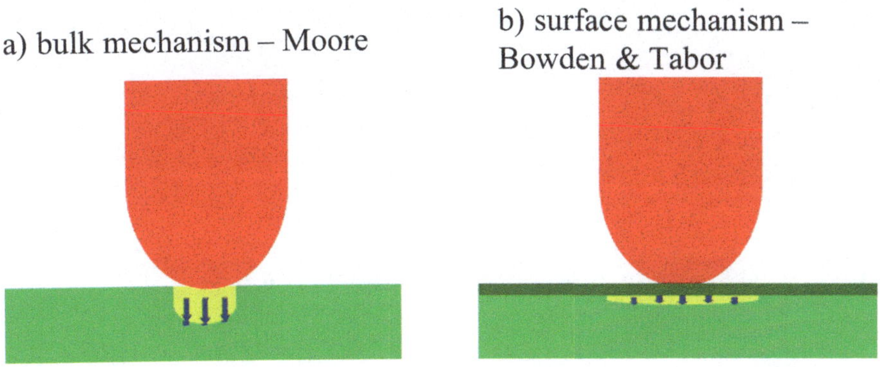

Fig. 27 Effect of ion beam bombardment on the friction mechanism of elastomers

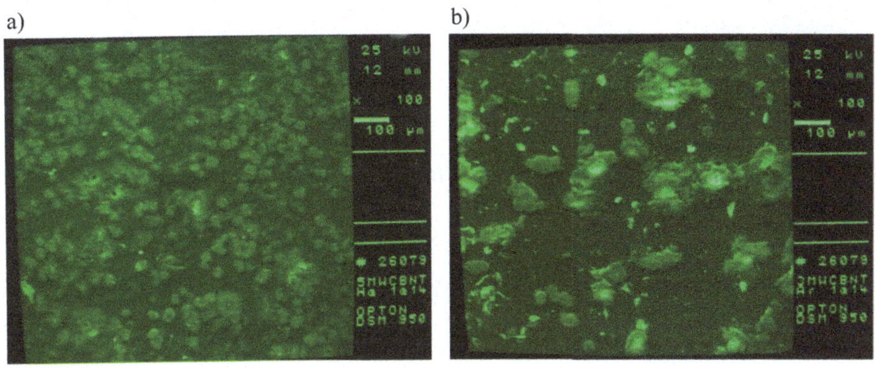

Fig. 28 Surface morphology of NBR/MWCNT containing 5 phr of MWCNT, subjected to ion etching: 1×10^{14} cm^{-2}/160 keV/He^{+} (**a**), or 1×10^{14} cm^{-2}/130 keV/Ar+ (**b**)

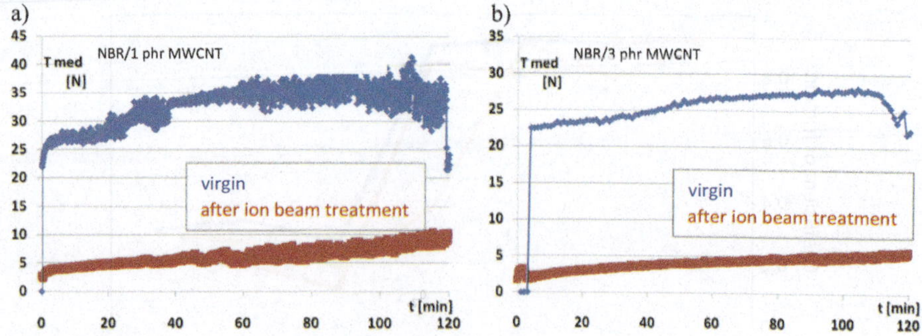

Fig. 29 Effect of Ar+ ion bombardment on the friction coefficient of NBR/MWCNT composites containing: 1 phr (**a**) or 3 phr (**b**) of MWCNT

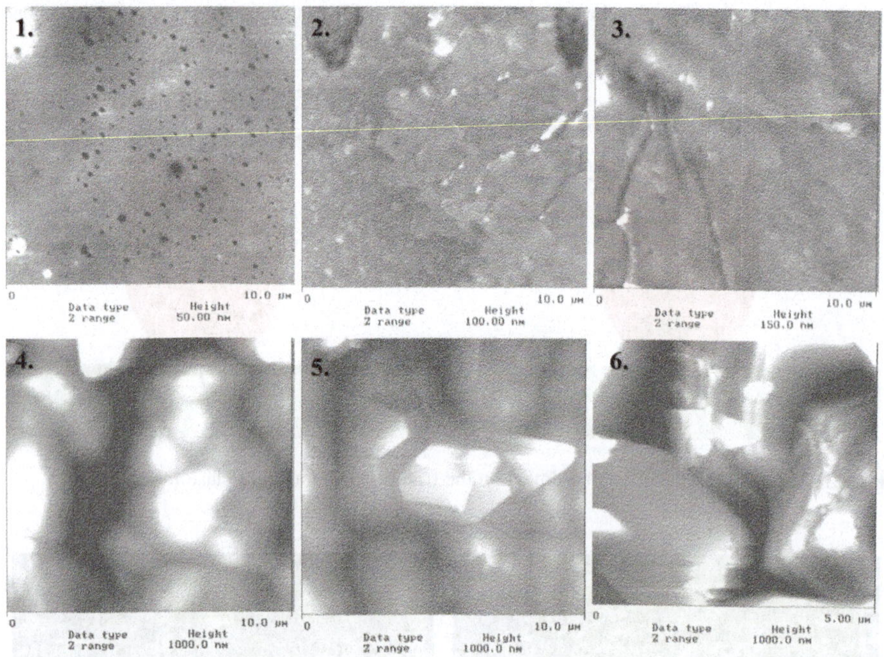

Fig. 30 Blooms of carboxylic acids on the surface of peroxide cured styrene-butadiene rubber (SBR) vulcanizates. *AFM—tapping mode*

to paraffin waxes, n-alcohols and n-carboxylic acids, with a backbone chain length of C11 to C18. A higher ratio of n-paraffins (of linear macromolecules) to iso-paraffins (of linearly branched macromolecules) increases the thickness of crystalline blooms, however, their adhesion to the substrate decreases, they break and tear off, losing their ability to further protect elastomers. From the point of view of friction and protection of the rubber against aging, the composition of the waxes should be optimized due

Table 2 Thickness, microroughness (R_a) and bearing area (B.A.) of the blooms of selected low-molecular weight additives (5 phr) present on the surface of SBR, and their influence on friction. *Microfriction: 5 μm stainless steel ball/ 0.1 μm/s*

Sample	Bloom thickness (μm)	Microroughness R_a (nm)	Bearing area B.A. (%)	Coefficient of micro-friction
SBR	–	6	98	0.18
SBR+SC11	0.02	8	71	0.37
SBR+SC12	0.20	35	25	0.22
SBR+SC16	4.74	227	52	0.27
SBR+SC17	4.47	129	44	0.16
SBR+SC18	3.42	256	71	0.24

SC11–SC18—low-molecular weight fatty acids containing from 11 (undecanoic acid) up to 18 (stearic acid) C atoms in a backbone chain

to the ratio of iso-paraffins to n-paraffins. An example of an amorphous bloom, that reduces the friction coefficient of rubber vulcanizates, is zinc stearate, which acts as a kind of grease. N-alcoholic blooms are not very effective from a tribological point of view, due to their low cohesive strength, derived from weak internal hydrogen bond interactions. N-carboxylic acid blooms are much more durable, with stronger internal interactions. They can reach very large thickness, even up to a few μm—Fig. 1. This results from the fact, that together with zinc stearate—always present on the surface of sulfur vulcanizates (product of the reaction between ZnO and stearic acid, formed during vulcanization with the sulfur system) (Jasiński 1988), they work to reduce the friction coefficient of the rubber, even on a macroscopic scale—Fig. 31.

In the contrast to thermoplastics or resins, the addition of solid lubricants, such as graphite or MoS_2, is less effective in reducing the friction coefficient of elastomers due to the simultaneous plasticization of the material, manifested by a significant increase in the deformation component of the friction force (Bieliński et al. 1993a, b). This is even more evident when trying to modify rubber with low molecular weight waxes, paraffin oil or silicone oil, acting even more effectively towards plastificizing the rubber, especially its surface layer—Fig. 32.

The increase in the deformation component of friction prevails over the decrease in its adhesion component, which is especially visible in the case of the addition of paraffin oil, being of a similar solubility parameter to hydrocarbon elastomers and therefore swelling them most effectively, while not manifesting surface migration. A better lubrication effect can be obtained by admixing of silicone oil, which is hardly soluble in NBR, but the lowest friction can be achieved by applying solid lubricants, the presence of which in the top layer of the rubber makes their addition most effective, which has been utilized in sealing technology.

The above observations are also very important from the point of view of car users, who could accidentally standing on one of the tires on an oil or a grease stain, or the safety of workers who have stepped into such spills of oil or grease. The danger of slipping is generally only present at the beginning and only if oil or grease does

a) n-alkanes *)

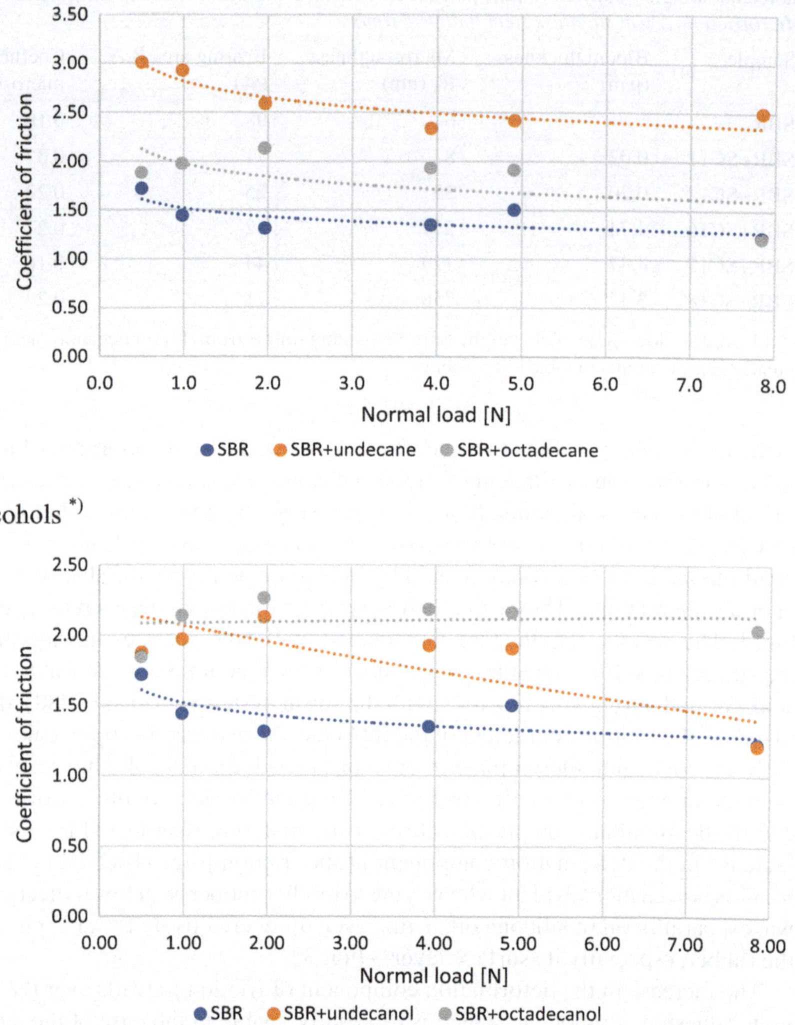

Fig. 31 Influence of the selected low-molecular weight additives (5 phr) on the coefficient of friction of SBR vulcanizates against steel. *Rubber pin-on-steel disc*

not swell the tread material of the tyre or the shoe sole. Due to the fact that most of the elastomers used in the above mentioned applications are non-polar by nature, slippage is expected mainly from polar oils and greases.

c) *n*-carboxylic acids and their esters

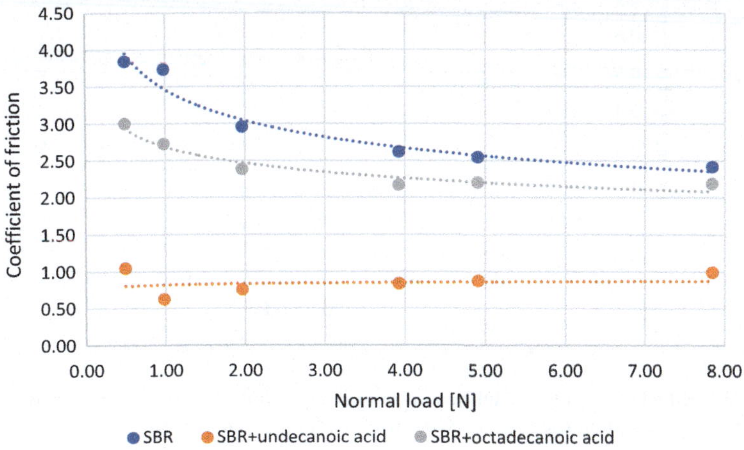

d) paraffin waxes

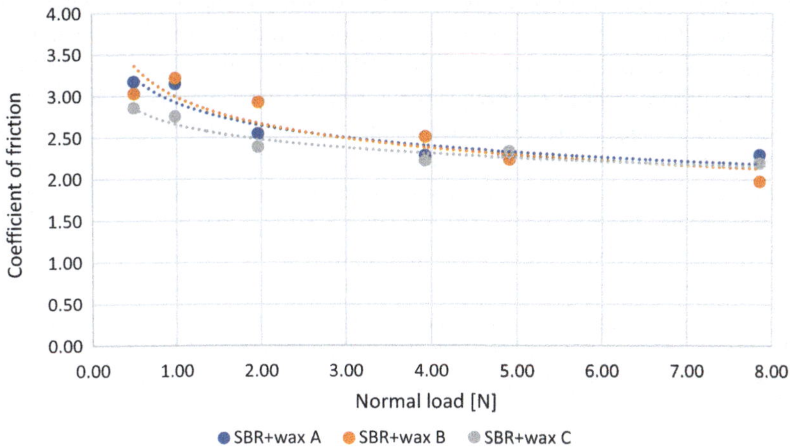

[*)] the tribological characteristic of the virgin elastomer is presented for comparison

Fig. 31 (continued)

4.2 *Surface Segregation in Polymer Blends*

The phenomenon of surface segregation is not only limited to low-molecular weight components of rubber mixes. Its existence was also confirmed in relation to elastomer-plastomer blends (Bieliński et al. 1997a, b; Bieliński 2004). This is most likely the result of surface migration of the low-molecular weight plastomer

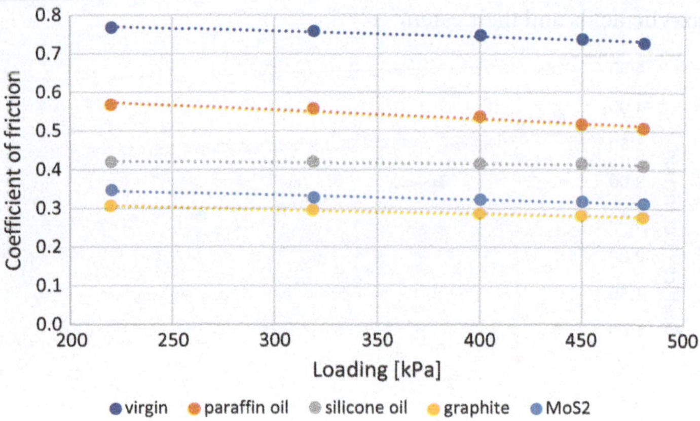

Fig. 32 Effect of the selected lubricating additives: 1. Oils—5 phr, 2. graphite and MoS_2—15 phr, on butadiene-acrylonitrile rubber (NBR) friction. *Rubber pin-on-steel disc*

fraction. The phenomenon is dependent on molecular weight, the difference in solubility parameters of both polymers and processing technology, allowing for surface segregation in polymer mixtures, as demonstrated for a system of natural rubber (NR)/trans-1,4-polypectenamer (TOR) (Bieliński et al. 1999a, b)—Fig. 33.

The surface layer of the blends (ca. 0.6–1.1 μm) is enriched with polyoctenamer compared to their bulk composition. The low-molecular weight cyclic fraction of TOR, segregating towards the surface, reduces the material microhardness. The effect is most visible for low polyoctenamer contents, not exceeding 20–25 phr—Fig. 34.

The effect described above is reflected by a significant decrease in the friction coefficient of the NR vulcanizates modified by TOR additive—Fig. 35.

An interesting phenomenon, known from industrial practice, is the protective effect of admixing diene terpolymer (EPDM) or ethylene-propylene copolymer (EPM) to styrene-butadiene rubber (SBR). Already added 5 phr of the former significantly improves the resistance of the latter to ozone aging. Infrared analysis (FTIR-ATR) indicates on the reversal of the ratio of the surface layer composition of the mixture compared to the EPM/SBR ratio by volume. Unlike styrene-butadiene rubber (SBR) filled with flexible polyolefines (EPDM or EPM), the system in which ethylene-propylene-diene rubber (EPDM), of similar chemical structure, forms a matrix for low molecular weight polyethylene (LDPE/EPDM), is characterized by the presence of polyethylene crystals on the surface of their mixtures. The crystalline phase of polyethylene is partially solvated by the amorphous matrix of elastomer during mixing above the melting temperature of LDPE. Most probably, the main driving force behind the surface migration of polyethylene in the system is its crystallization, which occurs more easily on the surface. The resulting crystallites form spherulitic structures only if the macromolecules of polyethylene exceed a certain molecular weight of $M_w = 1500$ g/mol—Fig. 36.

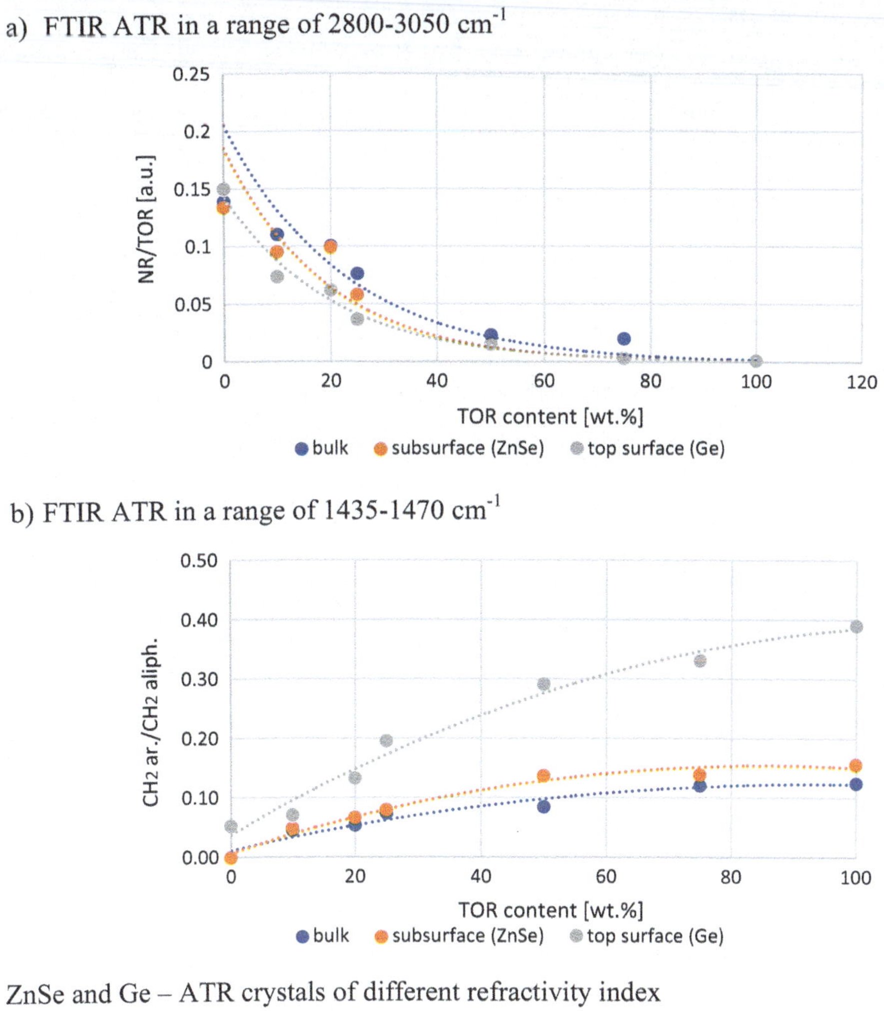

Fig. 33 Composition of the surface layer of NR/TOR blends, calculated based on FTIR analysis. *The arrow points towards the blend surface*

Surface segregation acts towards lowering of the coefficient of friction of polymer materials, especially under low load and the effect is durable—Fig. 37, and influences the mechanical properties of their surface layer—Fig. 38.

Morphology of the surface layer of polyolefine blends is also influenced by the macromolecular structure of the elastomers, which of course also affects the surface hardness profile and friction coefficient of the materials.

The largest surface segregation takes place in the system containing the elastomer of statistical structure (completely amorphous). Despite the lowest mechanical strength of the blend with its participation, the tribological effect achieved is the

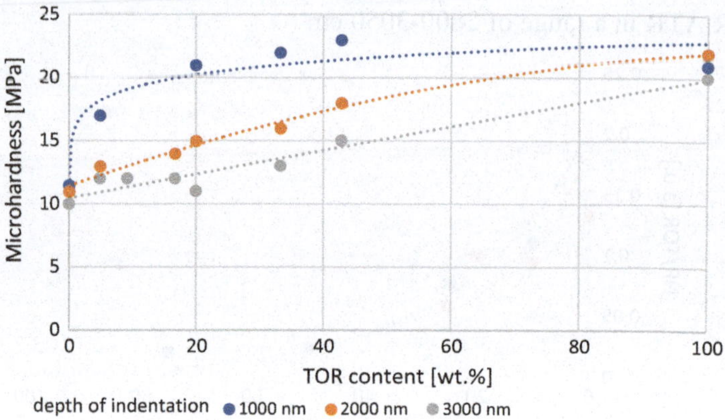

Fig. 34 Microhardness of NR/TOR blends. *Spherical microindentation*

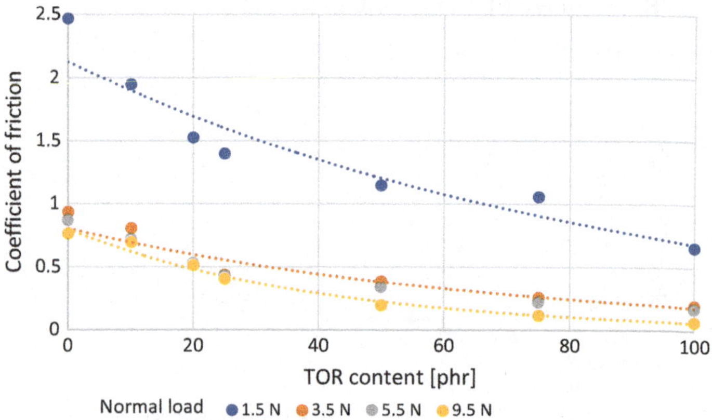

Fig. 35 Effect of the addition of polyoctenamer (TOR) on the friction of NR/TOR blends. *Steel block-on-elastomer sample ring*

greatest, which once again confirms the significant role of the surface layer in the friction of elastomers.

Molecular weight distribution (Mw/Mn) is also important. In the case of the polyethylene of high Mw/Mn value, its low-molecular weight fraction is solvated by a rubber matrix, limiting surface crystallization of the polyethylene. The application of LDPE with the so-called bimodal distribution of molecular weight allows, with appropriate mixing technology, to obtain an amorphous top layer consisting of a low-molecular-weight fraction of the polyethylene, which gradually fills the surface microroughnesses (Vijaybaskar and Bhowmick 2005)—Fig. 39.

In general, the addition of low-friction polymer additives like polytetrafluoroethylene (PTFE), isotactic polypropylene (iPP) or various kinds of polyethylene (PE) to

a) LDPE $M_w = 4000$ g/mol $M_w/M_n = 2.72$

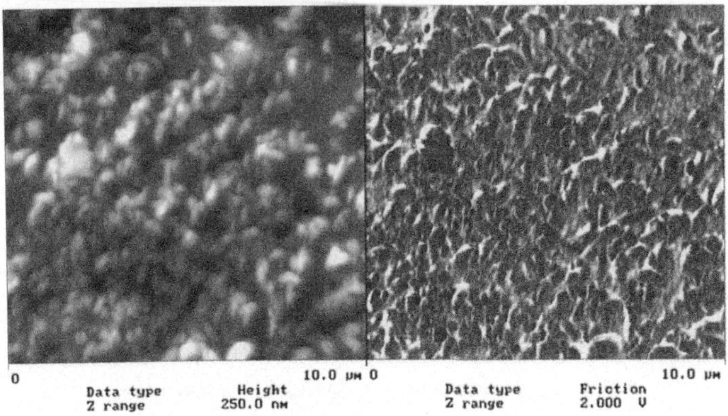

b) LDPE $M_w = 15000$ $M_w/M_n = 3.03$

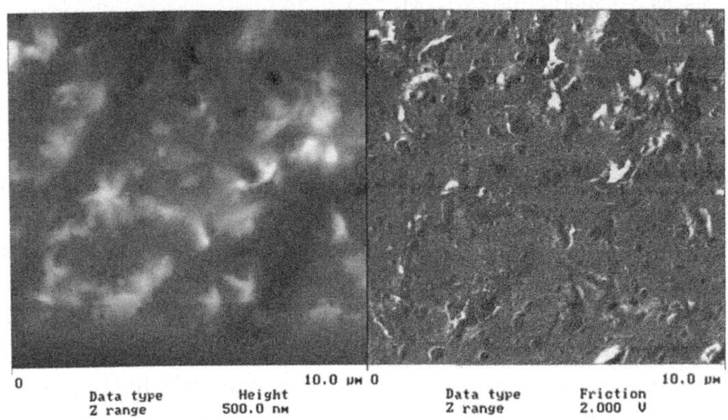

c) PE-LD $M_w = 35000$ $M_w/M_n = 2.73$

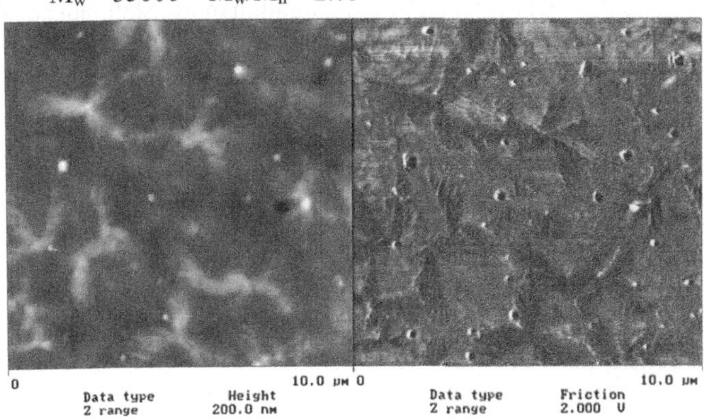

Fig. 36 Crystallization of low-density polyethylene (LDPE) on the surface of 5 phr LDPE/SBR blends. ***AFM – tapping mode***

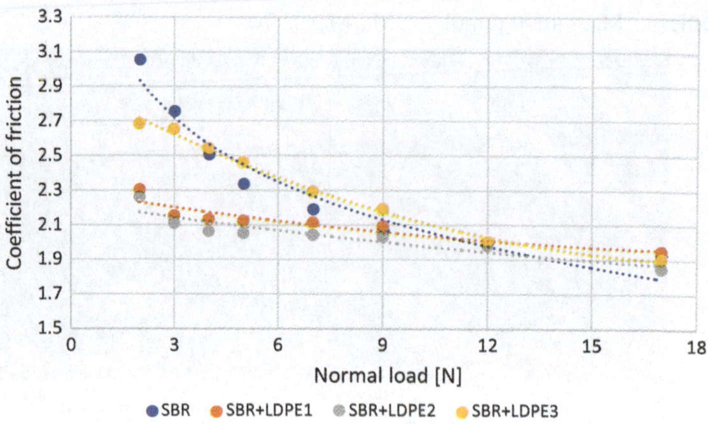

LDPE1 – M_w=4000 g/mol; M_w/M_n=2.72 LDPE2 – M_w=15000 g/mol; M_w/M_n=3.03
LDPE3 – M_w=35000 g/mol; M_w/M_n=2.73

Fig. 37 Effect of molecular weight and structure of polyolefines (LDPE) on the friction coefficient of 15 phr LDPE/SBR against steel. ***Steel block-on-elastomer sample ring***

elastomers, acts towards to reduce the friction coefficient of polymer blends—Fig. 40 (Bieliński et al. 1993a, b).

Introduction of iPP into EPDM at a temperature above melting point of the crystalline phase of the plastomer allows for better homogenization of the components, as a result of providing conditions for at least partial miscibility of the polymers at segmental level (polymers are not thermodynamically miscible) and/or for the co-crystallization of the propylene monomer units (Bielinski et al. 1997a, b). The observed increase in degree of crystallinity in the iPP/EPDM system results in decrease of friction coefficient of the elastomer. This can be explained, similar to thermoplastic block elastomers, by the limitation of translational and conformational mobility of flexible blocks. In contrast to iPP/EPDM blends, LDPE/EPDM systems are characterized by a lower content of the crystalline phase than would be the case with additive calculations. Most likely, an amorphous rubber matrix solvates the crystalline phase of the plastomer. It is highly likely, that the tribological effect achieved is due to the low molecular weight fraction that segregates to the surface, acting further as a kind of grease.

4.3 *Low-Friction Coatings and Solvent-Non Solvent Treatment*

Not every chemical or physical modification is effective in reducing friction of elastomers. Difficulties arise particularly with regard to specialty elastomers, such as

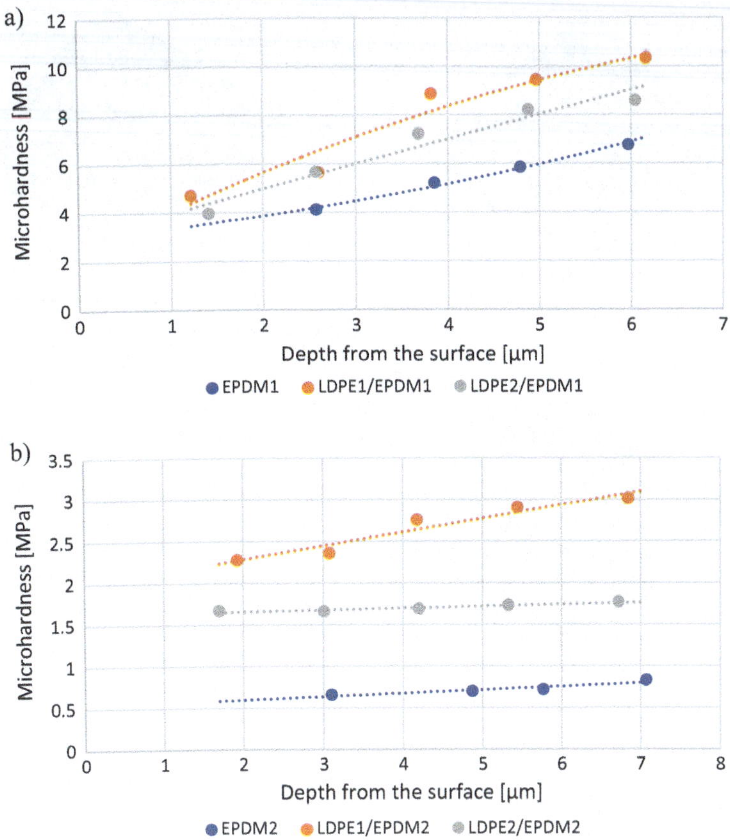

LDPE1 – M_w=4000 g/mol; M_w/M_n=2.72 LDPE2 – M_w=15000 g/mol; M_w/M_n=3.03
EPDM1 – block copolymer; EPDM2 – random copolymer

Fig. 38 Effect of the macromolecular structure of the elastomer matrix (EPDM) on the surface hardness profile of their blends with low-density polyethylenes (LDPE). ***Spherical microindentation***

ethylene-propylene or chloroprene rubbers, and specialty materials, e.g. fluorine and silicone elastomers. Alternatively, various types of coatings can be applied to the surface of materials, as required—containing pro-adhesive or lubricating substances (Lawson 1987; Bieliński et al. 2003). The morphology of coatings—Fig. 41, determining their tribological performance, can be controlled rheologically by monitoring their viscosity.

Particles of solid lubricants such as graphite or molybdenum disulphide can also be introduced directly into the surface layer of elastomer using the so-called "*solvent—non-solvent*" treatment (Chen and Ruckenstein 1992). The method is based on controlled swelling of the surface layer of elastomer by a suspension of solid lubricant in thermodynamically good *solvent* for it. The solvent reduces the degree

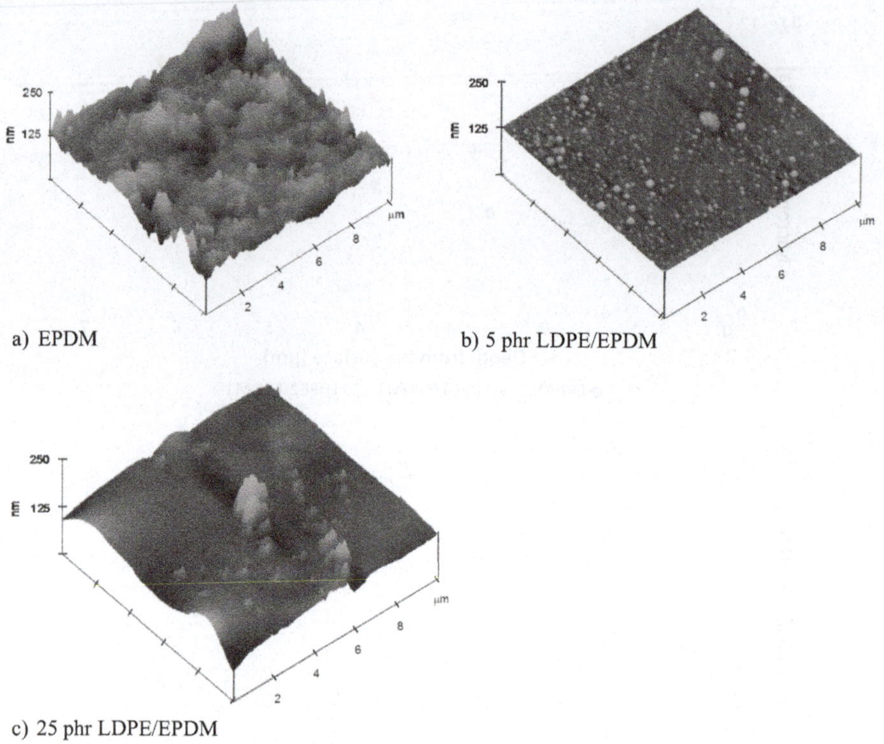

Fig. 39 Surface segregation in 5-25 phr LDPE of bimodal molecular weight distribution/EPDM blends. ***AFM—tapping mode***

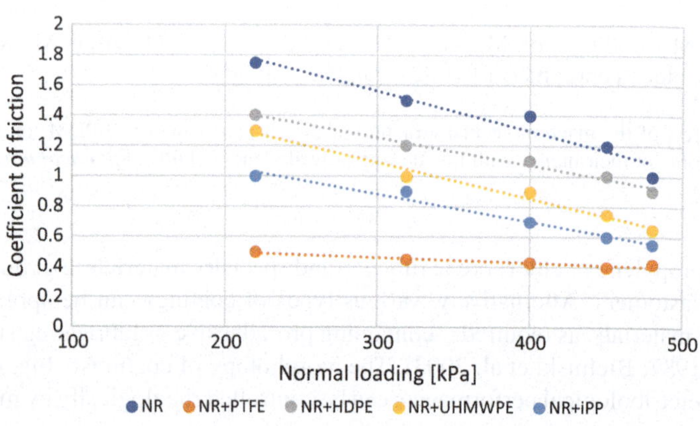

Fig. 40 Effect of admixing of 25 phr of selected polymers on friction of their blends with natural rubber. ***NR blend pin-on-steel disc***

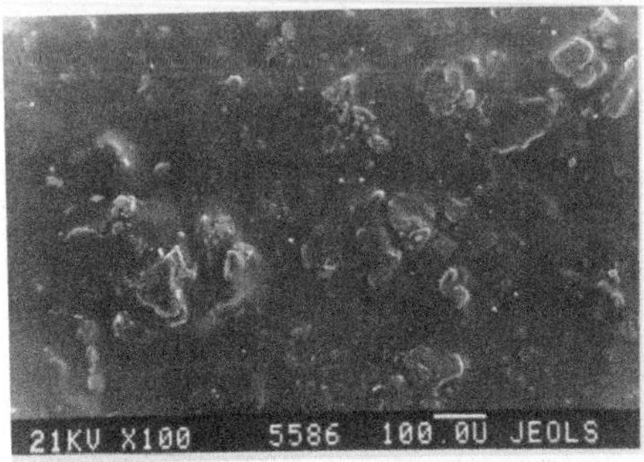

40 phr of graphite

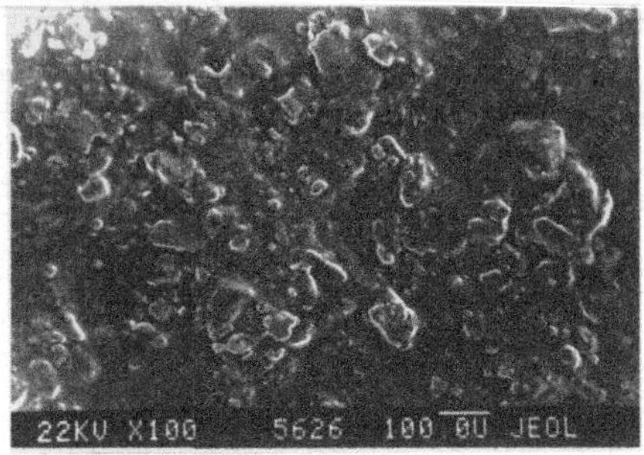

80 phr of graphite

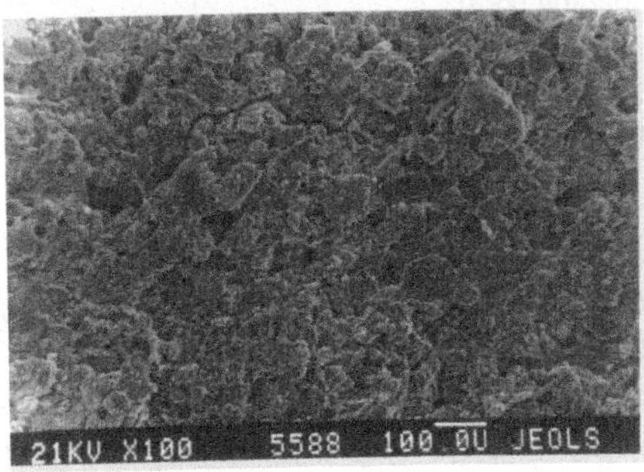

110 phr of graphite

Fig. 41 Influence of graphite content (related to rubber mass) on morphology of coatings made of the lubricant dispersion in toluene of SBR (SEM—100×)

of chain entanglement and packing, allowing the introduction of graphite or MoS_2 particles into the vulcanizate surface layer. After extraction with another solvent in which it dissolves well, but which does not swell rubber (*non-solvent*), the modifier particles anchor mechanically in the surface layer of the elastomer (substrate), providing a permanent effect of the modification—Fig. 42.

A coating with graphite, MoS_2, or their mixture, can significantly reduce the acrylonitrile-butadiene rubber vulcanizates friction against steel—Fig. 43. The modification was successfully verified as an effective method of modifying seals used in non-lubricated pneumatics (Ślusarski et al. 1996).

Similar modification can be made by applying to the surface of the vulcanizate a coating made of a mixture of solid lubricant (graphite or MoS_2) and rubber

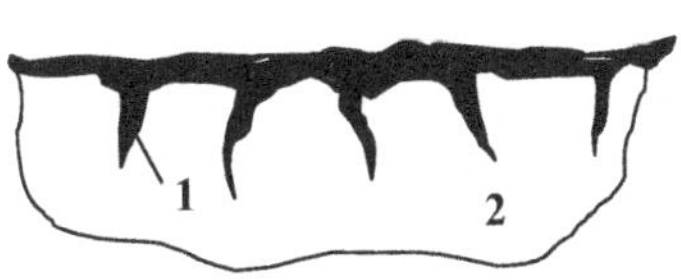

1 - solid lubricant (graphite or MoS_2)
2 - elastomer substrate

Fig. 42 Suggested morphology (cross-section) of low-friction coatings

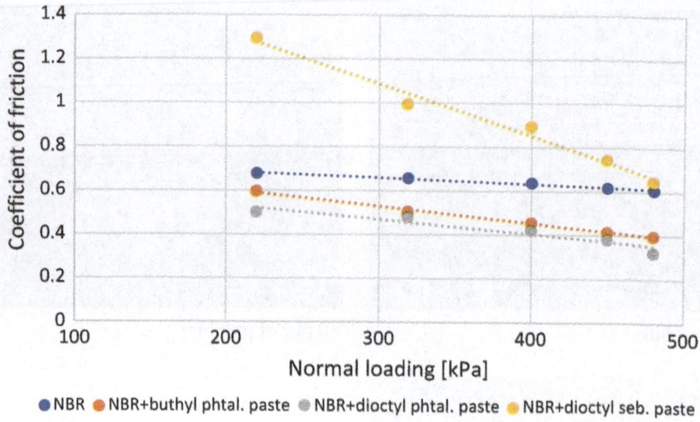

Fig. 43 Effect of solvent—non-solvent treatment on the friction characteristic of NBR vulcanizates. ***Rubber pin-on-steel disc***

latex or solution—preferably of chloroprene rubber, which has good adhesion and mechanical strength. In order for the coating to work effectively to reduce the friction coefficient, the solid lubricant particles must be distributed evenly, which can be tracked using a scanning electron microscope equipped with an X-ray analyser (SEM-EDX)—Fig. 44, or simply controlled rheologically.

The above coatings ("latex" coatings), contrary to the solvent—non-solvent ones (Fig. 42), have layered structure—Fig. 45.

Unlike in "*solvent—non-solvent*" coating, this time the outer layer is made of rubber and the friction occurs by slipping via a layer of lubricant (2), situated between the rubber substrate (3) and the outer layer (1).

5 Tribochemical Modification and the so-Called „Third Body" Formation

The least understandable aspects of friction are the phenomena occurring on the friction surface. The concept of the so-called "third body" (Heinicke 1984) is the basis on which the slip theories have been formulated—Fig. 46.

The so-called "third body" zone consists of wear and degradation products, originated from both counterfaces, impurities and lubricants, that change the friction conditions by changing the contact surface and/or acting as a kind of grease under the influence of plastic deformations, separating the rubbing bodies. Recently, the approach to friction has begun to change due to tribochemistry (Płaza 1997). Tribochemistry investigates chemical reactions taking place in the friction zone and their influence on friction conditions. It explains the process of selective transfer of the specimen material or its components to the counterface (Polak 1998), forming

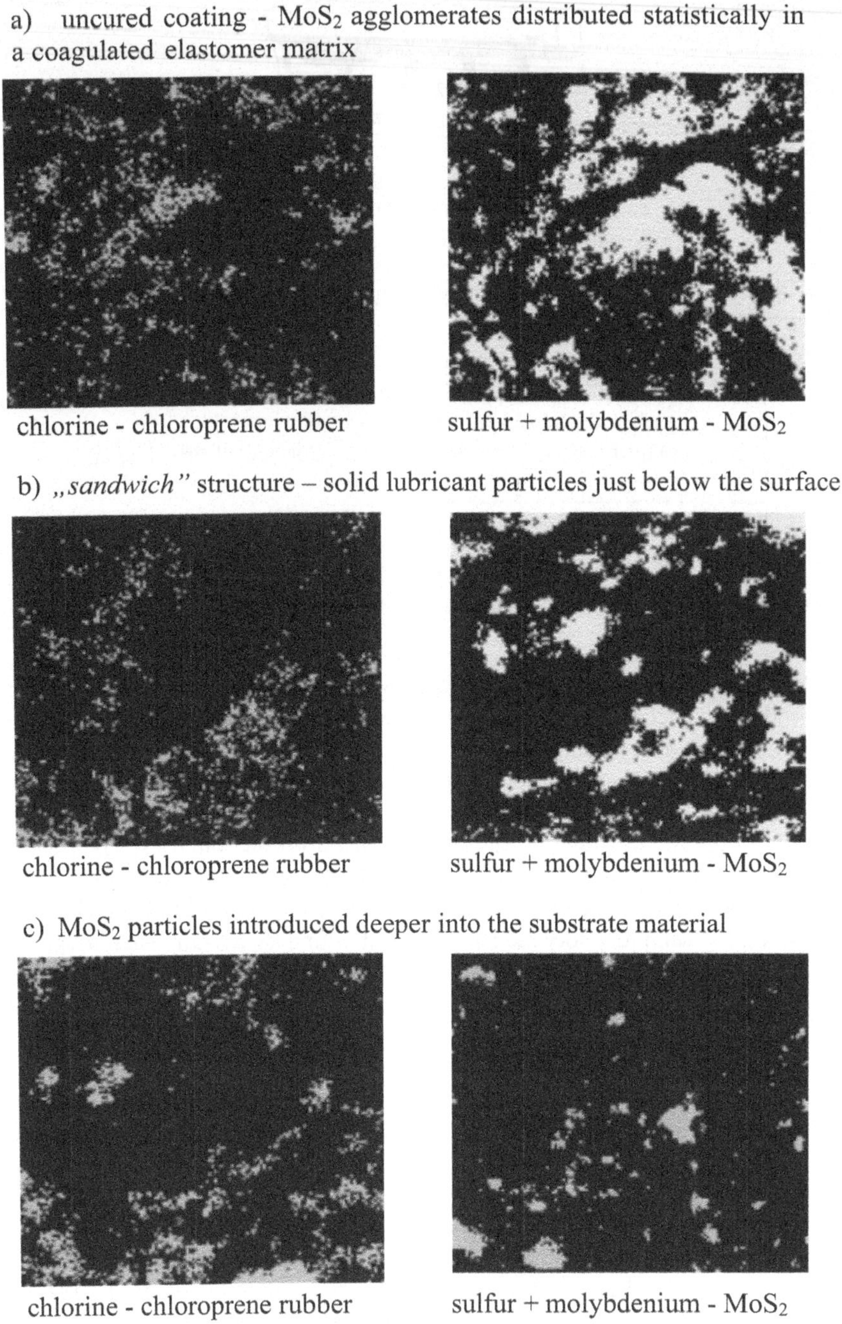

Fig. 44 The distribution of chemical elements—representing components, in coatings. ***SEM-EDX mapping***

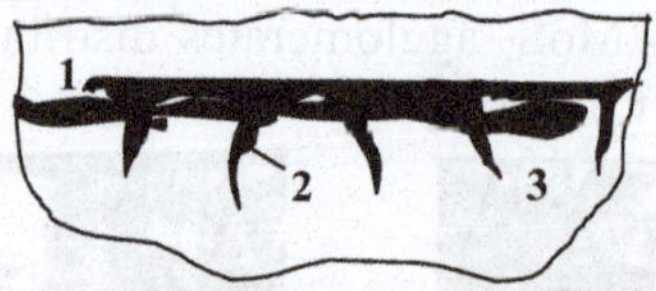

Fig. 45 Schematic representation of the „latex" coating morphology

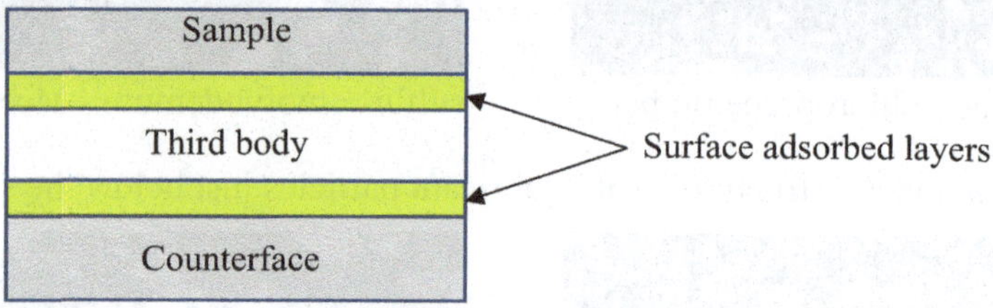

Fig. 46 The "five-zone" scheme, representing the "third body" model

boundary layers, protecting against further wear or reducing friction. Amorphous wear products, which are a consequence of abrasion and oxidation of "*soft*" polymer material, form an intermediate layer (i.e. "third body" layer), capable of dissipating mechanical energy (Myshkin 2000) and lubrication (Godet 1990). In the 1970s, the phenomenon of selective transfer, was the subject of numerous literature reports, especially in the Soviet Union, Central and East European countries. The selective transfer, also known as the Garkunov effect, is based on the formation of a layer of "*pure*" copper on the surface of the steel counterface being in frictional contact with the copper alloy sample, under the conditions of boundary friction when lubricating the contact with glycerine (Garkunov 1989). Significant reduction in the friction coefficient and practically no wear of the metal elements in some polymer-metal friction couples was also observed. The accompanying temperature rise in the friction contact zone favours the phenomenon of selective transfer of polymeric material components, which then enter into a chemical reactions with the surface layer of the metal counterface. So, the modification is not only limited to the polymer sample (Rymuza 1986), but also affects the surface layer of the metal. An example is the hydrogen wear of aluminium alloys, which cooperate intensively with a specimen made of polymer materials (Starczewski 2004). In the subject literature, only a few reports can be found on a similar modification, that would work towards reducing friction (Morrison and Porter 1983; Grossiord et al. 1998; Wang et al. 2005a, b; Bieliński et al. 2006a, b, 2007). It turns out, that as a result of intensive friction of sulfur vulcanizates of butadiene-styrene rubber against a sample made of Armco iron, a chemical modification of the surface layer of the metal occurs. Its range, except from the obvious dependence on the free sulfur and its compounds content

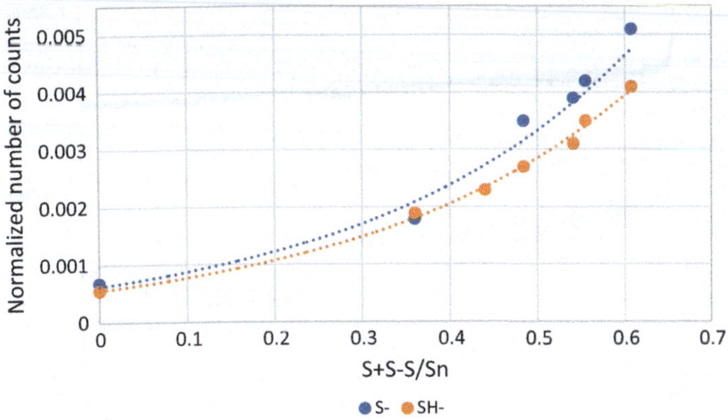

Fig. 47 Influence of the degree of crosslink sulfidity (S+S-S/S_n) of SBR vulcanizates on the relative content of sulfur containing ions (ToF-SIMS), present in the surface layer of Armco iron counterface after friction. *Metal block-on-SBR ring*

(originating from unreacted accelerators, and the products of vulcanization), which tend to surface migration, also depends on the type (sulfidity) and degree of crosslink density—Fig. 47.

From the results obtained it can be concluded, that the lower the sulfidity of crosslinks (S+S−S concentration), the higher the extent of the modification of iron counterface. The highest effect was detected in the case of iron counterace being in the frictional contact with SBR cured with a sulfur effective system of short (mono- and di-sulfidic) to long polysulfidic (S+S−S/S_n) crosslinks ratio equal to 0.55. The heat generated during friction facilitates the decomposition of polysulfidic crosslinks, which characterize themselves by the lowest energy among the sulfur crosslinks created during rubber vulcanization (Morrison and Porter 1983). Their decomposition, responsible for the liberation of sulfur ions being highly reactive towards iron, is most probable. The presence of 100–150 nm depth iron sulphide (FeS) layer was confirmed by the *depth profiling* SIMS studies of iron Armco sample after an extensive friction against SBR vulcanizates. FeS can effectively lubricate the metal surface, reducing its friction coefficient against the rubber vulcanizates—Fig. 48.

FeS, characterizing itself by good lubricating properties, becomes easily spread on the metal surface, filling its microroughnesses. Even a very thin layer of FeS can act effectively to reduce friction in the rubber-metal friction couple, due to the adhesion of the lubricating film to the metal surface. Metal oxides, being created simultaneously (mainly Fe_3O_4), act synergistically together with FeS, making a significant increase in the wear resistance of the metal counterface (Buckley 1981).

Attempts to modify the surface layer of light magnesium alloys, e.g. AZ 31 (96% Mg, 3% Al, 1% Zn), applying the "rubber friction" in order to achieve a lubricating effect, were also successful—Fig. 49 (Bieliński et al. 2006a, b, 2007).

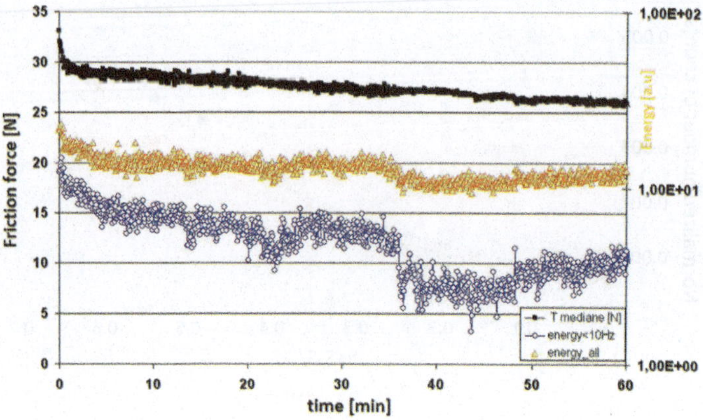

Fig. 48 Influence of the surface modification of Armco iron counterface on the friction coefficient and energy dissipation during friction of SBR vulcanizates against the metal. *Metal block-on-rubber ring*

Due to the modification the friction between MZ 31 and SBR could be reduced noticeably. The optimization of the efficiency of the metal counterface modification was tried, studying the effect of:

- crosslink density of rubber (conventional sulfur rubber vulcanizate vs. ebonite), and
- conformation of macromolecules, determining the stiffness of the polymer (polysulfone vs. polysulfidic rubber).

Analysis of the ToF-SIMS spectra obtained, revealed that the highest amount of S^- and SH^- ions was transferred to the surface layer of Armco iron subjected to extensive friction against ebonite—Fig. 50.

The modification with SBR and crosslinked polysulfide rubber is also effective (Siciński 2008). In the case of polysulfone, which characterizes itself by a strong binding of sulfur to the main chain and exhibits different from the other polymers studied mechanical degradation mechanism, the effect of sulfur transfer to the surface layer of the metal counterface is practically unnoticeable. The amount of iron sulfide produced depends on the reactivity and the concentration of the sulfur-containing polymer fragments present in the friction zone. Among the possible ingredients, leading to the formation of FeS, the greatest affinity for iron is shown by the polysulfide crosslinks and ionic products of their decomposition. They may be formed only in the case of styrene-butadiene rubber and ebonite vulcanizates, because of their structure. Distribution of all ions, present in the top layer of a metal sample subjected to friction against various polymers containing sulphur, is homogeneous. Again, apart from sulfuric ions, iron oxides are also formed on the metal surface, which presence promotes sliding (Wang et al. 2005a, b; Siciński 2008). Iron sulfide can be further oxidized during friction, leading to the formation of sulfones, that show even better lubricating properties. The complementary results obtained using

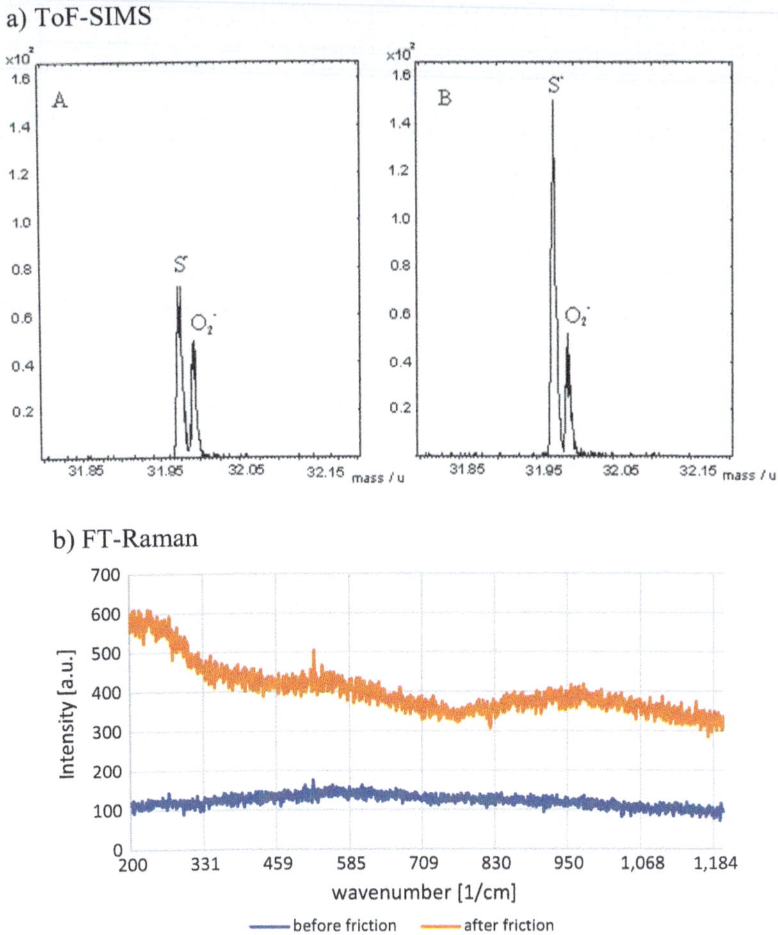

Fig. 49 The comparison between the ToF-SIMS (**a**) and confocal FT-Raman—from ca. 0.5 μm depth (**b**) spectra of AZ 31 alloy surface, collected before and after its friction against SBR. *Metal block-on-rubber ring*

Raman spectroscopy, confirm the metal modification data obtained by ToF-SIMS analysis—Fig. 51.

From the ToF-SIMS analysis it follows that only the spectra collected from the surface of the metal sample after friction with ebonite, SBR and cross-linked polysulfide rubber, indicate on a modification of the surface layer of iron. The FT-Ramana spectra of the metal counterface, obtained after friction with high-load ebonite or polysulphone, contain only the absorption bands that can be attributed to the degraded fragments of macromolecules or difficult to define compounds of carbon, hydrogen and oxygen. The example of ebonite additionally illustrates, that an increase in friction load does not always work towards an increase in the modification effect of the

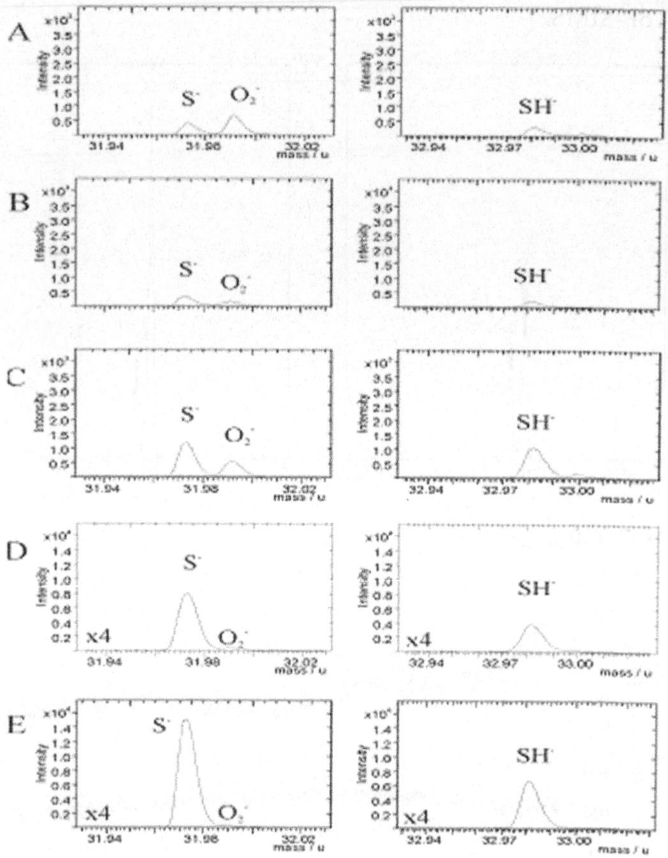

A - virgin iron sample (before friction); B - polysulfone under a normal load of 21.4 N; C - SBR under the normal load of 11.4 N; D - polysulfidic rubber under a normal load of 11.4 N; E - ebonite under a normal load of 21.4 N

Fig. 50 The specific ToF-SIMS spectra of the surface layer of Armco iron samples subjected to the friction against some polymer materials containing sulfur. **Metal block-on-polymer ring; v= 0,12 m/s; T = 23±15 °C**

surface layer of the metal. Under extreme conditions of friction, radical degradation of macromolecules occurs, associated with their strong oxidation. Under these conditions, the mechanism of ionic decomposition of crosslinks is not able to manifest itself.

Tribological characteristics of the polymer—ion Armco friction couples are presented in Fig. 52.

The tribological characteristics confirm the effect of the surface layer modification of iron samples with sulphur compounds, which occurs during friction of Armco iron against polymers containing sulfur. In the case of ebonite, the reduction of the friction

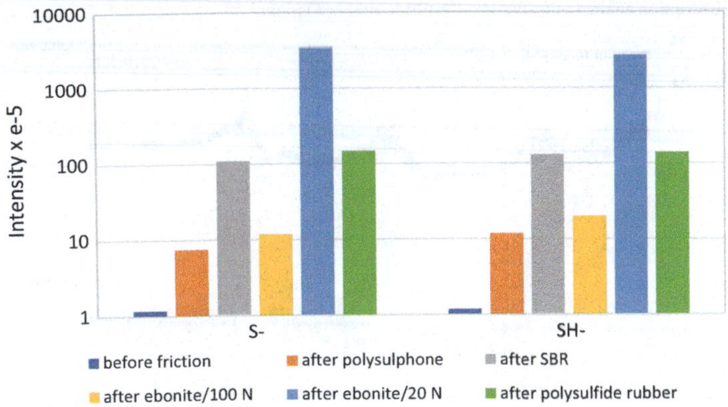

Fig. 51 Changes in the ToF-SIMS spectra of an Armco iron sample subjected to the friction against various polymers containing sulfur. **Metal block-on-polymer ring; v= 0.12 m/s; T = 23±15 °C**

coefficient is significant throughout the entire duration of the experiment, while its replacement with styrene-butadiene rubber or crosslinked polysulfide rubber is only effective for the first period of operation. Both the median friction force and the discrete friction energy levels have the most stable runs for ebonite. Approximately 30% reduction of friction is visible for SBR. However, an increase in the coefficient of friction occurs in tests with polysulphone, which end with abrasive wear of the metal surface.

The tribochemical phenomena accompanying friction do not only concern chemical reactions taking place in the surface layer of the friction couple elements of. Physical or chemical sorption of the additive particles present in the lubricating substances on the surfaces of the friction couple elements, is used in the synthesis of agents capable of reducing friction and wear (Heinicke 1984). The transfer of macromolecules from the lubricant to the surface of the metal counterface during friction was also observed, depending on the diffusion rate and magnitude of the interactions with the metal surface, which mechanism was proposed by Myshkin and Belyi (1983). This may eventually lead to the phenomenon of tribopolymerization, which can be described as the formation of oligomeric products called *friction polymers* on the counter-surface, characterized by a very effective lubricating action, also/especially under boundary friction conditions (Furey 1973). By the way, it is worth mentioning, that as a result of the transfer of polymer to the counterface sample, the microroughness of both elements of the polymer-metal friction pair is changed, which obviously affects the friction conditions.

b) friction against vulcanized polysulfide rubber (Thiokol A)

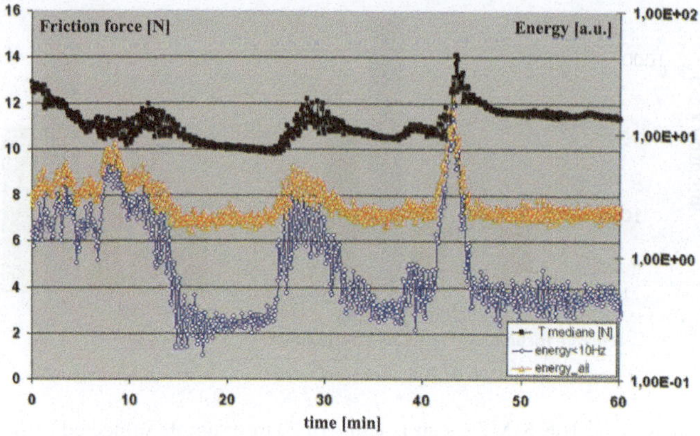

c) friction against ebonite

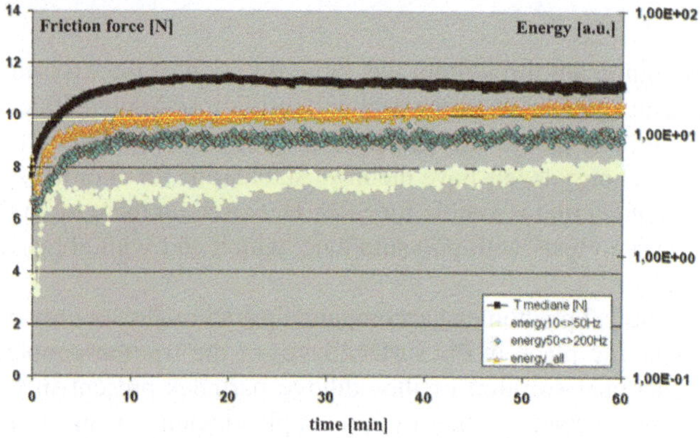

d) friction against polysulphone

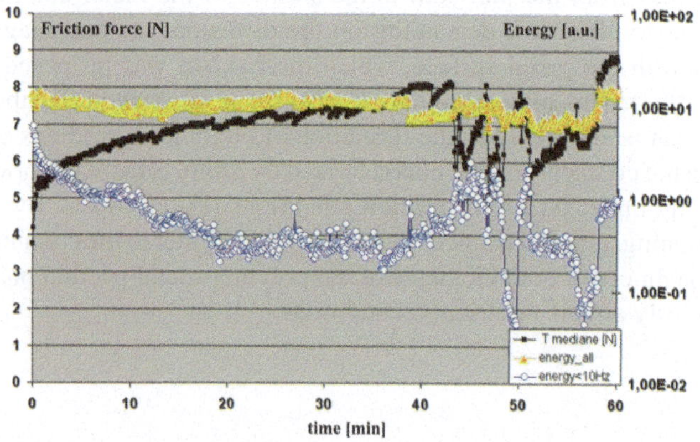

Fig. 52 Tribological characteristics of the polymer—iron Armco friction couples. *Metal block-on-polymer ring (normal load 11.4 N for SBR and polysulfide rubber, 100 N for polysulphone and ebonite); v= 0.12 m/s;* $T = 23^{\pm 15}$ *°C*

6 Summary and Conclusions

Tribological properties of elastomer materials, apart from the composition and structure of the polymer matrix, the kind of filler, filler loading and system morphology, characterizing their bulk, should also be considered from the point of view of their surface properties. The molecular weight distribution of polymers affects the morphology of their mixtures and blends. Similar to the migration of low molecular weight components (Bieliński et al. 2005a, b), also the surface segregation of components takes place in polymer systems (Bieliński et al. 1997a, b, 1999a, b; Bieliński 2004; Bieliński and Kaczmarek 2006). If the produced surface layer exhibits amorphous character and does not swell the polymer substrate, it is likely to serve as a some kind of lubricant (Bielinski et al. 2001a, b). However, if crystallization is the driving force for migration, the surface skin created is stiff and thin enough, it can withstand external loading, cooperating well with the elastic bulk of the polymer material (Bieliński 2000). The application of new high sensitivity analytical techniques, like AFM or nanoindentation enabled a deeper analysis of the surface structure of elastomer materials. It allows for a lot of phenomena accompanying their exploitation, initiated on a submicron scale or even at nanoscale, to be explained. The application of microindentation let the gradient character of vulcanization to be revealed (Bieliński et al. 2005a, b). The phenomenon can be controlled in order to optimize the characteristics and mechanical profile of the surface layer of rubber vulcanizates. The crosslink structure is another important factor, besides the filler loading and system morphology, influencing the tribological properties of the rubber. They can be modified either by changing the composition of crosslinking system or by applying appropriate treatment to the rubber products. The surface layer of polymer materials undergoes significant physical and chemical changes during processing (the so-called "active" processing), exploitation (aging) or is post-treated to obtain the defined properties (modification).

During friction of polymers against metals a mutual modification of frictional contact surfaces takes place. On the one hand, metal ions migrate to the surface layer of the polymers, on the other hand, polymer decomposition products react with the metal surface. Metal ions of variable valence facilitate the aging of polymers, contributing to the mechanism of their mechanical wear, while fragments of sulfuric crosslinks or free radicals of ruptured macromolecules can reacts with metals, producing lubrication (due to the creation of sulfides or oxides) or causing wear of a metal counterface respectively.

Unfortunately, up to now, the role of the surface layer is underestimated or even neglected in polymer technology. The approach to the problem from the point of view of material engineering and tribology reveals great potential, which has been described in this work on examples of chemical (halogenation) (Bieliński et al. 1995; 1997a, b, 2006a, b; Ślusarski et al. 1998) and physical modification (solvent—nonsolvent treatment (Chen and Ruckenstein 1992), plasma treatment (Egitto et al. 1990), ion bombardment (Turos et al. 2003; Pieczyńska et al. 2012; Bieliński et al. 2014) or laser patterning (Siciński et al. 2018). Due to the treatments it is possible to produce material with properties that precisely meet any requirements (the so-called "*tailored materials*"). It is a cheaper alternative to the usually complicated and more expensive synthesis of new materials.

The examples given in the work illustrate the issues presented above. They can be applied as a starting point for further tribological investigations, knowledge-based material design and modification from the point of view of controlling their friction.

References

Bieliński DM (1993) Modification of polymers and associated tribological properties. PhD Dissertation, University of Strathclyde

Bieliński DM (2000) Low friction coatings for rubber. Tribologia 169(1):63–75

Bieliński DM (2001a) Vulcanized rubbers of low friction characteristic. Archiv Mater Sci 22(4):241–260

Bieliński DM (2001b) The surface layer and friction of elastomers. Polimery 46(10):684–691

Bieliński DM (2004) On importance of the surface layer for exploitation of polymer materials. Kautsch Gummi Kunstst 57(1–2):13–21

Bieliński DM, Głąb P (2005) The way to improve the performance of finished rubber products. PL Patent 377,485, 6 Oct 2005

Bieliński DM, Kaczmarek Ł (2006) Surface segregation of polyethylene in low-density polyethylene/ethylene–propylene–diene rubber blends: aspects of component structure. J Appl Polym Sci 100(1):625–633

Bieliński D, Janczak KJ, Janczak T, Ślusarski L (1993a) Mechanical and tribological properties of elastomer-plastomer blends. Sci Prob Mach Oper Maintenance 28(4):383–396

Bieliński D, Janczak KJ, Ślusarski L, Loden AK (1993b) Physical modification of elastomers to improve their tribological properties. Wear 169(2):257–263

Bieliński DM, Ślusarski L, Affrossman S, Hartshorne M, Pethrick RA (1995) Influence of chemical modification on tribological properties of elastomers. J Appl Polym Sci 56(7):853–867

Bieliński DM, Ślusarski L, Affrossman S, O'Neil S, Pethrick RA (1997a) Influence of iodination on tribological properties of acrylonitrile-butadiene rubber. J Appl Polym Sci 64(10):1927–1936

Bieliński DM, Ślusarski L, Włochowicz A, Douillard A (1997b) Unusual behaviour of polyethylene in ethylene-propylene-diene matrix. Compos Interfaces 5(2):155–178

Bieliński DM, Ślusarski L, Włochowicz A, Cz Ślusarczyk (1998a) Structure and mechanical properties of nitrile rubbers modified with iodine. J Appl Polym Sci 67(3):501–512

Bieliński DM, Ślusarski L, Affrossman S, Pethrick RA (1998b) Surface modification of elastomers to improve their tribological properties. Kautsch Gummi Kunstst 51(6):429–438

Bieliński DM, Ślusarski L, Kleps T, Parasiewicz W (1999a) Surface structure and associated tribological properties of NR/TOR blends. Prog Rubber Plast Technol 15(3):123–131

Bieliński DM, Ślusarski L, Chapel J-P, Parasiewicz W (1999b) Constitution and structure of the surface layer and tribological properties of elastomers. Elastomery 14(3):3–10

Bieliński DM, Ślusarski L, Pyskło L, Dul J, Potocki K (2003) Modified starch as a complementary filler for rubber mixes. Przemysl Chem 82(8–9):1113–1116

Bieliński DM, Głąb P, Ślusarski L (2005a) Surface segregation of carboxylic acids in styrene-butadiene rubber. Compos Interfaces 12(5):445–457

Bieliński DM, Kajzer M, Ślusarski L, Kaczmarek Ł (2005b) Gradient structure of polymer materials. Polimery 50(4):298–304

Bieliński DM, Głąb P, Ślusarski L (2006) New approach to study tribological properties of polymer materials. A case of car windshield wipers. J Achiev Mater Manufact Eng 15(1–2):71–78

Bieliński DM, Grams J, Paryjczak T, Wiatrowski M (2006b) Tribological modification of metal counterface by rubber. Tribol Letters 24(2):115–118
Bieliński DM, Siciński M, Grams J, Wiatrowski M (2007) Influence of crosslink structure in rubber on the degree of modification of the surface layer of iron in elastomer—metal friction pair. Tribologia 212(2):55–64
Bieliński DM, Ostaszewska U, Jagielski J (2014) Application of ion bombardment to modify tribological properties of elastomers. Polimery 59(5):416–422
Boochathum P, Prajudtake W (2001) Vulcanization of *cis*- and *trans*-polyisoprene and their blends: cure characteristics and crosslink distribution. Eur Polym J 37(3):417–427
Borutto A, Crivellone G, Marani F (1998) Influence of surface wettability on friction and wear tests. Wear 222(1):57–65
Bowden FP, Tabor D (1980) Friction. An introduction to tribology (Polish trans). WNT, Warsaw
Briscoe BJ, Fiori L, Pelillo E (1998) Nano-indentation of polymeric surfaces. J Phys D Appl Phys 31(19):2395–2405
Buckley DH (1981) Surface effects in adhesion, friction, wear and lubrication. Elsevier, Amsterdam
Chen JH, Ruckenstein E (1992) Generation of porous polymer surface by solvent-nonsolvent treatment. J Appl Polym Sci 45(3):377–382
Dierkes W, Louis A, Noordermeer J, Blume A (2019) A novel approach of promoting adhesion of reinforcing cord to elastomers by plasma polymerization. Polymers 11(4):577–592. https://doi.org/10.3390/polym11040577
Egitto F, Vukanovic V, Taylor GN (1990) Plasma Etching of Organic Polymers. In: d'Agostino R (ed) Plasma deposition, treatment, and etching of polymers. Academic Press Inc, London
Extrand CW, Gent AN (1988) Contact angle and spectroscopic studies of chlorinated and unchlorinated natural rubber surfaces. Rubber Chem Technol 61(4):688–697
Feldman LC, Picraux ST (1978) Selected low energy nuclear reaction data. In: Mayer JW, Rimini E (eds) Ion beam handbook for material analysis. Plenum Press, New York
Ferrari AC, Robertson J (2000) Interpretation of Raman spectra of disordered and amorphous carbon. Phys Rev B 61(20):14095–14107
Furey MJ (1973) The formation of polymeric films directly on rubbing surfaces to reduce wear. Wear 26(3):369–392
Garkunov DN (1989) Tiboengineering. Mashinostroenie, Moscow
Garton A, Sturgeon PZ, Carlsson DJ, Wiles DM (1978) Plasma etching of polypropylene films and fibres. J Mater Sci 13(10):2205–2210
Gibson HW, Bailey FC (1980) Chemical modification of polymers. 13. Sulfonation of polystyrene surfaces. Macromolecules 13(1):34–41
Godet M (1990) Third-bodies in tribology. Wear 136(1):29–45
Grossiord C, Martin JM, Le Mogne Th, Palermo Th (1998) In situ MoS formation and selective transfer from MoDPT films. Surf Coat Technol 108–109:352–359
Hace D, Kovacevic V, Manojlovic D, Smit I (1990) The investigation of structural and morphological changes after the chlorination of rubber surfaces. Angew Makromol Chem 176–177:161–172
Heinicke G (1984) Tribochemistry. Academic Verlag, Berlin
Jagielski J, Grambole D, Jóźwik I, Bieliński DM, Ostaszewska U, Pieczyńska D (2011) Hydrogen loss from elastomers subjected to ion irradiation. Mater Chem Phys 126:262–271
Jasiński R (1988) Influence of stabilizer's type and quantity on rubber surface layer structure. J Polym Sci: Polym Lett Ed 26(11):473–480
Khorasani MT, Moemen Bellah S, Mirzadeh H, Sadatnia B (2006) Effect of surface charge and hydrophobicity of polyurethanes and silicone rubbers on L929 cells response. Colloids Surf B: Biointerfaces 51(2):112–119
Korzeniewska E, Walczak M, Pawlak R (2014) Laser modification of electrical properties of conductive layers made on teflon substrates. Przegląd Elektrotech 90:127–130
Lawson DF (1987) Corona discharge activation and reconstruction of elastomer surfaces. Rubber Chem Technol 60(1):102–110

Liu J, Chen F, Zheng H, Liu S, Sun J, Huang S, Song J, Jina Z, Liu X (2016) Adjusting the stability of plasma treated superhydrophobic surfaces by different modifications or microstructures. RSC Adv 83(6):79437–79447. https://doi.org/10.1039/C6RA14005J

Martin-Martinez JM, Fernandez-Garcia JC, Huerta F, Orgiles-Barcelo AC (1991a) Effect of different surface modifications on the adhesion of vulcanized styrene-butadiene rubber. Rubber Chem Technol 64(4):510–521

Martin-Martinez JM, Fernandez-Garcia JC, Orgiles-Barcelo AC (1991b) Halogenation of styrene-butadiene rubber to improve its adhesion to polyurethanes. J Adhes Sci Technol 5(12):1065–1080

Martin-Martinez JM, Fernandez-Garcia JC, Orgiles-Barcelo AC (1992) Contact angle measurements as a way to analyse synthetic rubber surfaces modified by chlorination. J Adhes Sci Technol 6(9):1091–1113

Moore DF (1975) The friction and lubrication of elastomers. Pergamon Press, Oxford

Moore DF, Geyer W (1974) A review of hysteresis theories for elastomers. Wear 30(1):1–34

Morra M, Occhiello E, Garbassi F (1989) Contact angle hysteresis in oxygen plasma treated poly(tetrafluoroethylene). Langmuir 5(3):872–876

Morrison NJ, Porter M (1983) Temperature effects on structure and properties during vulcanization and service of sulfur-crosslinked rubbers. Plast Rubber Proc Appl 3:295–304

Myshkin NK (2000) Friction transfer film formation in boundary lubrication. Wear 245(1–2):116–124

Myshkin NK, Belyi AV (1983) On the lubrication mechanism of substances with colloidal additives. ASLE Trans 26(3):405–410

Oldfield D, Symes TEF (1983) Surface modification of elastomers for bonding. J Adhesion 16(2):77–95

Pastor-Blas M, Fernandez-Gomez TP, Martin-Martinez JM (2000) Chlorination of vulcanized styrene-butadiene rubber using solutions of trichloroisocyanuric acid in different solvents. J Adhes Sci Technol 14(4):561–581

Pawlak R, Korzeniewska E, Koneczny C, Hałgas B (2017) Properties of thin metal layers deposited on textile composites by using the PVD method for textronic applications. AUTEX Res J 17(3):229–237

Perera MCS (1987) Surface modification of natural rubber laces. J Appl Polym Sci 34(7):2591–2600

Persson BNJ (1998) On the theory of rubber friction. Surf Sci 401(3):445–454

Pieczyńska D, Ostaszewska U, Bieliński DM, Jagielski J (2012) Modification of polymers with the application of ion beam bombardment. Part II. Modification of functional properties of rubber. Polimery 57(2):124–134

Płaza S (1997) Physico-chemistry of tribological processes. Lodz University Press, Łódź

Polak A (1998) Material transfer in polymer-steel sliding bearing. Mechanics Serie vol 233, Cracow Univ Technol Press, Cracow

Polish Ministry of Health and Social Care (1970) Polish Farmakopea vol 2, 4th edn. PWL, Warsaw, p 636

Puliyalil H, Cvelbar U (2016) Selective plasma etching of polymeric substrates for advanced applications. Nanomaterials 6(6):108. https://doi.org/10.3390/nano6060108

Roberts AD (1979) Looking at rubber adhesion. Rubber Chem Technol 52(1):23–42

Roberts AD, Brackley CA (1990) Surface treatments to reduce friction: rubber glove applications. Rubber ChemTechnol 63(5):722–733

Rymuza Z (1986) Tribology of sliding polymers. WNT, Warsaw

Rymuza Z (2004) Part Two: Reaction Processes and Mechanisms. 5. Plastics. In: Totten GE, Liang H (eds) Surface modification and mechanisms. Friction, stress and reaction engineering. Marcel Dekker, New York, p 99

Schallamach A (1957) Friction and abrasion of rubber. Wear 1(5):384–417

Schallamach A (1971) How does rubber slide? Wear 17(4):301–312

Schlögl S, Kramer R, Lenko D, Schröttner H, Schaller R, Holzner A, Kern W (2011) Fluorination of elastomer materials. Eur. Polym. J. 47(12):2321–2330

Siciński M (2008) Modification of the surface layer of the elements of polymer—metal friction couple. PhD Dissertation, Lodz University of Technology

Siciński M, Korzeniewska E, Tomczyk M, Pawlak M, Bieliński D, Gozdek T, Kałuzińska K, Walczak M (2018) Laser-Textured Rubbers with Carbon Nanotube Fillers. Polymers 10(10):1091. https://doi.org/10.3390/polym10101091

Ślusarski L, Bieliński D, Janczak T (1996) Application of solid lubricants to reduce the friction coefficient of the rubber. PL Patent 168,619, 29 March1996

Song-Hua G, Li-Hua G, Ke-Sheng Z (2011) Super-hydrophobicity and oleophobicity of silicone rubber modified by CF4 radio frequency plasma. Appl Surf Sci 257(11):4945–4950

Starczewski L (2004) Hydrogen wear of frictions' elements of machines. Problemy Eksploatacji 4:47–57

Sun DL, Hong RY, Wang F, Liu JY, Rajesh Kumar M (2016) Synthesis and modification of carbon nanomaterials via AC arc and dielectric barrier discharge plasma. Chem Eng J 283:9–20

Turos A, Jagielski J, Piątkowska A, Bieliński DM, Ślusarski L, Madi NK (2003) Ion beam modification of surface properties of polyethylene. Vacuum 70(2–3):201–206

Vega-Cantú Y, Hauge R, Norman L, Billups WE (2003) Enhancement of the chemical resistance of nitrile rubber by direct fluorination. J Appl Polym Sci 89(4):971–979

Vijaybaskar V, Bhowmick AK (2005) Dynamic mechanical analysis of electron beam irradiated sulphur vulcanized nitrile rubber network—some unique features. J Mater Sci 40:2823–2831

Wang H, Xu B, Liu J, Zhuang D (2005a) Investigation on friction and wear behaviors of FeS films on L6 steel surface. Appl Surf Sci 252(4):1084–1091

Wang H, Xu B, Liu J, Zhuang D (2005b) Characterization and tribological properties of plasma sprayed FeS solid lubrication coatings. Mater Characterization 55:43–49

Wolthuizen DJ, Martinez-Martinez D, Pei YT, De Hosson JThM (2012) Influence of plasma treatments on the frictional performance of rubbers. Tribol Lett 47:303–311. https://doi.org/10.1007/s11249-012-9985-9

Yasuda HK (1985) Plasma Polymerization, 1st edn. Academic Press Inc, London

Zaborski M, Ruciński J, Bieliński D (1991) Surface energy of vulcanizates differing in structure and density of crosslinks. Polimery 36(3):109–111

Tribological Study on Titanium Based Composite Materials in Biomedical Applications

S. Shankar, R. Nithyaprakash, and G. Abbas

Abstract Titanium (Ti) and titanium based alloy (Ti6Al4V) are known for its high strength, high stiffness, resistance to corrosion and biocompatibility. The mechanical properties as well as bio-compatibility of titanium lead to wide variety of applications in biomedical, environmental, aerospace, automotive and marine applications. The titanium based composites are widely preferred at present to enhance the wear, friction and corrosion behavior based on applications where it is used. Moreover, titanium based bio-composites are widely preferred in biomedical applications because of its improved friction and wear resistance properties. The present chapter emphasizes more on tribological study of titanium based composites in general and also titanium based bio-composites in hip joint replacement on biomedical applications. At the same time, the focus is also given to classify various coating techniques used to produce the titanium composites and parameters used to analyze the tribological behavior of titanium composites. This chapter also deals with approaches used to test with and without bio-lubricant, tribological and corrosion properties of titanium composites.

Keywords Titanium · Composite · Tribology · Biomedical · Techniques

1 Overview About Titanium Composites

Titanium composites are widely used in enhancing the tribological properties which are used in applications like biomedical, aerospace and sometimes nanocomposites are coated on titanium to treat waste water. Particular attention has been given to the

S. Shankar (✉) · R. Nithyaprakash
Department of Mechatronics Engineering, Kongu Engineering College, Erode, Tamil Nadu, India
e-mail: shankariitm@gmail.com

R. Nithyaprakash
e-mail: mtsprakash@gmail.com

G. Abbas
Department of Mechanical Engineering,, Kongu Engineering College, Erode, Tamil Nadu, India
e-mail: gabbas11897@gmail.com

M. T. Hameed Sultan et al. (eds.), *Tribological Applications of Composite Materials*, Composites Science and Technology, https://doi.org/10.1007/978-981-15-9635-3_8

tribological behavior of biomaterials. This chapter deals with various techniques used to fabricate titanium composites, tests carried out to study tribological properties and effective use of titanium composites in biomedical applications.

2 Introduction to Titanium Based Composites

Titanium (Ti) and Titanium-based materials are generally used in diverse areas due to its distinctive combination of excellent characteristics properties like strength and wear resistance. The stiffness of conventional titanium alloys could be further improved by using titanium-based matrix composites (TMCs) (Attar et al. 2018). Due to excellent bio-compatibility, stiffness and high corrosion resistance titanium based biocompatible materials are widely used in biomedical, chemical, aerospace and automotive applications (Geetha et al. 2009; Okulov et al. 2014).

The common reinforcements used in TMCs are continuously reinforced TMCs and discontinuously reinforced TMCs. For reinforcement of TMCs different particulates have been selected which include TiB_2, B_4C, TiN, ZrC, SiC, TiB, TiC, and Al_2O_3 (Geng et al. 2008). The formation of severe surface fractures were prevented by 20 vol% (TiB + TiC) reinforcement in titanium based composites. The wear rate of TMCs extremely decreased due to the reinforcements which reduced the surface damage against the abrasive wear. The TiB + TiC reinforced composites was fabricated by in-situ synthesis technique called investment casting process (Kim et al. 2013). The presence of hBN (hexagonal boron nitride) and titanium oxide (TiO_2) reinforcement coating in the Ti–6Al–4V alloy improves the antifriction property and this is achieved by techniques of micro arc oxidation (Lu et al. 2016). The reinforcing of Titanium Nitride (TiN) with titanium alloy provides higher electrochemical impedance, higher corrosion potential and lower corrosion density and this is produced by the techniques called laser irradiation (Zhao et al. 2018). There are several techniques which are used to produce titanium composites which include in situ method, micro arc oxidation, spark plasma sintering, plasma electrolytic oxidation, Plasma transferred arc (PTA) cladding process and laser processing techniques like laser cladding process, laser irradiation, laser melting deposition and laser alloying process. Among these methods, laser processing techniques is the most preferable (Yu et al. 2017). The titanium based bio-composites include titanium based hydroxyapatite (Ti6Al4V-HA), TiB-TiN reinforced Ti6Al4V, calcium phosphate based Ti composites which are developed to enhance the load bearing capacity of implants (Bandyopadhyay et al. 2016; Rahmati and Khodabakhshi 2018). Figure 8.1 shows the properties of biomaterials and Fig. 8.2 shows properties of Ti composites.

Some of the literatures reported the use of zirconium and ZrO_2 composites in Ti6Al4V with the help of coating techniques over the surface of Ti6Al4V to enhance surface properties (Li et al. 2012b). The tribological study of titanium and titanium alloy with composites include WC particle, TiO_2/hBN ceramic composites, TiCx reinforced metal matrix composite, TiB,TiC particles, NiTi, graphite, $TiNiZrO_2$

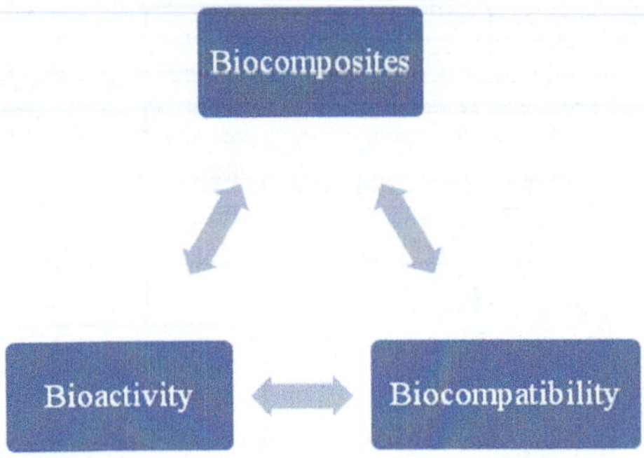

Fig. 8.1 Properties of biomaterials

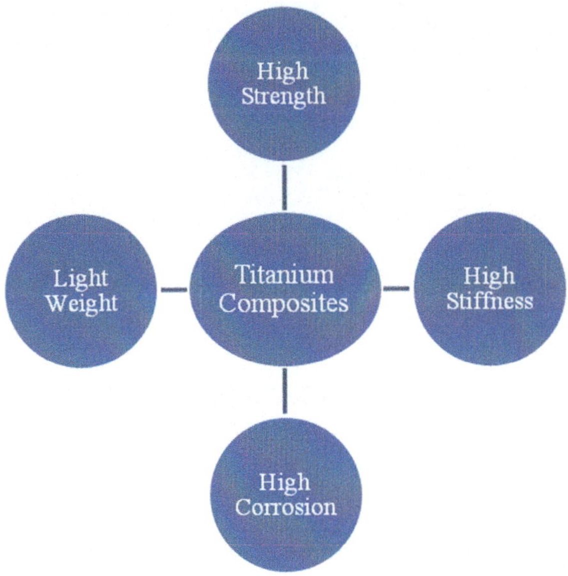

Fig. 8.2 Properties of titanium composites

composite, TiAlN (titanium aluminium nitride) nano composite, network structured CNT, nanocomposite $NiSi_2/Ti_5Si_3$, Ti_5Si_3 nanocomposite, gelatin functionalized graphene oxide composites, Ni60A/B_4C powder (Lu et al. 2016; Çelik 2013; Guo et al. 2007; Jiang et al. 2015; Li et al. 2012a; Li et al. 2011b; Li et al. 2010; Lin et al. 2015; Liu et al. 2014a; Mu et al. 2012; Obadele et al. 2015; Prem Ananth and Ramesh 2014; Umeda et al. 2015; Xu et al. 2013; Yan et al. 2015; Zhang et al. 2017). Few studies used carbon steel and stainless steel coated with Ti, TiN and TiC composites to investigate the tribological behaviour. The tribological studies considering biomedical applications were carried out with the bio-lubricants to identify better

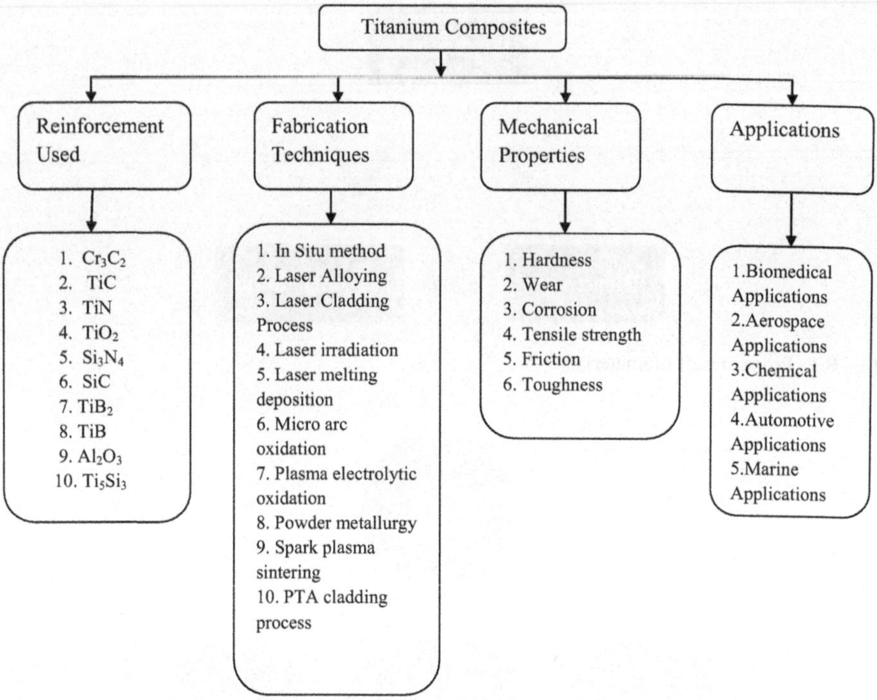

Fig. 8.3 Overview of titanium composites

wear resistant biomaterials (Jin et al. 2015; Peng et al. 2014; Wang et al. 2014). Figure 8.3 shows overview about Ti composites.

3 Types of Reinforcement Used

Reinforcement materials are added to the matrix materials to enhance the physical properties of the final composites. The composites have anisotropic properties due to the use of reinforcement in it. Generally, Composites are classified into three main categories based on the reinforcement characteristics,

- Particle Reinforced
- Fiber Reinforced
- Structural.

In particle reinforced composites, particles are suspended in a matrix. Particle reinforced composites are further classified into two types, i.e. Large particle and dispersion strengthened. In fiber reinforced composites, fibers (either natural fiber or man-made fibers) are reinforced in a matrix of composites. Fiber reinforced composites are classified into continuously reinforced and discontinuously reinforced composites. In continuously reinforced composites, fibers are aligned in a definite manner but in discontinuously reinforced, fibers are dispersed. Continuously reinforced composites show more strength when compared with discontinuously reinforced composites. In structural reinforcements, the classifications include laminated and sandwich panels. Usually, Titanium metal matrix composites (TMC's) can be classified into two main categories depending on the form of reinforcements, continuously reinforced TMC's and discontinuously reinforced TMC's. As discussed above, in continuously reinforced TMC's, the reinforcement materials diffuses into a matrix in a definite manner but in discontinuously reinforced TMC's, the reinforcing materials is spread into matrix of composites in a dispersed manner.

Commonly used reinforcements for the fabrication of TMC's are Cr_3C_2, TiC, TiN, Si_3N_4, SiC, TiB_2, TiB, Al_2O_3 and Ti_5Si_3. Some other composites include zirconium, ZrO_2, TiO_2/hBN Composite, calcium phosphate-based composite and Ti6A14V-HA biocomposites can also be used in biomedical field due to its improved wear resistance and high bio-compatibility. (Li et al. 2012a) used TiB and TiC as reinforcements on Ti composites to enhance mechanical and microstructral properties. In another study (Mu et al. 2012), TiO_2/graphite is used as reinforcements on Ti6A14V alloy which exhibits better self lubricating property. Xu et al. (2013) used Ti_5Si_3 as reinforcement on Ti6A14V substrate possess higher fracture toughness and hardness. Liu et al. (2014a) showed that TiC_x reinforcement on TC_4 titanium alloy has higher hardness. Jiang et al. (2015) synthesized TiB_2-TiN-(h-BN) reinforcement by in-situ approach for enhancing microhardness. For improving the mechanical properties and toughening mechanism (Lin et al. 2015) used TiB_2/NiTi reinforcement on TMC.

4 Techniques Used in Producing Titanium Composites

4.1 *In-Situ Approach*

In-situ technique refers to precipitation of reinforcing material as a result of melting of reinforcing material on substrate. The TMC of B_4C and pure Ti was fabricated using investment casting process also called lost wax. Three composite samples of TiB + TiC were prepared and cut into cubic shapes to perform fretting wear testing. The above process was carried out with vacuum induction melting (Kim et al. 2013). Another study (Yu et al. 2017) used induction cladding (IC) approach to form composite coating of TiC/Ti over Ti6A14V specimen. The advantages of this approach include reduced energy consumption, cost and highly reliable. The TiN coating on Ti-35Nb-7Zr-5Ta (TNZT) alloy having high bio-compatibility was developed through laser irradiation. The advantage of this process includes surface texturing and gas alloying using laser. The specimen used was 15 mm × 15 mm × 2 mm in size while laser source used was pulsed YAG: Nd laser (Zhao et al. 2018).

4.2 *Laser Alloying*

The laser alloying technique with help of CO_2 laser, composite coating of nitrogen gas and silicon powder on Ti6Al4V substrate, with sample size of 10 mm × 10 mm × 40 mm was developed (Tian et al. 2005a).

4.3 *Laser Cladding*

Energy from laser source is used to melt powder material to be coated over substrate to form metallurgical bonding on substrate. Ni-based powder is commonly used for titanium cladding. TiC_p composite coating was developed with help of Nd:YAG laser source (Candel et al. 2010). Another study (Dong and Wang 2009) reported intermetallic composite coating of TiC/Ti–Ni–Si pre-mixed powders fabricated on TA15 titanium alloy substrate with sample size of 10 mm × 30 mm × 50 mm. The laser source used was CO_2. Using TA15 substrate (Feng et al. 2012) with sample size of 50 mm × 20 mm × 10 mm another study developed the intermetallic composite coating of TiB-TiC/TiNi-Ti2Ni using Ti + Ni + B_4C powder. Ti-Al coating on pure titanium disc of diameter 31 and 10 mm in thickness was performed with help of CO_2 laser beam controlled by 4 axis computer numerical controlled (CNC) (Guo et al. 2007). With the help of TiB and TiC particles TMC coating over Ti6Al4V substrate composition of 6.5 wt% Al, 4.26 wt% V was performed. The substrate was made in the form of cylinder of 50 mm in diameter and 10 mm in length. Ti-B4C-Al or Ti-B4C-C-Al powders were used as the precursor materials (Li et al. 2010). Another study (Li et al. 2011a, 2012a) used TiB whiskers and TiC particles to develop composite coating on Ti6Al4V of 50 mm in diameter and 10 mm thickness using CO_2 laser. The coating of γ-NiCrAlTi/TiC + TiWC2/CrS + Ti_2CS on Ti6Al4V alloy substrate of 50 mm × 40 mm × 8 mm using CO_2 laser was performed and the volumetric wear rate was noted (Attar et al. 2014). Another literature (Lu et al. 2016) developed Ni60-hBN composite coating by varying hBN content developed on Ti6Al4V substrate with a dimension of 50 mm × 40 mm × 8 to evaluate its tribological properties using DILAS SD3000L–3 kW diode laser. Another study (Tian et al. 2005b) developed composite coating using graphite and silicon mixed powders on Ti6Al4V substrate with sample dimensions 10 mm × 10 mm × 40 mm using CO_2 laser. Table 8.1 establishes several laser techniques parameters.

4.4 *Micro-Arc Oxidation*

The composite coating of hBN and titanium oxide was formed with the help of micro-arc oxidation (MAO) on Ti6Al4V substrate made of round disc of (Φ 30 mm × 3 mm) and its tribological behaviour was investigated (Lu et al. 2016).

Table 8.1 Laser techniques parameters

Substrate material	Specimen Size	Laser used	Scanning Speed	Power output	References
Ti6Al4V	10 mm × 10 mm × 40 mm	1.5 kW CO_2 laser	3.5 mm/s	1200 W	Tian et al. (2005b)
Ti6Al4V	Φ50 × 5 mm	1 kW CW Nd:YAG laser	8–16 mm/s	400-800 W	Candel et al. (2010)
TA15 titanium alloy	10 mm × 30 mm × 50 mm	6 kW continuous wave CO_2 laser	600 mm/min	4 kW	Dong and Wang (2009)
TA15	50 mm × 20 mm × 10 mm	4 kW YLSí4000 fiber laser	500 mm/min	3.5 kW	Feng et al. (2012)
Pure titanium	31 mm in diameter, 10 mm in thickness	10-kW transverse-flow continuous-wave CO_2 laser	4 mm/s	5 kW	Guo et al. (2007)
Ti6Al4V	Ø 50 mm × 10 mm	5-kW CO_2 gas laser	5 mm/s	3.5 kW	Li et al. (2011a, 2012a)
Ti6Al4V	50 mm × 40 mm × 8 mm	10 kW CO_2 laser	6 mm/s	1.5 kW	Attar et al. (2014)
Ti6Al4V	50 mm × 40 mm × 8 mm	DILAS SD3000L–3 kW diode laser	4 mm/s	1.5 kW	Lu et al. (2016)

4.5 *Powder Metallurgy*

The powder metallurgy technique based on hot isostatic pressing (HIP), spark plasma sintering (SPS) and vacuum sintering was processed for titanium–titanium boride composites. Two different volume percentage namely titanium with 20% titanium boride and other with 40% titanium boride was investigated (Selva Kumar et al. 2012).

4.6 *Plasma Transferred Arc Cladding*

Ti_5Si_3 composite coating using Plasma transferred Arc (PTA) cladding was carried out over Ti6Al4V substrate using Ti53–Si32–Ni15, Ti43–Si26–Ni31 and Ti30–Si18–Ni52 (at.%) as precursor material (Liu et al. 2014c).

4.7 Coatings and Its Behaviors

4.7.1 Coating Techniques and Materials

In order to enhance the mechanical properties of substrate, surface of the substrate is coated with composite materials. In biomedical applications, coating is done for improving the corrosion and wear resistance, biocompatibility of metallic implants. There are several coating techniques for enhancing the physical properties of Titanium composites. Figure 8.4 shows the different types of coating techniques. Laser cladding of NiCrBSi/WC coating on titanium alloy showed reduced wear volume, coefficient of friction and also it decreased crack susceptibility (Attar et al. 2014). TC4 titanium alloy is used as a substrate which was coated by TiCx-$NiTi_2$/Ti cermet composite coatings which increased microhardness to clad layer (Liu et al. 2014a). In another study, carbon steel was coated with TiC composite coatings by laser cladding method which showed improved hardness, reduced wear rate and friction (Peng et al. 2014). TiB-TiC coatings was coated on titanium based composites which increased hardness and decreased the wear rate (Kim et al. 2013). TiAlN was coated on titanium alloy using the coating technique called cathodic arc PVD (plasma vapour deposition) which increased coating adherence and bonding strength (Prem Ananth and

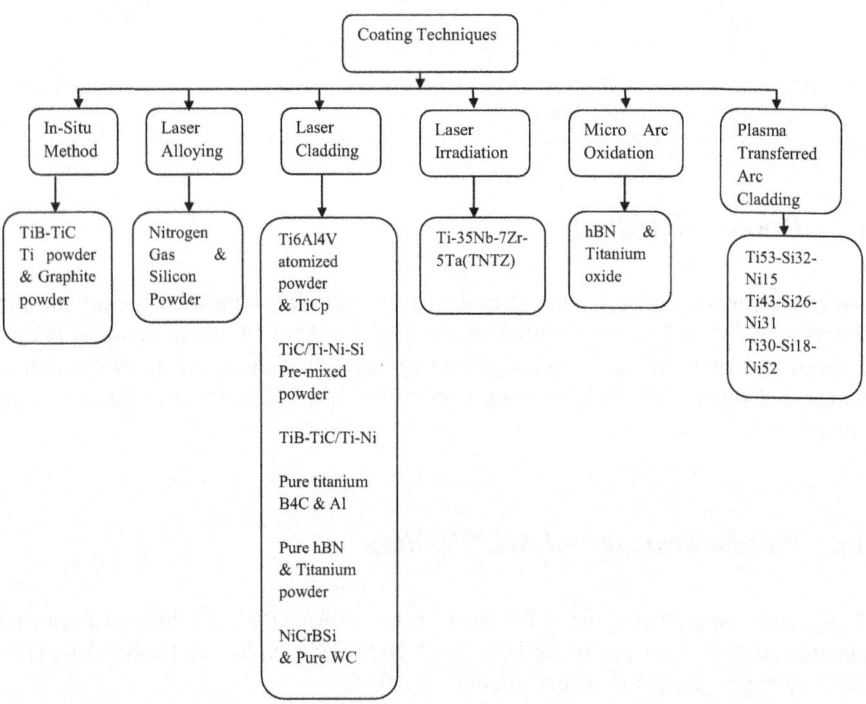

Fig. 8.4 Different types of coating techniques

Ramesh 2014). Ti-TiN was coated on stainless steel using plasma surface alloying which exhibited excellent wear resistance and chemical stability (Wang et al. 2014). TiB_2-TiN-(h-BN) coating on Q235 steel substrate improved microstructural uniformity and compactness and also enhanced interfacial bonds of coating using plasma cladding process (Jiang et al. 2015). The higher corrosion potential as well as electrochemical impedance and lower corrosion current density was obtained by laser irradiation method in which TiN was coated on Ti substrate (Zhao et al. 2018). Table 8.2 shows various coating and its coating techniques.

It is well known that, bio implants must have higher corrosion properties to withstand the human environment and it should not be reacting with body fluids. If the metallic implants reacts with inner human environment, it causes severe bone cell damages, corrode the implants and many side effects would be induced. For this cause, the study of corrosion performance of the composites is essential. One of the studies (Saba et al. 2018) showed that the corrosion resistance of Ti nano-composites was enhanced with increase in nano-reinforcement up to 0.25 wt%. (Xu et al. 2012) revealed that the nanocomposite $NiSi_2/Ti_5Si_3$ bilayer film had higher corrosion resistance than the Ti–6Al–4V alloy. Lee (2012) showed that addition of CNTs into Ni coatings on Ti–6Al–4V improved the electrochemical resistance in hank solution. Zhao et al. (2018) reported that the surface with TiN coating showed improved corrosion resistance of composites. Hu et al. (2016), revealed that β-Ta_2O_5 nanoceramic coating on Ti–6Al–4V alloy showed excellent corrosion resistance.

5 Tribological Study of Titanium Composites

The tribological behaviour of pure titanium as well as titanium alloys against composite materials were investigated in many previous literatures. Tkachenko et al. (2013) which studied the tribological performance of Ti-Si based composites sliding against silicon nitride balls which showed good wear resistance and less coefficient of friction. In another study (Lee et al. 2008), alumina ceramic ball were used as counter body material which sliding against TiB_2-TiB composites with light minerals as lubricants with normal load of 49 N, exhibited excellent hardness and wear resistance and also better corrosion resistance. In another study, Si_3N_4 ceramic ball was sliding against TiAl matrix self lubricating composites with Graphite as a lubricant under conditions of 10 N load showed better wear resistance and lubrication (Shi et al. 2013). Yang et al. (2017) studied the tribological behaviour of TiAl composites against Si_3N_4 balls with multilayer graphene, silver (Ag) as lubricant under 12 N load, 0.8 m/s sliding speed which showed excellent wear properties. In another study, Li et al. (2018) investigated the tribological performance of TiAl composites using pin-on-disk tribometer with serpentine (solid lubricant) which exhibited good hardness and wear resistance properties. Suthar and Patel (2018) performed the tribological test under different loads with sliding time duration of 30 min in which TiB-TiC composites was used against 52,100 bearing steel. The result showed better hardness and reduced friction. The friction and wear behavior of TMCs using pin-on-disk

Table 8.2 Composite coatings and techniques

S. No.	Substrate	Coatings	Coating technique	Advantages	References
1	Titanium alloy	NiCrBSi/WC coating	Laser cladding	Reduced wear volume, coefficient of friction, decreased crack susceptibility	Attar et al. (2014)
2	TC4 titanium alloy	TiCx-NiTi2/Ti cermet composite coatings	Laser cladding	Increased micro hardness to clad layer	Liu et al. (2014a)
3	Carbon steel	TiC composite coating	Laser cladding	Increased hardness, reduced wear rate and friction	Peng et al. (2014)
4	Titanium alloy	TiAlN coating	Cathodic arc PVD	Improved coating adherence, increased bonding strength of coating	Prem Ananth and Ramesh (2014)
5	Stainless steel	Ti-TiN coating	Plasma surface alloying	Excellent wear resistance and chemical stability	Wang et al. (2014)
6	Q235 steel	Fe-based alloy coatings containing TiB2–TiN–(h-BN)	Plasma cladding process	Improved microstrutural uniformity and compactness, enhanced interfacial bonds of coating	Jiang et al. (2015)

(continued)

Table 8.2 (continued)

S. No.	Substrate	Coatings	Coating technique	Advantages	References
7	316L stainless steel	Ti-Cu coating	Closed field unbalanced magnetron sputtering (CFUBMS)	Improved antibacterial activity, corrosion and tribological properties	Jin et al. (2015)
8	Ti–6Al–4V alloy	TiB2/NiTi reinforced TMC coating	Laser cladding	Improved fracture toughness	Lin et al. (2015)
9	Pure Ti plate	Carbon nano tubes(CNTs) coating	Surface texturing	Reduced friction coefficient, improved abrasive wear	Umeda et al. (2015)
10	Titanium based composites	TiB-TiC	In situ approach	hardness increased, wear rate dropped	Kim et al. (2013)
11	Ti–6Al–4V alloy	TiC/Ti composite coating	In situ approach(high frequency indution cladding method)	Hardness was increased more than two times that of Ti-6Al-4 V alloy	Yu et al. (2017)
12	Ti–6Al–4V alloy	Si_3N_4, Ti_5Si_3, Ti_2N coatings	Lase alloying	Better wear resistance and oxidation resistance, reduction of crack formation	Tian et al. (2005a)

(continued)

Table 8.2 (continued)

S. No.	Substrate	Coatings	Coating technique	Advantages	References
13	Ti–6Al–4V hot rolled samples	TiC particle reinforced Ti6Al4V MMC coatings	Laser cladding	Increased micro hardness, decreased coefficient of friction and wear rate	Candel et al. (2010)
14	Ti–6Al–4V	Alpha-NiCrAlTi/TiC + TiWC2/CrS + Ti2CS coatings	Laser cladding	Enhanced wear resistance and friction reducing capability	Liu et al. (2014b)
15	Ti–6Al–4V alloy	Ni60-hBN high temperature self-lubricating anti-wear composite coatings	Laser Cladding	Increased micro hardness and improved wear resistance	Lu et al. (2016)
16	Biomedical titanium alloy	TiN coating	Laser irradiation	Higher electrochemical impedance, higher corrosion potential and lower corrosion current density	Zhao et al. (2018)

(POD) wear tester under various conditions like sliding speed of 125 mm/s, load of 0.35 N and sliding time of 30 min was investigated by Kim et al. (2011). The findings showed enhanced mechanical properties for TMCs under above parameters. Table 8.3 shows tribological overview of various Titanium composites.

6 Titanium Based Bio-Composites in Hip Joint Replacement

Biocompatibility refers to ability of an artificial material to medically interact with human body without causing any side effects. Titanium based biocomposites are mainly used in artificial joint replacements owing to its excellent biocompatibility, high strength and corrosion resistance properties. The reinforcement mainly used is zirconia, zirconium, hydroxyapatite and calcium phosphate. Sometimes coating techniques are also used to deposit composite layer over titanium alloy to enhance the tribological property to reduce wear debris being released into the human body. This helps in reducing revision surgery and also enhances the life of implants.

6.1 Hydroxyapatite Based Titanium Composites

The applications of biomaterials are extended day by day. The introduction of hydroxyapatite (HA) is a scientific milestone in biomedical field because hydroxyapatite and natural bone has identical chemical structure and it improves the on-growth of natural bone (Berndt et al. 2014). HA is a hydrated calcium phosphate material. Hydroxyapatite (HA) ($Ca_{10}(PO_4)_6(OH)_2$) is the foremost promising inorganic element with excellent biocompatibility. It is mainly utilized for dental and bone implants (Fathi and Zahrani 2009; Liang et al. 2004; Que et al. 2008). Except bone tissue related applications, HA is also used for tissue repair and replacement. HA exhibits better biocompatibility due to its chemical and crystallographic structure being similar to that of natural human bone. HA is porous in nature, bioactive and due to that, it is partially resorbed and replaced by naturally bone after some time (Williams 1987; Niespodziana et al. 2010). Rao and Kannan (2002), Silva et al. (2001), Mishra et al. (2017) suggested that, HA when reinforced with elements like Al_2O_3, ZrO_2, TiO_2, showed reduction in brittleness without affecting its biocompatibility. Ti is bio-active and bio-inert because of mechanical and chemical bonding with natural bone (Khorasani et al. 2015). Most of the implants are prepared using titanium or its alloy due to the outstanding mechanical properties, chemical stability, endurance and low density in body fluid. Metal implants have good hardness and mechanical strength but their biocompatibility is less. To enhance the properties of these metal implants, HA is widely used because of its better biocompatibility and chemical stability (Szcześ et al. 2017). Rathore (2015) showed 10% TiO_2 with HA

Table 8.3 Tribological overview of titanium composites

Equipment	Composites	Counter body material	Lubricant used	Sliding distance	Sliding speed	Load	Temperature	References
Ball-on-disc system	Ti–Si based composites	Silicon nitride (6 mm balls)	25% fetal serum solution + sodium azide and ethylenediaminetetraacetic acid	1024 m	0.04 m/s	Normal load of 0.5 N	Room temperature	Tkachenko et al. (2013)
Ball-on-disk tester	Ti–Si based composites	Zirconia balls			0.04 m/s	3 N		Tkachenko et al. (2014)
Ball-on-disk type friction tester	(TiB–TiC) particulate-reinforced Titanium Matrix Composites	52,100 bearing steel				1, 3.5, and 10 N		Suthar and Patel (2018)
Ball-on-disk type tribometer	TA15 composite	GCr15			224 rpm	0.98, 2.94, and 4.9 N	Room temperature and 600 °C	Guo et al. (2016)
Ball-on-disk type friction tester	Ti/QC composites	100Cr6 steel ball		166.3 m	2 cm/s	1 N		Khun et al. (2015)
Ball-on-disk micro-tribometer	Titanium composites	100Cr6 steel ball			3 cm/s	1 N		Khun et al. (2016)
Ball-on-disk friction tester	Titanium matrix (TiB + TiC) composites	52,100 bearing steel			120 rpm (125 mm/s)	0.35 N		Kim et al. (2011)
Ball-on-disk	TiB_2-TiB composites	Alumina ceramic ball	Light mineral oil	1000 m	0.12 m/s	5.0 kg (49 N)	Room temperature	Lee et al. (2008)
HT-1000 ball-ondisk high temperature tribometer	TiAl matrix self-lubricating composites	Si_3N_4 ceramic ball	Graphite, MoS_2		0.23 m/s	10 N load	200, 400, 600 and 800 °C	Shi et al. (2013)

(continued)

Table 8.3 (continued)

Equipment	Composites	Counter body material	Lubricant used	Sliding distance	Sliding speed	Load	Temperature	References
HT-1000 ball-on-disk high temperature tribometer	TiAl matrix self-lubricating composites	GCr15(6 mm dia) balls	TiB_2, MoS_2, multilayer graphene (MLG), graphite and boron nitride		0.2 to 1.1 m/s	5 N	room temperature	Xu et al. (2015a)
Ball-on-disk tribometer of HT-1000, unidirectional sliding	TiAl composites materials	Si_3N_4 balls	Multilayer graphene, graphite, AgVo3, silver (Ag)		0.8 m/s	12 N		Wang et al. (2017)
Ball on disc type HF-1000 friction and wear tester	TiBw/Ti6Al4V composites	Si_3N_4 ceramic ball			400 and 600 rpm	5, 10 and 15 N		An et al. (2018)
Ball on plate type	Ti–6Al–4V/TiC composites	Si_3N_4 ceramic balls	Graphite	54 m		0.5 N	20–25 °C	Bai et al. (2019)
Pin-on-disk tribometer	TiB reinforced Ti composite	tungsten carbide			1.5 m/s	10 N		Bao et al. (2018)
Pin-on-disk tribometer	TiCp reinforced titanium composite	5 mm WC-6Co ball		500 m		10 N		Candel et al. (2010)
Block-on-wheel dry sliding wear tester	TiC reinforced Ti–Ni–Si intermetallic composite	Hardened GCr15 bearing steel		3024 m	0.84 m/s	98 N, 147 N and 196 N	Room temperature	Dong and Wang (2009)
Pin-on-disc	(TiB + TiC)/Ti composite	45 steel (HRC 32)			0.176 m/s	17.5 N		Li et al. (2011a)
Pin-on-disk tribometer	TiAl composites		Serpentine (Solid Lubricant)		0.2 m/s	12 N	25–800 °C	Li et al. (2018)

(continued)

Table 8.3 (continued)

Equipment	Composites	Counter body material	Lubricant used	Sliding distance	Sliding speed	Load	Temperature	References
MXP-2000 ring-on-ring dry sliding wear tester	Ti5Si3 reinforced intermetallic composite	hardened GCr15 bearing steel		576 m	0.13 m/s	49 N, 98 N and 196 N,	Room temperature	Liu et al. (2014c)
Pin-on-disk machine	titanium matrix composite		No lubrication	212 m	0.54 m/s	40 to 100 N	Room temperature	Qin et al. (2011)
Ball-on-plate	Ti-TiB-TiN_x in-situ composite	Alumina ball				10 N		Silva et al. (2016)
Ball-on-disk high-temperature tribometer	Ti2AlN/TiAl Composite	Si_3N_4 balls (6 mm of diameter)			600 m/min	4 N	25,600 °C	Wang et al. (2018)
Block-on-wheel dry sliding wear tester	Ti_2Ni_3Si reinforced intermetallic composite			2939 m	0.816 m/s	49, 98 and 147 N	Room temperature	Wang and Wang (2004)
Ball-on-disk high-temperature tribometer	Ti_2AlN/TiAl Composites	Si_3N_4 ceramic ball			600 rpm	10 N	25, 200, 400, 600 and 800 °C	Li et al. (2018)
Pin-on-disk high-temperature wear tester	(TiB + TiC)/Ti–6Al–4V matrix composite			1200 m	0.25, 0.5 and 1 m/s	10–200 N	25 °C (room temperature, RT), 200 °C and 400 °C	Zi-Run et al. (2017)
Pin-on-disk machine	Ti_3SiC_2-based composites	Ti_3SiC_2/SiC pin	Distilled water + anhydrous C_2H_5OH	288 m	40 mm/s	0.49 N		Hibi et al. (2006)

(continued)

Table 8.3 (continued)

Equipment	Composites	Counter body material	Lubricant used	Sliding distance	Sliding speed	Load	Temperature	References
Ball-on-disk high-temperature tribometer	Ti_3SiC_2/TiAl Composite	GCr15 steel ball			0.2–0.8 m/s	2 to 8 N	20–25 °C	Xu et al. (2015b)
HT-500 ball-on-disc tribometer	Nanocomposite $NiSi_2/Ti_5Si_3$	Ceramic ball made of ZrO_2		792 m	22 cm/s	2.8, 3.3 and 3.8 N,	Ambient temperature and 500 °C	Xu et al. (2012)
Ball-on-disk tribo-meter	Ti_3SiC_2/Cu/Al/SiC composite	Si_3N_4 ball			0.188 m/s	5 N	RT, 200, 400, 600, and 800 °C	Dang et al. (2016)
Reciprocating ball on disk type	(Ti, Nb)Cp/Fe-based laser composite				480 rpm (rotation of wheel)	100 N		Wang et al. (2018)
Ball-on-disk tribo-meter	Ti–6Al–4V	Si_3N_4 ceramic bal (4 mm dia and 1700HV)	Hexagonal boron nitride (hBN) has a graphite		16.89 m/min	5 N	20 °C, 300 °C and 600 °C	Lu et al. (2016)
HT-1000 ball-ondisk high temperature tribometer	TiAl matrix self-lubricating composites containing	Si_3N_4 ceramic ball (6 mm of diameter)	Ag and Ti_2AlC lubricants (solid lubricants)		0.23 m/s	10 N	25 (RT), 200, 400, 600 and 800 1C	Shi et al. (2013)
HT ball-on-disk wear tester	TiO_2/Al_2O_3 composite	Si_3N_4 balls			0.15 m/s	5 N	300 °C	Wang et al. (2015)
Ball-on-disk high-temperature tribometer	Ti_2AlN/TiAl composite	Si_3N_4 ball (6 mm of diameter)			600 rpm	1.5–20 N	25–1000 °C	Wang et al. (2018)

(continued)

Table 8.3 (continued)

Equipment	Composites	Counter body material	Lubricant used	Sliding distance	Sliding speed	Load	Temperature	References
HT-1000 rotational ball-on-disk high-temperature tribometer	TiAl–TiB_2 composites	Si_3N_4ball is 6.35 mm in diameter			0.188 m/s	10 N	(RT, 200, 400, 600 and 800 C	Wang et al. (2017)
HT-1000 ball-on-disk high-temperature tribometer	Ti_2AlN/TiAl composites	Si_3N_4ceramic ball (6 mm dia)			600 rpm	10 N	25, 200, 400, 600 and 800 °C	Liu et al. (2014b)
Block-on-ring wear-corrosion testing apparatus	TiAl matrix composites	sintered Al_2O_3 ceramic ring		216 m	0.12 m/s or 100 rpm	5gf		Li et al. (2018)

had excellent compressive strength of 59.8 MPa. Chen et al. (2012) showed hardness and relative density of Ti–35Nb–2.5Sn/15HA was increased and wear rate was decreased with the increase of HA powders. In another study, Rosu et al. (2012) combination of TiN with HA was used because of its better mechanical and bone tissue interactions.

6.2 Calcium Based Titanium Composites

Calcium phosphate (CaP) is a biomaterial and it is widely used in biological applications because of its excellent biocompatibility, bioactivity and osteoconductivity properties, Calcium phosphate is a best artificial bone substitute (Adeleke et al. 2017). Narayanan et al. (2008) reported that the electrochemical properties of Ti–6Al–6Nb and Ti–6Al–4V were influenced by structure, composition and surface roughness of CaP coatings. Titanium implants show good fixation to the bone when calcium phosphate coatings on their surface (Narayanan et al. 2008). In another study, titanium was coated by Tricalcium phosphate ceramics to improve the bone cell-material interactions and load bearing characteristics using Laser engineering net shaping process (Roy et al. 2008). In order to produce a physically textured surface and enhancing osseointegration, CaP coatings was deposited on a Ti–6Al–4V alloy by using laser deposition method (Nag et al. 2013). In some of the literatures, adding CaP to titanium increased hardness and strength. Also the addition of 5% wt. CaP to Ti64 decreased the wear rate by 70% and its raises 0.2% offset yield strength, ultimate compressive strength. So, the addition of calcium phosphate with titanium made the implants to be applicable for load bearing applications (Bandyopadhyay et al. 2016).

6.3 Approaches Used to Test with and Without Bio-lubricant

Bio-lubricant is a type of lubricant which is non toxic to human beings. Bio-lubricants in hip joint replacement are mainly used for improving the wear resistance and lifetime of implants. Bio-lubricants play an important role in reducing friction and wear rate of bio-implants. Natural synovial joints, such as hip and knee joints, are lubricated with synovial fluid present in the joints and thus prevents the friction between bones while movements (Scholes et al. 2016). There are many bio-lubricants used in hip implants (i.e., bovine serum solution, saline solutions, hank solution, solid bio-lubricants and custom-made lubricants) (Brown and Clarke 2006). Luo et al. (2013) investigated the wear rate, coefficient of friction, worn surface and the wear mechanism of Ti6Al4V alloy under dry and three different lubrication conditions (deionized water, physiological saline and bovine serum lubrications). The results revealed that Ti6Al4V alloy had high CoF, high wear rate under dry friction but showed lower wear rate and CoF under three different lubrications. Among three lubricants, bovine serum solution showed better tribological behavior. Razak et al. (2015) explored the use of alternative bio-lubricants such as palm olein, palm kernel oil and palm fatty acid distillate for improving tribological behavior in metal-on-metal contact. Results revealed that palm oil could optimize the friction and wear rate on metallic acetabular cup due to its unique physical and chemical properties. Scholes et al. (2016) studied the potential synthetic bio-lubricant as an alternative to bovine serum and their results reveals that the novel 0.5% gellan gum fluid gel lubricant has lower wear when compared to bovine serum lubricant. The addition of self lubricant CaF_2 with anti-wear composite coating possessed superior in friction reducing and anti-wear properties than no lubrication condition (Xiang et al. 2014). In another study, (Yang et al. 2017) reported that the Ti–48Al–2Nb–2Cr self-lubricating materials with 10 wt% solid lubricant produced lesser wear. Farnoush et al. (2013) studied the Ti–6Al–4V samples in both dry and simulated body fluid (SBF) solution. The samples studied under SBF solution showed better corrosion and wear resistance than in dry conditions. In another study, Tkachenko et al. (2013) studied Ti–Si based composites with 25% fetal serum solution + sodium azide and ethylene diamine tetra acetic acid exhibited good wear and corrosion resistance and also showed higher hardness.

7 Approaches to Test the Corrosion Properties

Corrosion is one of the serious issues resulting in the failure of biomedical implants. Corrosion resistance is one of the important properties for implants in biomedical applications. In tribological perspective, tribocorrosion refers to "tribological contact of materials leading to irreversible transformation caused by mechanical surface interactions and simultaneous physicochemical reactions". Corrosion in metal implants mainly occur due to electrochemical reaction between implant surface and body fluids. In general corrosion is an electrochemical process. There are several approaches used to evaluate the corrosion resistance properties of implant materials. Some approaches are CHI660D electrochemical analyzer, Potentiodynamic polarization and electrochemical impedance spectroscopy. Figure 8.5 shows some testing

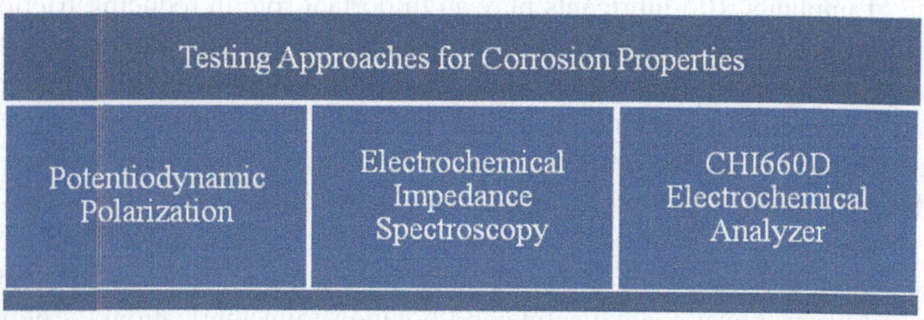

Fig. 8.5 Some testing approaches for corrosion properties

approaches for corrosion properties. Saba et al. (2018) investigated the corrosion properties of Ti/nanodiamonds (NDs) nanocomposites by using CHI660D electrochemical analyzer. Corrosion test results showed that corrosion resistance of titanium increased with increase in nano-reinforcements up to 0.25%wt of ND's. To improve the corrosion resistance of Ti6Al4V alloy, the novel $NiSi_2/Ti_5Si_3$ bilayer film was coated on to it by double cathode glow discharge method. The findings showed that the electrochemical resistance of coated alloy was superior to that of Ti6Al4V alloy. The above mentioned novel coated alloy's electrochemical behavior was characterized by potentiodynamic polarization and electrochemical impedance spectroscopy (Xu et al. 2012). Lee (2012) investigated the corrosion behavior of electrodeposited Ni-CNT composite coating on Ti6Al4V alloy using Hank's solution. The results suggested that the adding CNTs (carbon nanotubes) to the nickel coating increased the pitting corrosion resistance and it is verified by potentiodynamic polarization.

A novel β-Ta_2O_5 nanoceramic coatings was used to improve the corrosion resistance of Ti6Al4V and the results showed that the corrosion resistance of β-Ta_2O_5 nanoceramic coatings was superior than that of Ti6Al4V alloy in 3.5%NaCl solution. The electrochemical behavior of novel composite coatings was examined by potentiodynamic polarization and electrochemical impedance spectroscopy (Hu et al. 2016).

8 Advancements in Development of Titanium Based Composite Materials

The recent advancement in titanium composites called Polyaniline (PANI)-Titanium-supported nanocomposites were used as ion-exchanger material for waste water decontamination. The above mentioned titanium composite was developed using sol-gel method (Shahadat et al. 2020). The development of titanium dioxide-coated biochar composites was used to remove organic pollutants in aqueous medium. The composite was prepared using sol-gel method (Cai et al. 2018). One of the recent literatures showed yttria stabilized ZrO_2 nano coating was performed on titanium substrate to enhance biocompatiblity, bioactivity and hardness. The coating was done using electrode deposition (Yin et al. 2018). Self healing of materials was investigated for SS304 modified with boron (B), cerium (Ce) and titanium (Ti) and it showed improved creep resistance (Kilicli et al. 2018). The recent study reported the fabrication of novel Ti–6Al–4V/TiB composites by hot pressing as well as spark plasma sintering (SPS) showed improved tensile strength than wrought Ti–6Al–4V (Luo et al. 2019). Another recent study revealed that the mechanical and biological properties of the Ti6Al4V/HA composite porous scaffolds could be easily adjusted by the 3D printing techniques and it is a promising one in bone tissue regeneration (Yi et al. 2020). In the twenty first century, developing renewable energy is one of the major scientific challenges. Recently various advanced TiO_2 nanostructures have been developed for photo catalytic fuel production (H_2 generation, CO_2

reduction, etc.) due to their low cost, abundance and environmental friendly in nature. Zhang et al. (2020) discussed in detail about the several applications of hierarchically mesoporous TiO_2 materials in energy and environmental related areas, such as photo electrochemical water splitting, chemical catalysis, photo catalytic degradation of pollutants, lithium-ion batteries and sodium- ion batteries. The $Ti_3C_2T_x$@$NiCO_2O_4$ composite materials could be used for high performance microwave absorber due to its good conductivity, high dielectric constant, and dielectric loss matching with ohmic loss, magnetic loss and interface polarization. The above mentioned composites were prepared by simple hydrothermal method and annealing treatment (Hou et al. 2020).

9 Conclusions

Titanium as well as titanium alloy along with coating of different materials are discussed in this chapter. The different reinforcements and coating techniques used for titanium composites help readers in choosing better technique for producing these composites. Also titanium and its alloy are widely used for implants because of its enhanced mechanical properties and its biocompatibility. Due to its excellent corrosion properties, it is widely preferred in biomedical applications. Also, Titanium has better tribological properties and its mechanical characteristics could be enhanced by several coating techniques. Hydroxyapatite with titanium exhibits excellent biocompatibility and bioactivity. Also calcium with titanium composites possess excellent load bearing characteristics which could be used in joint replacements. The recent development in titanium composite also showed it could be used in effective treatment of waste water and self healing of materials. Moreover, the recently developed titanium nanocomposites are used in energy conservation and degradation of pollutants. This helps in maintaining ecosystem in balanced manner. Thus it is evident from this chapter that, titanium and its composites have endless applications due to its diversified properties.

References

Adeleke S, Bushroa A, Sopyan I (2017) Recent development of calcium phosphate-based coatings on titanium alloy implants. Surf Eng Appl Electrochem 53(5):419–433

An Q, Huang L, Bao Y, Zhang R, Jiang S, Geng L, Xiao M (2018) Dry sliding wear characteristics of in-situ TiBw/Ti6Al4V composites with different network parameters. Tribol Int 121:252–259

Attar H, Calin M, Zhang L, Scudino S, Eckert J (2014) Manufacture by selective laser melting and mechanical behavior of commercially pure titanium. Mater Sci Eng, A 593:170–177

Attar H, Ehtemam-Haghighi S, Kent D, Dargusch MS (2018) Recent developments and opportunities in additive manufacturing of titanium-based matrix composites: a review. Int J Mach Tools Manuf 133:85–102

Bai M, Namus R, Xu Y, Guan D, Rainforth MW, Inkson BJ (2019) In-situ Ti-6Al-4V/TiC composites synthesized by reactive spark plasma sintering: Processing, microstructure, and dry sliding wear behaviour. Wear 202944

Bandyopadhyay A, Dittrick S, Gualtieri T, Wu J, Bose S (2016) Calcium phosphate–titanium composites for articulating surfaces of load-bearing implants. J Mech Behav Biomed Mater 57:280–288

Bao Y, Huang L, An Q, Jiang S, Geng L, Ma X (2018) Wire-feed deposition TiB reinforced Ti composite coating: formation mechanism and tribological properties. Mater Lett 229:221–224

Berndt C, Hasan F, Tietz U, Schmitz K-P (2014) A review of hydroxyapatite coatings manufactured by thermal spray. In: Advances in calcium phosphate biomaterials. Springer, Berlin, pp 267–329

Brown SS, Clarke IC (2006) A review of lubrication conditions for wear simulation in artificial hip replacements. Tribol Trans 49(1):72–78

Cai X, Li J, Liu Y, Yan Z, Tan X, Liu S, Zeng G, Gu Y, Hu X, Jiang L (2018) Titanium dioxide-coated biochar composites as adsorptive and photocatalytic degradation materials for the removal of aqueous organic pollutants. J Chem Technol Biotechnol 93(3):783–791

Candel J, Amigó V, Ramos J, Busquets D (2010) Sliding wear resistance of TiCp reinforced titanium composite coating produced by laser cladding. Surf Coat Technol 204(20):3161–3166

Çelik ON (2013) Microstructure and wear properties of WC particle reinforced composite coating on Ti6Al4V alloy produced by the plasma transferred arc method. Appl Surf Sci 274:334–340

Chen Y, Wang X, Xu L, Liu Z, Do Woo K (2012) Tribological behavior study on Ti–Nb–Sn/hydroxyapatite composites in simulated body fluid solution. J Mech Behav Biomed Mater 10:97–107

Dang W, Ren S, Zhou J, Yu Y, Wang L (2016) The tribological properties of Ti3SiC2/Cu/Al/SiC composite at elevated temperatures. Tribol Int 104:294–302

Dong Y, Wang H (2009) Microstructure and dry sliding wear resistance of laser clad TiC reinforced Ti–Ni–Si intermetallic composite coating. Surf Coat Technol 204(5):731–735

Farnoush H, Bastami AA, Sadeghi A, Mohandesi JA, Moztarzadeh F (2013) Tribological and corrosion behavior of friction stir processed Ti-CaP nanocomposites in simulated body fluid solution. J Mech Behav Biomed Mater 20:90–97

Fathi M, Zahrani EM (2009) Fabrication and characterization of fluoridated hydroxyapatite nanopowders via mechanical alloying. J Alloy Compd 475(1–2):408–414

Feng S, Tang H, Zhang S, Wang H (2012) Microstructure and wear resistance of laser clad TiB–TiC/TiNi–Ti2Ni intermetallic coating on titanium alloy. Trans Nonferrous Met Soc China 22(7):1667–1673

Geetha M, Singh AK, Asokamani R, Gogia AK (2009) Ti based biomaterials, the ultimate choice for orthopaedic implants—a review. Prog Mater Sci 54(3):397–425

Geng L, Ni D, Zhang J, Zheng Z (2008) Hybrid effect of TiBw and TiCp on tensile properties of in situ titanium matrix composites. J Alloy Compd 463(1–2):488–492

Guo B, Zhou J, Zhang S, Zhou H, Pu Y, Chen J (2007) Phase composition and tribological properties of Ti–Al coatings produced on pure Ti by laser cladding. Appl Surf Sci 253(24):9301–9310

Guo B, Wang Z, Li H (2016) Study on the friction and wear behavior of a TA15 alloy and Its Ni-SiC composite coating. J Mater Eng Perform 25(5):1763–1772

Hibi Y, Miyakc K, Murakami T, Sasaki S (2006) Tribological behavior of SiC-reinforced Ti3SiC2-based composites under dry condition and under lubricated condition with water and ethanol. J Am Ceram Soc 89(9):2983–2985

Hou T, Wang B, Ma M, Feng A, Huang Z, Zhang Y, Jia Z, Tan G, Cao H, Wu G (2020) Preparation of two-dimensional titanium carbide (Ti3C2Tx) and NiCo2O4 composites to achieve excellent microwave absorption properties. Compos B Eng 180:107577

Hu W, Xu J, Lu X, Hu D, Tao H, Munroe P, Xie Z-H (2016) Corrosion and wear behaviours of a reactive-sputter-deposited Ta2O5 nanoceramic coating. Appl Surf Sci 368:177–190

Jiang S, Wang G, Ren Q, Yang C, Wang Z, Zhou Z (2015) In situ synthesis of Fe-based alloy clad coatings containing TiB 2–TiN–(h-BN). Int J Miner Metall Mater 22(6):613–619

Jin X, Gao L, Liu E, Yu F, Shu X, Wang H (2015) Microstructure, corrosion and tribological and antibacterial properties of Ti–Cu coated stainless steel. J Mech Behav Biomed Mater 50:23–32

Khorasani AM, Goldberg M, Doeven EH, Littlefair G (2015) Titanium in biomedical applications—properties and fabrication: a review. J Biomater Tissue Eng 5(8):593–619

Khun N, Li R, Khor K (2015) Mechanical and tribological properties of spark plasma–sintered titanium composites filled with different Al-Cr-Fe quasicrystal contents. Tribol Trans 58(5):859–866

Khun NW, Tan AWY, Liu E (2016) Mechanical and tribological properties of cold-sprayed Ti coatings on Ti-6Al-4V substrates. J Therm Spray Technol 25(4):715–724

Kilicli V, Yan X, Salowitz N, Rohatgi PK (2018) Recent advancements in self-healing metallic materials and self-healing metal matrix composites. JOM 70(6):846–854

Kim I, Choi B, Kim Y, Lee Y (2011) Friction and wear behavior of titanium matrix (TiB+ TiC) composites. Wear 271(9–10):1962–1965

Kim J-S, Lee K-M, Cho D-H, Lee Y-Z (2013) Fretting wear characteristics of titanium matrix composites reinforced by titanium boride and titanium carbide particulates. Wear 301(1–2):562–568

Lee C (2012) Wear and corrosion behavior of electrodeposited nickel–carbon nanotube composite coatings on Ti–6Al–4V alloy in Hanks' solution. Tribol Int 55:7–14

Lee C, Sanders A, Tikekar N, Chandran KR (2008) Tribology of titanium boride-coated titanium balls against alumina ceramic: wear, friction, and micromechanisms. Wear 265(3–4):375–386

Li J, Yu Z, Wang H, Li M (2010) Microstructural evolution of titanium matrix composite coatings reinforced by in situ synthesized TiB and TiC by laser cladding. Int J Miner Metall Mater 17(4):481–488

Li J, Yu Z, Wang H (2011) Wear behaviors of an (TiB+ TiC)/Ti composite coating fabricated on Ti6Al4V by laser cladding. Thin Solid Films 519(15):4804–4808

Li J, Yu ZS, Wang HP (2011b) Effects of Y2O3 on microstructure and mechanical properties of laser clad coatings reinforced by in situ synthesized TiB and TiC. Mater Sci For 589–592

Li J, Yu Z, Wang H, Li M (2012) Microstructure and mechanical properties of an in situ synthesized TiB and TiC reinforced titanium matrix composite coating. J Wuhan Univ Technol-Mater Sci Ed 27(1):1–8

Li X, Tang B, Ye J (2012) Fabrication of Zr and Zr–N surface alloying layers and hardness improvement of Ti–6Al–4V alloy by plasma surface alloying technique. Appl Surf Sci 258(6):1981–1984

Li X, Shi T, Zhang C, Li S, Zhang J, Zhang L-C (2018) Improved wear resistance and mechanism of titanium aluminum based alloys reinforced by solid lubricant materials. Mater Res Exp 5(8):086502

Liang H, Shi B, Fairchild A, Cale T (2004) Applications of plasma coatings in artificial joints: an overview. Vacuum 73(3–4):317–326

Lin Y, Lei Y, Fu H, Lin J (2015) Mechanical properties and toughening mechanism of TiB2/NiTi reinforced titanium matrix composite coating by laser cladding. Mater Des 80:82–88

Liu S, Liu Z, Wang Y, Yue P (2014) Ti-based composite coatings with gradient TiCx reinforcements on TC4 titanium alloy prepared by laser cladding. Sci China Technol Sci 57(7):1454–1461

Liu X-B, Meng X-J, Liu H-Q, Shi G-L, Wu S-H, Sun C-F, Wang M-D, Qi L-H (2014) Development and characterization of laser clad high temperature self-lubricating wear resistant composite coatings on Ti–6Al–4V alloy. Mater Des 55:404–409

Liu Y-F, Zhou Y-L, Zhang Q, Pu F, Li R-H, Yang S-Z (2014) Microstructure and dry sliding wear behavior of plasma transferred arc clad Ti5Si3 reinforced intermetallic composite coatings. J Alloy Compd 591:251–258

Lu X-L, Liu X-B, Yu P-C, Qiao S-J, Zhai Y-J, Wang M-D, Chen Y, Xu D (2016) Synthesis and characterization of Ni60-hBN high temperature self-lubricating anti-wear composite coatings on Ti6Al4V alloy by laser cladding. Opt Laser Technol 78:87–94

Luo Y, Yang L, Tian M (2013) Influence of bio-lubricants on the tribological properties of Ti6Al4V alloy. J Bionic Eng 10(1):84–89

Luo S, Song T, Liu B, Tian J, Qian M (2019) Recent advances in the design and fabrication of strong and ductile (tensile) titanium metal matrix composites. Adv Eng Mater 1801331

Mishra A, Khobragade N, Sikdar K, Chakraborty S, Kumar SB, Roy D (2017) Study of mechanical and tribological properties of nanomica dispersed hydroxyapatite based composites for biomedical applications. Adv Mater Sci Eng 2017

Mu M, Zhou X, Xiao Q, Liang J, Huo X (2012) Preparation and tribological properties of self-lubricating TiO2/graphite composite coating on Ti6Al4V alloy. Appl Surf Sci 258(22):8570–8576

Nag S, Paital SR, Nandawana P, Mahdak K, Ho YH, Vora HD, Banerjee R, Dahotre NB (2013) Laser deposited biocompatible Ca–P coatings on Ti–6Al–4V: microstructural evolution and thermal modeling. Mater Sci Eng, C 33(1):165–173

Narayanan R, Seshadri S, Kwon T, Kim K (2008) Calcium phosphate-based coatings on titanium and its alloys. J Biomed Mater Res Part B: Appl Biomater: Off J Soc Biomater Jpn Soc Biomater Austr Soc Biomater Korean Soc Biomater 85(1):279–299

Niespodziana K, Jurczyk K, Jakubowicz J, Jurczyk M (2010) Fabrication and properties of titanium–hydroxyapatite nanocomposites. Mater Chem Phys 123(1):160–165

Obadele BA, Andrews A, Olubambi PA, Mathew MT, Pityana S (2015) Tribocorrosion response of laser clad TiNiZrO2 composite coatings on Ti6Al4V. Mater Sci For 345–350

Okulov I, Kühn U, Marr T, Freudenberger J, Soldatov I, Schultz L, Oertel C-G, Skrotzki W, Eckert J (2014) Microstructure and mechanical properties of new composite structured Ti–V–Al–Cu–Ni alloys for spring applications. Mater Sci Eng, A 603:76–83

Peng D-X, Kang Y, Huang Y-J (2014) Microstructure and tribological properties of gas tungsten arc clad TiC composite coatings on carbon steel. Ind Lubr Tribol 66(5):609–617

Prem Ananth M, Ramesh R (2014) Tribological improvement of titanium alloy surfaces through texturing and TiAlN coating. Surf Eng 30(10):758–762

Qin Y, Geng L, Ni D (2011) Dry sliding wear behavior of extruded titanium matrix composite reinforced by in situ TiB whisker and TiC particle. J Mater Sci 46(14):4980–4985

Que W, Khor KA, Xu J, Yu L (2008) Hydroxyapatite/titania nanocomposites derived by combining high-energy ball milling with spark plasma sintering processes. J Eur Ceram Soc 28(16):3083–3090

Rahmati R, Khodabakhshi F (2018) Microstructural evolution and mechanical properties of a friction-stir processed Ti-hydroxyapatite (HA) nanocomposite. J Mech Behav Biomed Mater 88:127–139

Rao RR, Kannan T (2002) Synthesis and sintering of hydroxyapatite–zirconia composites. Mater Sci Eng, C 20(1–2):187–193

Rathore A (2015) Development of hydroxyapatite-titania composite for implant application

Razak D, Syahrullail S, Sapawe N, Azli Y, Nuraliza N (2015) A new approach using palm olein, palm kernel oil, and palm fatty acid distillate as alternative biolubricants: improving tribology in metal-on-metal contact. Tribol Trans 58(3):511–517

Rosu RA, Serban V-A, Bucur AI, Dragos U (2012) Deposition of titanium nitride and hydroxyapatite-based biocompatible composite by reactive plasma spraying. Appl Surf Sci 258(8):3871–3876

Roy M, Krishna BV, Bandyopadhyay A, Bose S (2008) Laser processing of bioactive tricalcium phosphate coating on titanium for load-bearing implants. Acta Biomater 4(2):324–333
Saba F, Zhang F, Liu S, Liu T (2018) Tribological properties, thermal conductivity and corrosion resistance of titanium/nanodiamond nanocomposites. Compos Commun 10:57–63
Scholes SC, Colledge CJ, Naylor A, Mahdi MH, Smith AM, Joyce TJ (2016) Potential synthetic biolubricant as an alternative to bovine serum. Lubricants 4(4):38
Selva Kumar M, Chandrasekar P, Chandramohan P, Mohanraj M (2012) Characterisation of titanium-titanium boride composites processed by powder metallurgy techniques. Mater Charact 73 (Complete)
Shahadat M, Ahmad A, Bushra R, Ismail S, Ahammad SZ, Ali SW, Rafatullah M (2020) Recent advancement in wastewater decontamination technology. In: Modern age waste water problems. Springer, Berlin, pp 1–22
Shi X, Xu Z, Wang M, Zhai W, Yao J, Song S, ud Din AQ, Zhang Q (2013) Tribological behavior of TiAl matrix self-lubricating composites containing silver from 25 to 800 C. Wear 303(1-2):486–494
Silva VV, Domingues RZ, Lameiras FS (2001) Microstructural and mechanical study of zirconia-hydroxyapatite (ZH) composite ceramics for biomedical applications. Compos Sci Technol 61(2):301–310
Silva J, Alves AMVCP, Pinto A, Silva FS, Toptan F (2016) Dry sliding wear behaviour of Ti-TiB-TiNx in-situ composite synthesised by reactive hot pressing. Int J Surf Sci Eng 10(4):317–329
Suthar J, Patel K (2018) Processing issues, machining, and applications of aluminum metal matrix composites. Mater Manuf Processes 33(5):499–527
Szcześ A, Hołysz L, Chibowski E (2017) Synthesis of hydroxyapatite for biomedical applications. Adv Coll Interface Sci 249:321–330
Tian Y, Chen C, Chen L, Huo Q (2005) Wear and oxidation resistance of composite coatings fabricated on Ti-6Al-4V by laser alloying with nitrogen and silicon. J Phys D Appl Phys 38(23):4217
Tian Y, Chen C, Wang D, Huo Q, Lei T (2005) Microstructure and wear properties of composite coatings produced by laser cladding of Ti-6Al-4V with graphite and silicon mixed powders. Surf Rev Lett 12(02):161–165
Tkachenko S, Datskevich O, Kulak L, Engqvist H, Persson C (2013) Tribological properties of Ti-Si-Zr alloys. In: 22nd conference on metallurgy and materials, METAL 2013, pp 15–17
Tkachenko S, Nečas D, Datskevich O, Čupera J, Spotz Z, Vrbka M, Kulak L, Foret R (2014) Tribological behavior of Ti–Si based in situ composites under sliding
Umeda J, Fugetsu B, Nishida E, Miyaji H, Kondoh K (2015) Friction behavior of network-structured CNT coating on pure titanium plate. Appl Surf Sci 357:721–727
Wang Y, Wang H (2004) Wear resistance of laser clad Ti2Ni3Si reinforced intermetallic composite coatings on titanium alloy. Appl Surf Sci 229(1–4):81–86
Wang H, Shu X, Liu E, Han Z, Li X, Tang B (2014) Assessments on corrosion, tribological and impact fatigue performance of Ti-and TiN-coated stainless steels by plasma surface alloying technique. Surf Coat Technol 239:123–131
Wang S, Zhao Q, Liu D, Du N (2015) Microstructure and elevated temperature tribological behavior of TiO2/Al2O3 composite ceramic coating formed by microarc oxidation of Ti6Al4V alloy. Surf Coat Technol 272:343–349
Wang L, Cheng J, Zhu S, Yu Y, Qiao Z, Yang J, Liu W (2017) High temperature wear behaviors of TiAl–TiB2 composites. Tribol Lett 65(4):144
Wang D, Sun D, Han X, Wang Q, Wang G (2018) Investigation on tribological properties of the pre-oxidized Ti2 AlN/TiAl composite. J Mater Eng Perform 27(4):1973–1986
Williams D (1987) Tissue-biomaterial interactions. J Mater Sci 22(10):3421–3445
Xiang Z-F, Liu X-B, Ren J, Luo J, Shi S-H, Chen Y, Shi G-L, Wu S-H (2014) Investigation of laser cladding high temperature anti-wear composite coatings on Ti6Al4V alloy with the addition of self-lubricant CaF2. Appl Surf Sci 313:243–250

Xu J, Liu L, Xie Z-H, Munroe P (2012) Nanocomposite bilayer film for resisting wear and corrosion damage of a Ti–6Al–4V alloy. Surf Coat Technol 206(19–20):4156–4165

Xu J, Liu L, Jiang L, Munroe P, Xie Z-H (2013) Unraveling the mechanical and tribological properties of a novel Ti5Si3/TiC nanocomposite coating synthesized by a double glow discharge plasma technique. Ceram Int 39(8):9471–9481

Xu Z, Chen L, Shi X, Zhang Q, Ibrahim AMM, Zhai W, Yao J, Zhu Q, Xiao Y (2015) Formation of friction layers in graphene-reinforced TiAl matrix self-lubricating composites. Tribol Trans 58(4):668–678

Xu Z, Xue B, Shi X, Zhang Q, Zhai W, Yao J, Wang Y (2015) Sliding speed and load dependence of tribological properties of Ti3SiC2/TiAl composite. Tribol Trans 58(1):87–96

Yan Y, Zhang X, Mao H, Huang Y, Ding Q, Pang X (2015) Hydroxyapatite/gelatin functionalized graphene oxide composite coatings deposited on TiO2 nanotube by electrochemical deposition for biomedical applications. Appl Surf Sci 329:76–82

Yang K, Shi X, Huang Y, Wang Z, Wang Y, Zhang A, Zhang Q (2017) The research on the sliding friction and wear behaviors of TiAl-10 wt% Ag at elevated temperatures. Mater Chem Phys 186:317–326

Yi T, Zhou C, Ma L, Wu L, Xu X, Gu L, Fan Y, Xian G, Fan H, Zhang X (2020) Direct 3-D printing of Ti-6Al-4V/HA composite porous scaffolds for customized mechanical properties and biological functions. J Tissue Eng Regen Med

Yin X, Liang C, Ge F (2018) Electrodeposition of a YSZ–Yttria stabilized zirconia composite coating on a titanium bone implant. Int J Electrochem Sci 13:822–831

Yu H, Zhang W, Wang H, Ji X, Song Z, Li X, Xu B (2017) In-situ synthesis of TiC/Ti composite coating by high frequency induction cladding. J Alloy Compd 701:244–255

Zhang H, Yu H, Chen C (2017) Microstructure and wear resistance of composite coating by laser cladding Ni60A/B4C pre-placed powders on Ti-6Al-4V substrate. Sci Eng Compos Mater 24(4):541–546

Zhang W, Tian Y, He H, Xu L, Li W, Zhao D (2020) Recent advances in synthesis of hierarchically mesoporous TiO2 materials for energy and environmental applications. Natl Sci Rev

Zhao X, Zhang P, Wang X, Chen Y, Liu H, Chen L, Sheng Y, Li W (2018) In-situ formation of textured TiN coatings on biomedical titanium alloy by laser irradiation. J Mech Behav Biomed Mater 78:143–153

Zi-Run Y, Hai-Xiang H, Jiang C-F, Li W-M, Liu X-R, Lyu S (2017) Evaluation on dry sliding wear behavior of (TiB+ TiC)/Ti-6Al-4V matrix composite. Int J Precis Eng Manuf 18(8):1139–1146

The Effect of Fillers on the Tribological Properties of Composites

R. Muraliraja, T. R. Tamilarasan, Sanjith Udayakumar, and C. K. Arvinda Pandian

Abstract Recent advances in the sphere of materials engineering have seen the advent of a new generation of composite materials that have become a replacement for the traditional materials used in different industries over the years. The proven potential of composite technology to deliver in line with critical global trends in the aerospace, automobile, and marine industries has placed it on top of other materials in the market. As composite materials can be lightweight and durable, they score very highly on efficiency measures with desired engineering properties required. A careful and wise combination of matrices, reinforcements, and additives can be tailored for specific applications of the end product. Traditionally, most of the fillers were considered as additives, limiting their contribution to a composite only on reducing their cost. However, the diversity of applications and a broad spectrum of their usage has led to high demand for incorporating fillers in composite technology. In this perspective, the objective of this chapter is to explore the works of literature for providing information about the fillers concerning processing, functions, mechanical and tribological characteristics, environmental impact. Moreover, the recent advances and challenges in employing different types of fillers in different classes of composites have been briefly discussed.

Keywords Fillers · Composites · Wear · Friction

R. Muraliraja (✉)
Vels Institute of Science Technology and Advanced Studies, Chennai, India
e-mail: muralimechraja@gmail.com

T. R. Tamilarasan · C. K. Arvinda Pandian
B.S. Abdur Rahman Crescent Institute of Science and Technology, Chennai, India
e-mail: tamilarasan.tr@gmail.com

C. K. Arvinda Pandian
e-mail: doctratearvind@gmail.com

S. Udayakumar
Universiti Sains Malaysia-Engineering Campus, Nibong Tebal, Pulau Pinang, Malaysia
e-mail: sanjithudayakumar@gmail.com

M. T. Hameed Sultan et al. (eds.), *Tribological Applications of Composite Materials*, Composites Science and Technology, https://doi.org/10.1007/978-981-15-9635-3_9

1 Introduction

"Filler is a solid material capable of changing the physical and chemical properties of materials by surface interaction or its lack thereof and by its own physical characteristics" (Wypych 2016). Fillers are used for diverse industrial applications all over the world based on the nature and needs of the application. These additive components are incorporated in a mixture to elevate the properties and reduce the cost of the composites. The fillers are commercially available in the form of gases, solids, and liquids. Based on the requirements of the material's applications, the fillers are selected among the various available categories. These fillers not only reduce the cost of the composite but also improve the processing and mechanical behaviour of the composite. They serve for a variety of purposes in different industrial applications such as adhesives, agriculture, aerospace, appliances, automotive materials, bottles and containers, building components, business machines, cables and wires, coated fabrics, coatings, paints, face creams and powders and health care medicines, composites used for dental applications, fibres, film, foam, food feed, friction materials, geosynthetics, hoses and pipes, magnetic devices, medical applications, membranes, noise dampening, optical devices, paper, railway transportation, roofing, telecommunication, tires, sealants, sports equipment, waterproofing and windows (Wypych 2016).

Initially, fillers were mainly used to reduce the cost of the material, but it is also subjected to particle size considerations. The effectiveness of a filler addition is influenced by several factors such as purity, surface preparation, shape, and distribution. Fillers can also be used to vary the density of a material. High-density polymer material up to 2 g/cm^3 can be obtained with the help of fillers. Similarly, low-density polymers with lighter materials like foam as a filler can also be obtained (Chiellini and Solaro 1996). The colour of a material can be altered with the help of metallic powder fillers wherein the final polymer gazes like a composite metal. The surface properties of a material can be improved using graphite, PTFE, MOS_2 as fillers to reduce the coefficient of friction, as these types of fillers possess self-lubrication properties (Kano and Akiyama 1996). The shape of a material can be retained in polymer foams with the help of fillers. Besides, the use of hollow spherical particles as a filler provides the best insulation to electric and thermal conduction. The porosity of a material can also be influenced by the right choice of fillers (Srirangan and Paulraj 2016). Fillers are known to significantly affect most of the mechanical properties of materials (Asadi et al. 2011). Fillers contribute towards reducing fire accidents by reducing the auto-ignition temperature, and it decreases the smoke formation during hazardous conditions; also, it increases the char formation. The fillers are also capable of reducing the heat transmission rate and prevent dripping. Some of the critical properties of a material can be improved through the proper and right use of fillers.

2 Fillers in Composites

2.1 Role of Fillers in Composites

Additives, fillers, and reinforcements are mostly used to improve or change the properties of a composite. These days, fillers are used in the composites to improve the properties such as wear, friction, corrosion, heat conductivity, degradation of the material, decrease the thermal expansion, electrical properties, reduce the amount of shrinkage, and swelling. Incorporating fillers such as graphite and SiC particles in the glass-epoxy composite systems significantly improve the friction behaviour and exhibits superior wear-resistant properties (Suresha et al. 2006). For tribological applications, short fibres are used as reinforcement in polymer composites. Fillers in the polyetheretherketone (PEEK) composite have been used to reduce the wear rate, but its level of influence lies on the capability of the composite layer to form transfer films (Bahadur and Gong 1992). Lignin fillers are used in natural rubbers to protect it from thermo-oxidative degradation in the air (Košíková et al. 2007). Fillers that possess high thermal conductivity are added to epoxy composites to increase the breakdown power (Li et al. 2011). The composites used in dental applications also have fillers for several purposes, such as colour, strength, and bonding.

2.2 Classification of Fillers

Fillers are classified mainly by their material, size, and shape. Fibres and particles are commonly used fillers in the polymer matrix and metal matrix composites, respectively. They are either incorporated to serve as reinforcements or fillers. Particulate fillers in the composites are classified based on its size are macrofillers, midifillers, minifillers, microfillers, and nanofillers. The size limits of the particulate fillers are prescribed as 10–100 μm, 1–10 μm, 0.1–1 μm, 0.01–0.1 μm, –0.005 to 0.01 μm, respectively. The characteristics of the fillers are different from one another in the various aspects such as surface properties of the filler, shape, size, distribution of the particles, and impurities. From the literature survey, it was observed that finer particle sized fillers produced composites of better mechanical properties, whereas the coarser particles showed a declining trend of the properties in the composite. Impurities may create serious problems such as unwanted chemical reactions with the matrix materials during elevated temperatures (Nassar et al. 2017; Muraliraja et al. 2018). The toughness of the thermoplastics can be improved by adding rubbery kind of fillers. Layers of fibrous filler materials are implanted in the plastic or polymer material to increase the strength (Raja and Retnam 2019). Fibres are mainly classified into two categories, namely natural and manmade. Natural fibres are extracted from animals, plants, and naturally available minerals. The artificially synt6hesized organic and inorganic are the manmade fibres used in composites (Nassar et al. 2017) Fig. 1.

Fig. 1 Classification of fillers (Nassar et al. 2017)

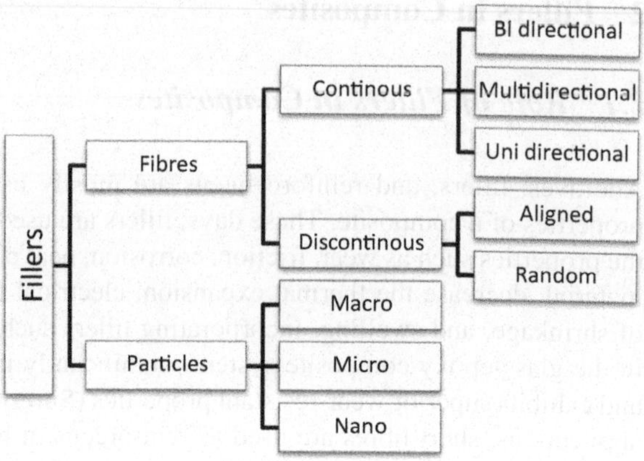

2.3 Fillers in Different Processing Methods *Wypych (2016)*

The preparations composites based on different fillers are mentioned in the table given below. The goal of this section is to identify the different types of processes or methods used to incorporate fillers in newly formulated composite materials Table 1.

Table 1 Different processing methods used to incorporate fillers in composites

No	Fillers	Process/methods	References
1	Glass fibre	Blow moulding	Palutkiewicz et al. (2019)
2	Carbon fibre sheet	Compression moulding	Wulfsberg et al. (2014)
3	Particulate filler (nano-YSZ)	Dip coating	Bakhsheshi-Rad et al. (2016)
4	Particulate fillers	Dispersion	Li et al. (2012)
5	Biomass residues as fibre fillers	Extrusion	Bajwa et al. (2011)
6	Wood fibre	Foaming	Matuana et al. (1998)
7	Cellulose fibre	Injection moulding	Graupner et al. (2016)
8	Glass fibre (treated)	Pultrusion	Chen et al. (2017)
9	Silane treated glass fibre	Reaction injection moulding	Yoo et al. (2017)
10	Natural fibre (Jute)	Resin transfer moulding	Ashworth et al. (2016)
11	Natural fibres (agave, coir, and pine)	Rotational molding	Cisneros-López et al. (2017)
12	Chopped carbon fibre	Sheet molding	Tang et al. (2019)
13	Multi-walled carbon nanotubes	Spinning	Lai et al. (2015)
14	Hemp fibres	Thermoforming	Ciupan et al. (2017)

2.4 *Factors Affecting the Functions of Fillers*

Fillers generally reduce the weight of the composite without compromising its strength. They also reduce the cost of the matrix material by replacing the specific content of the matrix. The main factors influencing the properties of the composites are the shape and the size of the fillers. Fillers are generally in the shapes of spheres and flakes. Typically, fillers that are sized in between of micrometers and nanometers are preferred, as they have more surface to the contact area. Based on the size and the shape, different fillers tend to exhibit different effects on the properties and behaviour of the composites. In particular, the fillers are selected as per the application for which the composites are developed.

2.5 *Effect of Fillers on Mechanical Properties of the Composites*

The addition of fillers to a base matrix tends to enhance the mechanical properties of the composite. Factors like, "dispersion, distribution and adhesion of fillers along with the interface between filler and matrix" have an influential role over the mechanical behaviour of the composites. Polymer-based composites showed enhancement of the mechanical properties like ultimate tensile strength, impact strength and hardness. Both natural, as well as artificially prepared fillers, have demonstrated a significant contribution in improving the properties of the base material. Composites with filler provide additional energy-absorbing damage modes, effective stress transfer between the particles and the matrix with combined advantages of the constituent phases.

In an experimental study by Naidu et al. (2019), the influence of graphitic carbon nitride (g-C(g-C_3N_4) nanofillers on the mechanical properties of epoxy-glass fibre composites was investigated. Various proportions of the filler (1, 1.5, 2, 2.5, and 3%) were added to the epoxy matrix and mechanical characterization of the filled composites revelated that for 2 wt% of g-C_3N_4 filling, the tensile and flexural strength was enhanced by 11% and 13% respectively (Naidu et al. 2019). Pandian and Jailani (2019) performed a comparative study on the effect of silica fumes on the mechanical attributes of jute-linen composites. The composites were prepared with and without incorporating silica fumes (1 wt%). The results inferred that the tensile strength of the filler added composite had increased by 7% and the flexural property too had improved by 5.2%, which was more than that of the composite without filler (Pandian and Jailani 2019). The researchers also explored the effect of industrial waste silica fumes as fillers on polymer composites. The dynamic mechanical properties of industrial waste silica fumes (1, 2 and 3 wt%) incorporated natural fibre fabrics reinforced polymer composites were evaluated. The incorporation of a small quantity of silica fumes had significantly increased the composite's dynamic mechanical properties ("storage modulus, loss modulus, and glass transition temperature"). However, for a higher weight fraction (above 2%) of silica fume addition, no further improvement in the mechanical properties of the composite was observed, which could be ascribed to the agglomeration of silica fumes (Arvinda Pandian and Siddhi Jailani 2019).

Nano silicon dioxide of different concentrations (0, 5 and 10%) were incorporated into jute-epoxy composites, and their influence on the fatigue, flexural and tensile attributes were studied. The Young's modulus, peak load, ultimate tensile and flexural attributes were found to be the highest in case of 5% filler loaded composite. Similarly, in the fatigue test, the composite with 5% nano silicon dioxide withstood the highest number of cycles and exhibited good results compared to the other loaded composites (Ashik et al. 2017). The potential of molybdenum disulphide or shungite, organo-modified montmorillonite, graphite nanoplates as fillers in Ultra-high-molecular-weight polyethylene (UHMWPE) was investigated. The deformation and strength characteristics of UHMWPE composites for filler content of up to 0.06 vol% were studied. This study showed that filler addition had a significant effect on the deformation and strength characteristics. Under the filler contents of up to 0.06 vol%, the highest increase in the elastic modulus of composites was provided by the lamellar nanofillers, namely graphite nanoplates and montmorillonite (Grinev et al. 2018). Mohanty et al. (2014) studied the effect of alumina nanoparticles, glass fibre and carbon fibre on the tensile properties of the incorporated epoxy composites. Four different types of composites were prepared with different concentrations of the additives mentioned above and their properties were evaluated. Alumina particles, short glass and carbon fibres were added separately in the range of 1–5 wt% in the first three types of composites, while the fourth type was synthesized by combining both alumina particles (2 wt%) and fibres. Of all the synthesized composites, the combination of alumina particles and fibres in the epoxy composites exhibited excellent tensile strength and modulus.

An interesting study by Swain and Biswas (2017) on the influence of moisture on different mechanical attributes of jute/epoxy with Al_2O_3 fillers at wet and dry conditions was conducted. The maximum impact and flexural attributes were 1.902 J and 72.94 MPa, respectively. The observations from the experimental results indicated that the mechanical properties began to decrease on absorbing water (Kane et al. 2016). A similar study by Sideridis et al. (2017) to examine the effect of water absorption on the flexural attributes of low-content iron particle–epoxy composites was performed. The flexural strength and strain witnessed a decrease with an increase in the filler content in the presence of water (Sideridis et al. 2017).

Aveen et al. (2019) developed the glass epoxy composite with aluminium powder, mother of pearl and fly ash powder as fillers. The study was conducted to determine the mechanical properties of fabricated composites by conducting flexural and tensile tests. In the total volume, the filler material was varied in percentage composition by 3, 6 and 9%. From the tests, it was found that the composite with aluminium filler material exhibited better tensile property while the fly ash filler composite exhibited better flexural property (Aveen et al. 2019). Sudheer et al. (2014) investigated the mechanical characteristics of potassium titanate whisker (PTW) incorporated epoxy composites. PTW inclusions showed a positive effect on hardness, density and stiffness properties of the composites (Sudheer et al. 2014). Investigation on the effects of polymer–filler and filler–filler interactions on mechanical as well as dynamic rheological properties of High-density polyethylene (HDPE)–wood composites were conducted. The results indicated that enhanced filler–filler interaction increased the complex viscosity and storage modulus of composites and decreased the mechanical properties. Enhanced polymer–filler interaction increased the complex viscosity and storage modulus and also improved the mechanical properties (Yang et al. 2010).

2.6 Environmental Benefits and Health Hazards

In modern days, there are hundreds of sources of industrial waste polluting the environment, consequently having damaging effects on the earth and its inhabitants. Also, owing to "ecological necessities and strict regulations", incorporating natural fibres in the place of synthetic fibres has become inevitable "for the manufacturers to accomplish new composite materials originated from renewable sources" (Arvinda Pandian et al. 2017). Hence the "whole gamut of engineering sector" has started shifting towards natural fibre-oriented materials from synthetic materials (Arvinda Pandian and Siddhi Jailani 2018).

Several natural fillers such as coconut coir, rice husk and wheat husk fillers were incorporated as fillers, and the mechanical attributes of filled glass/epoxy composites were characterized. Amongst the different composites, the composite with coconut coir had exhibited excellent mechanical attributes (Dhawan et al. 2013). In another study, ground walnut shells, organic waste fillers were used to modify the properties of epoxy composites. Composites with 20, 30, 40, and 50 wt% of walnut shell were prepared and tested. The composites containing the natural filler exhibited a considerable degree of improvement in hardness and stiffness, yet witnessed a decrease in the impact, and tensile properties (Salasinska et al. 2018). Bamboo, E-glass, and coconut shells were also used in polyester composites, and a significant improvement in the fatigue life and tensile strength of the composite was observed with the addition of coconut shell powder as a filler (Raja and Retnam 2019).

3 Effect of Fillers on Tribological Properties of Composites

3.1 Tribology

The understanding of the design, friction, wear, and lubrication of interacting surfaces in relative motion are described in the scope of tribological science. Almost every industrial part or component is subjected to different types of direct and indirect tribological loadings, namely adhesive and abrasive, during their service. Therefore, the tribological behaviour of materials becomes an essential criterion to be considered in the design of any mechanical part. Wear is the dominant reason behind material wastage and loss of mechanical, thermal, or electrical performance of the related application. Friction is the primary reason behind wear and energy loss because it affects the period of complete utilisation of the material. The two influential tribological properties of a mechanical system, i.e., friction and wear, are neither mutually exclusive nor completely inclusive. It is hardly discerned that tribology affects the life of people to a much greater extent than what is actually realized. Any reduction in wear or improvement friction control in an application can result in a considerable amount of savings. Several measures have been developed to reduce the coefficient of friction as much as possible by either eliminating or at least controlling the factors such as surface finish, temperature, operational load, relative speed, nature of relative motion between the surfaces, and lubrication characteristics. Friction in a mechanical system can generally be reduced by the use of sacrificial bearing surfaces made of wear-resistant or low shear materials, modifying the surface of the moving or stationary component by coating the surface, replace sliding friction by rolling friction, or improve the lubricity between the sliding surfaces by changing the viscosity, use of improved additives or suitable lubricants (Kumar and Srivastava 2016; Vinayagamoorthy 2018).

The huge amount of direct and indirect costs incurred by tribological deficiencies and failures dramatically affects the economy of a country. The expenditure mostly incurs due to the simultaneous loss of material and energy on every mechanical component in operation. Among the two characteristic properties in the tribological behaviour of a system, wear is a more critical factor compared to friction as excessive wear of a component in a machine may result in catastrophic failures and operational breakdowns, which can adversely impact productivity and hence, cost (Holmberg and Erdemir 2017; Tzanakis et al. 2012). There is a need to emphasize the importance of sustainable tribology in our era to showing that tribology is frequently the primary cause and, simultaneously, the solution for most of the mechanical maintenance problems. The development and deployment of modern tribological solutions for engineering systems can provide measurable beneficial, financial, and environmental outcomes to society and the industry.

Studies have revealed that 90% of failures in mechanical parts and components occur as a result of tribological loadings (Jost 2006). Therefore, a substantial sum of expenditure involved in repair costs can be lowered if a proper understanding of tribological principles and its applications are made. In tribology, the friction and wear of a material depend on some of the essential parameters such as surface roughness, relative motion, velocity, type of material, load, temperature, stick-slip, relative humidity, lubrication and vibration (Blanchet 2012). Nonetheless, effective modern methodologies for good tribological design can be very costly, yet efficient. An understanding of the nature of wear and friction in the system is essential to formulate the right mechanism to control their tribological behaviour. The criteria by which the wear life or frictional behaviour of a product is regulated may vary strongly across different segments of application. In some applications, the function is far more critical than manufacturing costs. One effective way of reducing wear and controlling friction is by the use of lubricants, which comprises of dry and wet lubricants. A wide range of liquids and soft solids have been studied to exhibit effective lubricant properties and have also proven to be cost-effective solutions. The apparent advantage of solid lubrication over oil lubricants is their superior cleanliness.

Composite materials emerged as a potential alternative for several classes of other materials in engineering technology, wherein two or more natural or artificial elements are combined by a series of chemical or physical processes to obtain a final product with added strength, efficiency or durability. The technical advantage of employing composites for manufacturing mechanical components or devices is that they can be prepared with the desired engineering properties by a careful selection of matrix and a compatible reinforcement. In the scope of tribology, composite materials are expected to provide a viable support structure to support the load and frictional heat. The inherent nature of the composite system exhibits mechanical stability and also fashions the attractions of means of dissipating frictional work. The continuous progress and innovations in modern technology are laying out newer rheological demands on lubricants, which are not often met by conventional lubricants (Dorri Moghadam et al. 2015; Omrani et al. 2016). On the other hand, self-lubricating composites have gained importance, as they can be tailored by using a base matrix, reinforcements, solid lubricants, or additives that may be required for the typical application. Advancements in the field of tribo-engineering have offered attractive solutions as self-lubricating composites, which are mostly compositions of metallic, ceramic or polymeric matrices added with functional fillers providing the desired tribological functionality (Erdemir 2005).

In general, the role of a filler in a composite material is versatile, and the nature of role purely is influenced by the composite matrix and the filler material itself. Some of the unique advantages of using fillers are to strengthen the matrix (load carrying capacity), improve the sub-surface crack arresting ability, enhancement of the thermal or electrical conductivity of the material, and providing a lubricating effect at the interface by decreasing the shear stress. Although fillers demonstrate superior behaviour in terms of the mechanical and tribological properties, the use of a specific type of fillers (e.g., particulate type) may somewhat affect the other properties; therefore, proper optimization of the mechanical and tribological properties of the composite has to be carried to avoid this disadvantageous compromise of the behaviour. Occasionally, in some cases, fillers are also used as cheap additives, as they reduce the material costs in polymer applications besides enhancing the tribological behaviour of the material (Bobby and Samad 2017; Saba et al. 2014; Senbet 2008).

3.2 *Common Testing Methods for Tribology Tests*

There is an endless number of methods that can be employed to characterize the tribological behaviour of a particular material. In order to make meaningful interpretations of the tribological outcomes, the most suitable test for the particular purpose needs to be selected. The nature of results obtained from a tribo test does not merely relate to the behavioural characteristics and properties of the composite but represents the mechanical system as well. Proper testing of composites using standardised methods can reveal information on the likely product life of the system. In the tribological characterisation of composites, the experimental attributes such as the matrix, type and structure of reinforcement, filler-matrix interface, and internal lubricants usually influence the measurement of wear and friction. The testing of wear and friction differ for different tribological systems, and therefore, composites need to be tested under standardised procedures before application of the actual results. Standard testing methods and devices have been elucidated in several references (Nirmal et al. 2011). Most of the methods have been exclusively published by the Society of Automotive Engineers (SAE) and American Society of Standards and Materials (ASTM) as standardised procedures for testing and evaluating material properties. Following the standard guidelines will enable the use of identical devices to obtain very nearly identical results facilitating comparability and correlatability.

Some of the commonly used tribological testing machines are dry sand rubber wheel (based on ASTM G65), pin on drum (based on ASTM A514), pin on disc (ASTM G99), Linear tribo-machine, Block on ring (ASTM G77, G137-95), Block on disc (ASTM G99), and 4-ball test (ASTM D2266) (Bhushan 2000; Singh et al. 2016). One of the most widely used methods for evaluating the tribological characteristics is the Pin-on-disc wear and friction testing that follows the methods as per the ASTM G99 standard. The specimen (pin) is held vertically or horizontally under loading against a rotating counterface (disc). The counterface exhibits a constant area of contact throughout the test as a result of which the method is more suitable for application involving sliding wear. The distinct operating parameters include applied load, sliding distance, wet or dry sliding condition, sliding velocity while the specimen contact area is maintained constant with resOpect to sliding time. The Pin on drum wear test rig is built based on ASTM A514 standard. The specimen travels horizontally (linearly) against a rotating drum that rotates at the desired speed using a drive chain. While the wear test can be conducted for both abrasive or adhesive conditions, wear simulates the applications of goods on rotating rollers or conveyor belts.

The Block on ring wear testing, which is based on ASTM G77 standard has a working principle very similar to that of Pin on disc testing method. The specimen is held against a rotating wheel or ring at 90° to the wheel or ring axis of rotation. The characteristic operating parameters include applied load, sliding velocity, temperature, sliding distance, wet or dry sliding condition but the specimen contact area is varied with respect to sliding time. This testing method simulates applications such as pulleys and camshafts. The Dry sand rubber wheel wear testing is based on the ASTM G65 standard. The specimen is held against a rubber wheel, while sand is introduced to the rubber interface to simulate an abrasive testing condition. The testing method can also be used for adhesive testing in the absence of sand. This kind of wear test typically simulates the applications such as tyres, bushes, bearings and rollers.

Although the standard methods of testing can produce adequate results to predict the failure mode or lifetime of a tribological system, the standard tests may not simulate the experience of the composite in the practical system. Although simple bench model tests can easily, quickly, and cheaply evaluate an extensive range of materials under well-controlled and simulated test conditions, they do carry some limitations. Generally, the practical representation and realism of the data and features of surface damage, and the prospects of making reliable inferences about the performance or usability from the tests to an application is somewhat decreased as we move from a field test to a simpler model test. The primary criterion for confirming whether a chosen model test is applicable or not is based on to what extent the wear mechanisms appearing in the application during the actual service conditions are being reproduced. Moreover, it is critical to comprehend the actual mechanism of wear for accurately characterizing the tribological behaviour of the tested materials.

3.3 Wear and Frictional Behaviour of Filled Composites

The addition of filler materials to composites is mainly targeted towards improving the thermal, mechanical and tribological properties. In the scope of improving the tribological properties, the ability of the fillers lies mainly in influencing the development of transfer film and counterface adhesion. Composites are classified based on the type of the primary matrix, or reinforcement or the nature of the interface while the fillers used in composites are generally categorised into metallic fillers, ceramic fillers, polymeric fillers, and mineral fillers. Most of the time, micro- and nano-sized inorganic fillers are used for modifying the tribological behaviour of composites. The systemic effect of a filler on the tribological behaviour of a composite only becomes noticeable when the filler is at the surface. The ability of fillers to reduce friction depends on the degree of fineness of the filler particles, as they need to be fine enough to effectively increase the deflection temperature of the matrix near the surface because frictional heating and matrix softening are two factors responsible for high friction. In terms of the contribution of fillers towards wear resistance, the proportion of the filler in the matrix will determine the property. In general, for composites, the best abrasion resistance and low coefficient of friction will be obtained with those fillers providing the highest packing fraction (P_f) or at concentrations approaching P_f. With the development of mature technology, industries have begun using fillers in diverse applications as multifunctional additives rather than just cost-reducing expedients. Powdered Teflon, graphite, micro-size CuO, silica, $CaCO_3$, PTFE and molybdenum sulphide (MoS2) and so on, are some of the prominent fillers used in composites for improving the tribological behaviour. A considerable amount of work on the evaluation of the tribological behaviour of composites has been carried out previously. The scope of this review is limited to briefly examining the various kinds of fillers used in composites for improving their tribological behaviour.

3.3.1 Ceramic Fillers

Ceramics, such as metallic oxides, carbides, borides, nitrides, silicides, etc., are known to retain their mechanical properties at higher temperatures. This inherent property of ceramics has been exploited in ceramic fillers which are primarily compounds of the aforementioned inorganic groups. The properties of the filler and their functionality comes from the inherent nature of the base ceramic itself. The inherent advantages of ceramics in terms of their high hardness, strength, melting point, and abrasion resistance make them a much-preferred variety of fillers in composites operating in high-temperature applications. While the most usual type of bonding observed in ceramics is a combination of ionic and covalent, Van-der-Waals forces or a metallic component may also be present depending on the type of ceramic. Some of the commonly used ceramic fillers in composites are AlN, Al_2O_3, SiC, Si_3N_4, $Sr_2Ce_2Ti_5O_{16}$, zirconium silicate ($ZrSiO_4$), wollastonite ($CaSiO_3$), silicon dioxide (SiO_2), beryllium oxide (BeO), CeO_2, boron nitride (BN), and ZnO.

Addition of ceramic particulate, whisker or microfibre fillers to composites or nanocomposites have revealed a dramatic improvement in the wear resistance of composites as much as three times the magnitude of the unfilled composites. Nanometer-sized particles have been extensively used as fillers for obtaining superior tribological characteristics. Hong-Bin Qiao et al. (2007) investigated the wear and friction properties of the Al_2O_3 particles (5%) filled polyetheretherketone (PEEK) and polytetrafluoroethylene (10%) PEEK against medium carbon steel (AISI 1045 steel) ring under dry sliding conditions. The PEEK composites exhibited a notable decrease in the wear rate with the addition of the nanometer and micron-sized Al_2O_3 (in the absence of PTFE) but had less effect on the friction coefficient. In contrast, the wear rate, as well as friction coefficient, had significantly lowered for the pure PEEK composite filled with 10 mass % of PTFE. But, when 10 mass % PTFE was filled into the Al_2O_3/PEEK composites, the behaviour contradicted the expectation wherein the coefficient of friction decreased while the wear rate increased (Qiao et al. 2007).

A comparative evaluation of using SiC and Al_2O_3 on the wear behaviour of jute/epoxy composite was conducted by varying the weight percent of fillers with respect to the resin. In the absence of fillers, the jute–epoxy composites were easily subjected to wear under higher normal loads and sliding velocities whereas the addition of ceramic fillers displayed considerably lowered the wear rate of jute epoxy composites. The lowest coefficient of friction was observed in 15 wt% filled jute–epoxy composites which complemented its higher wear resistance. Furthermore, Al_2O_3 filled jute epoxy composites exhibited a lower coefficient of friction and wear loss compared to SiC filled composites for all compositions. The wear mechanism of the in-filled was characterized by microcracking, pit and debris formation in contrast to the unfilled composite, which was dominated by fibre breakage and plastic deformation (Sabeel Ahmed et al. 2012).

In another study, the effect of different loads and abrasion distance was investigated for Polyamide 66/Polypropylene (PA66/PP) blend, nano clay filled PA66/PP and short carbon fiber reinforced PA66/PP nanocomposites based on the three-body abrasive wear mechanism. The addition of nano clay/short carbon fiber and PA66/PP had exhibited a sound effect on the wear rate for different abrading distance and loads. On comparison of the effects of their addition, nano clay/short carbon fibre composites demonstrated a lower wear rate than short carbon fiber filled PA66/PP composites (Ravi Kumar et al. 2009). The possibility of using Rice bran ceramics (RBC) as fillers were explored, as the hard-porous carbon material, made from rice bran provide low friction and high wear resistance characteristics (Shibata et al. 2014). The tribological properties of the thermoplastic resin-based RBC composites were experimentally determined for polyamide 66 (PA66), polyamide 11, polyoxymethylene, polybutylene terephthalate, and polypropylene matrix resins. Higher wear resistance and lower friction levels were observed for thermoplastic-based RBC compared to their pure resins. Also, the RBC fillers had the upper hand over Glass fibre (GF) fillers, as a substantial improvement in the strength of the composites was noticed whereas the friction coefficient and wear did not witness any effect. The RBC particles, in addition to improving the fracture toughness of the composite, also contributed to a decrease in the friction coefficient, resulting in mild wear. Thus, the RBC based composited indicated their great potential as an anti-wear hard particulate filler in the industry (Shibata et al. 2014, 2012). So far, four different forms of glasses, namely plain GFs, hollow glass beads (GBs), solid GBs, and glass flakes have been incorporated in glass-filled thermoplastic composites. Of all the forms, the lowest wear rate was observed for GF-filled composites and solid GB-filled composites, whereas the hollow GB-filled composite showed the highest wear (Klaas et al. 2005).

Metallic compounds such as ZnO, TiO_2, CuO have been exploited as filler owing to their desirable tribological properties. Addition of ceramic fillers in polymers had improved the performance during abrasive wear. The effectiveness of the filler to function as a wear-resisting material is determined by factors such as the content of filler added, the interaction of filler matrix and the type of the matrix used (Suresha et al. 2010). Nanosized CuO filled and short carbon (CF) and aramid (Kevlar) fibres-reinforced polyphenylene sulfide (PPS) composites prepared by compression moulding were investigated for their tribological behaviour using a pin-on-disc apparatus. In case of the filled composite, a steady-state wear rate was observed while it was reduced to half of that in the case of filled CF reinforcement composite. However, the addition of CuO filler had not contributed to improvement in the wear resistance of the fabric-reinforced composites owing to the poorly developed transfer film in the presence of fibres and as a result, the composites became fragile (Bahadur and Polineni 1996). Li et al. (2002) reported the use of nano ZnO as filling material to PTFE to reduce the wear rate of the polymer. The wear rate obtained for the composite containing 15 vol% nano ZnO was determined as the optimum level of nano ZnO to be incorporated, however a higher coefficient of friction was observed for the nanocomposite than that of the unfilled PTFE (Li et al. 2001). Short carbon fibre (SCF), graphite flakes, and microparticles of TiO_2 and ZnS were used as fillers in thermoplastic composites, such as polyetheretherketone (PEEK) and polyetherimide (PEI). The tribological characteristics of the two types of high-temperature-resistant filled thermoplastic composites were evaluated under dry sliding conditions against steel counterparts. The addition of fillers like SCF and graphite flakes demonstrated considerable improvement in the wear resistance and the load-carrying capacity of the base polymers.

Nonetheless, with the addition of microparticles of TiO_2 and ZnS, the coefficient of friction and wear rate of the composites had further reduced especially at high temperatures (Chang et al. 2007). TiO_2 and ZrO_2 were incorporated as ceramic fillers in the reinforcement of bamboo-glass hybrid polymer composites. These compounds have identical morphology and were incorporated as fillers with particular emphasis on exploring their effect on the tribological behaviour of reinforced thermosets (including micro-particle). Among the two inorganic fillers, the comparison of the wear characteristics of bamboo-glass-epoxy hybrid composites revealed that ZrO_2 fillers gave better strength and wear-resistant properties when compared to the TiO_2 filled composite (Latha and Rao 2018).

The potential of Silicon carbide (SiC) or boron carbide (B_4C) ceramic fillers was investigated by introducing them into three-dimensional needled carbon fibre to prepare ceramic modified carbon/carbon (C/C) composites. On observing their morphology, the pore size distribution was found to be uniform on the addition of the ceramic fillers in the C/C composites. Further, the ceramic modification considerably lowered the COF fade in seawater conditions for C/C composites. C/C-SiC_f displayed superiority over C/SiC for its excellent stable friction behaviour without any fade in seawater conditions. Also, a cumulative effect of lubrication from water film and SiO_2 film seemed to increase the COF fade for C/SiC compared to that of C/B_4C (Cai et al. 2013). A combination of nanostructured fillers based on BN and SiO_2 micro powders was used to reinforce $AK_{12}M_2MgN$ alloy. The structure of the fillers led to a uniform dispersion of the structural components of the alloy and resulted in a considerable increase in its wear resistance and a decrease in the coefficient of friction (Komarov et al. 2013). A synergistic effect was observed for epoxy based composited filled with varying concentrations of short carbon fibre (SCF) and solid lubricants, i.e. PTFE and graphite. These composited were also supplemented with the addition of varying amounts of sub-micron sized TiO_2 (300 nm). In the investigation by Zhang et al. (2004), it was inferred that the synergistic effect was subjected to tribological characteristics of the composites in comparison with monolithic systems (Zhang et al. 2004).

Nanoparticles of SiO_2, SiC, Si_3N_4 and ZrO_2 used as fillers have demonstrated to be very effective in lowering the coefficient of friction and specific wear rate of PEEK composites when sliding against the steel counter surfaces (Wang et al. 2000, 1996). Apart from these, nano laminated Ti_3SiC_2, granite dust and powder, porcelain waste, nanodiamond and Ti_3AlC_2TiC particles and fly ash have also been as fillers in metal matrix and polymer matrix composites.

3.3.2 Carbon and Organic Fillers

Carbon black and carbon fibres have been used as additives in a wide range of thermoplastic, and thermoset resins. The layered structure of graphite, which is characterized by a weak Van-Der Waals force between the layers possess a high electrical conductivity. The nature of the graphite structure also helps in reducing friction and wear by forming a lubricant film between the mating surfaces, which is utilized in its application as a filler. Carbon black, on being incorporated as fillers in composites showed attractive tribological characteristics. The addition of carbon black fillers (at 1–5% filling) along with PTFE composites exhibited a tremendous increase in the wear resistance of up to 700 times. The excellent dispersion of fillers in the PTFE matrix enhanced the interaction effect of carbon black in the dry conditions of testing against smooth metal surfaces.

Moreover, the wear rate of a PTFE composite mainly depends on the properties of an ultra-disperse filler, such as its specific surface area. With an increase in the specific surface area of the added filler, the wear rate witnesses a proportional increase due to the increase in the area of interphase interaction of the components per unit mass of the filler and high activity at its surface centres (Aderikha and Shapovalov 2010). Carbon black derived from wood apple shell, obtained by pyrolysis at 400 °C was also used as a filler in an epoxy resin. Carbon black particulates composite showed minimum wear on comparison with raw particulate composite (Ojha et al. 2014). Hybrid fillers containing multi-walled CNTs and carbon black in natural rubber were also investigated for dry friction and wear behaviour. It was inferred that a considerable increase in the wear resistance and unexpected reduction in the friction coefficient resulted from the increase in the applied load for every fixed sliding speed (Ojha et al. 2014). The investigations on the mechanism of wear in the low-filled PTFE-CB composites revealed a delamination mode of wear wherein the changes in the wear resistance was relatable with the structural transformations of the composite.

The exceptional properties of the single and multi-walled carbon nanotubes (CNTs) favour them as a recognized filler for epoxy-based composites and coatings. Various boating and automotive applications utilised a wide range of reinforced polymer–matrix composites because of the capacity of these composites to withstand longer durations of sliding contact-based wear conditions. The tribological response of an epoxy matrix of polyamide filled with graphite and/or carbon nanotubes was highlighted by the enhanced wear resistance which was owing to the carbon fillers dispersed in the epoxy matrix. Of all the carbon-based fillers, the most desirable result was obtained for the composited filled with TCNTs. The TCNTs were well dispersed in the epoxy matrix due to the presence of the NH_2 groups. Despite the enhanced wear resistance exhibited by EpCNTs and Ep-Graphite, Ep-Hybrid demonstrated relatively much inferior tribological properties. Subsequently, even the combination of graphite and CNT fillers did not witness any synergistic effect (Sakka et al. 2017). Sam-Daliri et al. (2019) found a novel method to improve the tribological behaviour of an unsaturated polyester matrix wherein a relatively small amount (0.2 wt%) of well-dispersed MWCNTs within the wood flour polyester composite was used a filler in the polyester matrix. The enhanced tribological behaviour was attributed to the transfer of a soft layer of wood flour on the worn-out area during the wear process, which acted as a self-lubricating material (Nabinejad et al. 2019). Composites were developed with a novel combination of multiwalled carbon nanotubes (MWCNTs) and short carbon fibres (SCF) as fillers for an automotive brake system. All the combinations of the composite with the carbon fiber had shown superior properties of lower wear rate and coefficient of friction (Gbadeyan et al. 2018).

Researchers have extensively used carbon fibres for their desirable mechanical properties. Surface modified carbon nanofibres (CNF) filled PTFE composites were subjected to tribological characterization under dry sliding conditions. The CNF was treated with HNO_3 followed by coupling agent treatment before incorporation to the PTFE matrix. The optimum content of CNF in the PTFE resulted in reduction of the wear rate, almost 30% lower than that of untreated CNF filled PTFE for an applied load of 200 N (Shi et al. 2008). Suresha et al. (2010) compared the wear characteristics of carbon–epoxy and graphite filled carbon–epoxy composites at different loads in abrasive condition using different grades of SiC abrasive paper (150 and 320 grit size). Improved abrasion resistance was witnessed for graphite filled carbon-epoxy composites investigated for different loads and distances of abrasion. The improvement in the properties corresponds to the filler to filler interaction and uniformity in the distribution of the added fillers in the carbon-epoxy matrix (Suresha et al. 2010). The abrasion and attrition wear of experimental composites related to dental application with samples of different resin viscosities was investigated. Although raising the resin viscosity lowered the wear resistance, it had minimal influence on composites holding nonbonded nanofiller. However, an increase in resin viscosity increased abrasion and attrition in composites containing silanated nanofiller, with equivocal effects in composites containing unsilanated nanofiller (Musanje et al. 2006).

Any discussion on carbon-based fillers is incomplete without highlighting the use of graphene/graphene oxide fillers. The tribological properties of metal matrix composites containing graphene were reported in several papers earlier. The functional advantage of using graphene fillers is that multilayer graphene and reduced graphene oxide can not only enhance the strength drastically but also reduce the friction coefficient and wear rate of composites (Li et al. 2017a, b; Llorente et al. 2019). Carbon fibre composites have gained much importance in marine, sports, construction industries besides aerospace applications.

3.3.3 Metallic Fillers

Mild steel substrates coated with metallic fillers such as micro-nickel, aluminium, silver and zinc powders have demonstrated remarkable tribological properties and surface energy characteristics. The attractive results exhibited by the metallic filler-based composites has encouraged several other metallic powders to be explored as a viable possibility. Three types of metallic fillers, namely steel fibers, brass fibers and copper powder, were used in the preparation of non-asbestos organic (NAO) composites. The tribological properties of the composites were evaluated for different loads and speeds. Although the addition of metallic fillers led to enhancement in friction performance of the composites, for every subsequent increase in the amount of the metallic filler, the wear resistance exhibited an increase. The higher wear resistance due to increasing metallic contents resulted in an increase in the thermal conductivity

(TC). Especially, copper filler (10%) based composites showed significant performance from both wear and friction properties followed brass, while iron powder based composite showed moderate behaviour (Kumar and Bijwe 2010).

Researchers attempted to investigate the influence of various metallic fillers (Cu, steel, or Al) on the friction and wear performance of brake pad composites. The experiments were conducted using a small-scale friction tester against two counter disks (grey cast iron and aluminium metal matrix composite (Al-MMC)) at ambient and elevated temperature ranges. The ambient temperature tests against grey cast iron disc revealed that the composites with Cu fibers showed a noticeable negative $\mu-\nu$ (friction coefficient versus sliding velocity) inferring that stick-slip may occur at low speeds. At higher temperatures, wear tests showed that the Cu-fiber composites exhibited better wear resistance than the other composites. The tests with steel fibers revealed that they were not well-matched with Al-MMC disks due to substantial material loss and irregular friction behaviour during sliding at elevated temperatures (Jang et al. 2004). The use of copper and its alloys to increase the thermal diffusivity at the friction interface has become a common practice. The natural tendency of copper to endure the high temperatures attained at the interface for higher levels of the friction coefficient is taken as an advantage. During high temperatures, the copper oxide formed at the interfacial layer dissipates the frictional heat effectively. Hence, copper and its alloys are added as fillers to regulate the friction level while avoiding the fierceness against the counterpart. Likewise, aluminium fibers are also added to the composites in applications where aluminium metal matrix composite (Al-MMC) brake rotors are used (Wilson and Alpas 1996; Urquhart 1991).

Steel fibres have also been presented as a potential replacement for other fillers that lack the mechanical properties as that of steel. Qu et al. (2004) examined the wear and frictional characteristics of composites filled with continuous steel fibres. The fibre orientations are hosted concerning the sliding direction, namely parallel (P) normal (N) and antiparallel (AP) along the fibre direction. The wear rates were found to increase with changing sliding directions and exhibited a dependency on the stability of the film. The friction coefficient (range 0.49 to 0.54) also remained dependent on the fibre alignment directions. At higher temperatures, the iron-rich transfer film formed on the specimen promotes adhesive interaction, and steady-state friction is observed (Qu et al. 2004). The wear and friction characteristics of the friction materials reinforced by brass fibres against grey cast iron demonstrated superior wear resistance of the composite. The interacting elements showed fatigue wear mechanism which was confirmed based on the following inferences: (i) formation of a copper transfer film on the friction surface of the grey cast iron counterpart (ii) the worn surface of the counterpart revealed fatigue cracks. The wear loss and coefficient of friction increased slightly when the mass fraction of brass fibres was over 19% (Xian and Xiaomei 2004). Compression moulding technique to add the carbon fibres and nanofillers like Al and Zn as reinforcements to epoxy hybrid composites proved to be effective in reducing the specific wear rate up to 0.5 wt. % of Al/Zn filler loading (Divya and Suresha 2018).

Apart from the above mentioned, steel wool, nickel, silver, brass fibres, copper powder, sodium are the other types of metallic fillers used in the composites. Besides, mineral silicates (Vasilev et al. 2019), talc (Zhao et al. 2012), feldspar (Cai et al. 2015), nano clays (Bobby and Samad 2017), calcium carbonate are also being used as fillers for many polymer applications. Polyvinyl chloride, polyolefins, phenolics, polyesters, and epoxies are all resins compatible with $CaCO_3$ as fillers, which is not only attributed to economic but performance considerations as well (Palanikumar et al. 2019).

3.3.4 Polymeric Fillers

Polymers have become more of a commodity and relatively less expensive these days. Addition of fillers to reduce cost has become a less significant factor because by judiciously combining the filler with the resin, one can achieve a spectrum of materials with properties intermediate between those of the two ingredients. Polymeric fillers are generally classified into natural and synthetic varieties, wherein natural polymers consist of cellulose fibres, graphite fibres, wood flour, flax, cotton and starch. Synthetic polymers are not available in the natural environment, so these are tailored polymer materials catering to different applications, namely polyamide, polyethene, polytetrafluoroethylene, polyester, aramid, and polyvinyl alcohol fibres. Polymeric fillers usually limit their usage to controlling permeability, instilling softness to the matrix, damping control, and imparting the desired tribological behaviour to the composite system.

Polymer-based fillers such as fluorinated polyether ketone (aryl ether ketone) (FPEK) and polytetrafluoroethylene (PTFE) have also been used to modify the properties of epoxy-based composites. The nature of FPEK tends to render the composite extremely resistant to scratch and as a result, its presence in the corresponding matrix, the composite resists even higher degree of abrasion (Saba et al. 2014). The friction coefficient of the composite is considerably lowered with the addition of PTFE fillers to the composite. PTFE is a commercial filler that could reduce the frictional coefficient, and, due to this fact, sometimes also the wear rate of polymeric composites is reduced. The unique molecular and morphological structure of PTFE renders the polymer its extraordinary tribological properties. A third-body transfer film is formed when the polymer slides against the steel counterparts, which is the mechanism behind the lowering of the frictional coefficient (Zhang et al. 2004; Şahin 2018).

4 Recent Advances, Challenges and Future Trends

Composite materials have become a sustainable alternative to conventional materials used in automotive, aerospace and other industries in the last few years. The speciality of this technology is that the end product material can be engineered with desired specific properties by a careful combination of matrix and additives. Although the composites offer many attractive properties, it is unquestionable that the manufacturing process involved in commercializing these materials present new challenges day by day. On the other hand, the technological capabilities and diversity of applications of different reinforced composites have steadily increased. The broadening spectrum of applications of composites is driven by the need to replace an existing expensive material, usually a metal. While most of the metals are facing pressing demand and acute shortage in metal processing industries, traditional composites have been facing a need for up-gradation to fulfil the demand for superior properties and high performance. In this aspect, nanocomposites are evolving as a promising solution to cater to the needs and requirements of modern industries. The advantages of the newer generation of composites are their ability to improve two or more desired properties simultaneously. The tribological research on the nanocomposites, particularly hybrid nanocomposites and polymer nanocomposites is still at an early stage and has great potential to cater to many domestic and industrial applications.

Traditionally, most fillers were considered as additives, considering their contribution to a composite only focused on reducing their cost. However, the diversity of applications and a broad spectrum of their usage has led to high demand for fillers or reinforcing fillers in composite technology. Although the diversity of fillers has crossed countable limits, the challenge is to select the most suitable and appropriate filler by understanding the performance criteria of the composite being developed. Several commercial polymer composites survived the rigorous market demands, because of the right blending of ingredients to maximize the magnitude of the desired property and simultaneously mitigating the loss of others. Apart from these, filler manufacturers have to necessarily take environmental concerns, such as recycling, sustainability and life cycle impact into consideration. Natural, preferably biological and organic fillers are gaining more dominance over synthetic owing to their environmentally friendly properties. There is no doubt that more emphasis will be imposed to improve the durability and viability of the fillers used in the future. Many reasons such as lack of familiarity, limited availability have so far prevented large scale penetration of natural fillers. However, tremendous growth is still anticipated with large companies entering the market and technological advances being made.

Thus, a stable and healthy advancement in tribological research of nanocomposites is anticipated to take over the current scenario as the trend is moving towards developing novel, green measures, chemical or physical alterations of improving the tribological behaviour of polymer composites (pristine, hybrid and nanocomposites).

5 Conclusion

In this chapter, the different fillers used in the composite have been reviewed broadly for the first time. The most important contribution of the current work is identifying various fillers used in the composite materials and its influence on the tribological behaviour of the composite. It is expected that the review would be beneficial to the researchers seeking guidance on different types of fillers for the preparation of newly formulated composite materials. The roles, classifications, processing methods and the influences of different categories of fillers in the composites are discussed intricately. The challenges faced during the fabrication of composited are highlighted. The recent advancements in the processes of incorporating the fillers in composites and the future possibilities of the research have also been briefly reviewed.

References

Aderikha VN, Shapovalov VA (2010) Effect of filler surface properties on structure, mechanical and tribological behavior of PTFE-carbon black composites. Wear 268(11–12):1455–1464

Arvinda Pandian CK, Siddhi Jailani H (2018) Investigation of viscoelastic attributes and vibrational characteristics of natural fabrics-incorporated hybrid laminate beams. Polym Bull 75(5):1997–2014

Arvinda Pandian CK, Siddhi Jailani H (2019) Dynamic and vibrational characterization of natural fabrics incorporated hybrid composites using industrial waste silica fumes. Int J Polym Anal Charact 24(8):721–730

Arvinda Pandian CK, Siddhi Jailani H, Rajadurai A (2017) Natural fabric sandwich laminate composites: development and investigation. Bull Mater Sci 40(1):139–146

Asadi P, Givi MKB, Abrinia K, Taherishargh M, Salekrostam R (2011) Effects of SiC particle size and process parameters on the microstructure and hardness of AZ91/SiC composite layer fabricated by FSP. J Mater Eng Perform 20(9):1554–1562

Ashik KP, Sharma RS, Raghavendra N (2017) Effect of filler on mechanical properties of natural fiber reinforced composites. Asian J Chem 29(8):1697–1701

Ashworth S, Rongong J, Wilson P, Meredith J (2016) Mechanical and damping properties of resin transfer moulded jute-carbon hybrid composites. Compos Part B Eng 105:60–66

Aveen KP, Bhajantri V, D'Souza R, Londe NV, Jambagi S (2019) Experimental analysis on effect of various fillers on mechanical properties of glass fiber reinforced polymer composites. AIP Conf Proc 2057(January)

Bahadur S, Gong D (1992) The role of copper compounds as fillers in the transfer and wear behavior of polyetheretherketone. Wear 154(1):151–165

Bahadur S, Polineni VK (1996) Tribological studies of glass fabric-reinforced polyamide composites filled with CuO and PTFE. Wear 200(1–2):95–104

Bajwa SG, Bajwa DS, Holt G, Coffelt T, Nakayama F (2011) Properties of thermoplastic composites with cotton and guayule biomass residues as fiber fillers. Ind Crops Prod 33(3):747–755

Bakhsheshi-Rad HR, Hamzah E, Ismail AF, Daroonparvar M, Yajid MAM, Medraj M (2016) Preparation and characterization of NiCrAlY/nano-YSZ/PCL composite coatings obtained by combination of atmospheric plasma spraying and dip coating on Mg-Ca alloy. J Alloys Compd 658:440–452

Bhushan B (2000) Modern tribology handbook: volume one: principles of tribology. Mod Tribol Handb Vol One Princ Tribol 1–1697

Blanchet TA (2012) Friction and wear of polymer materials. Handb Lubr Tribol II Theory Des Second Ed 34-1–34-14

Bobby S, Samad MA (2017) Enhancement of tribological performance of epoxy bulk composites and composite coatings using micro/nano fillers: a review. Polym Adv Technol 28(6):633–644

Cai Y, Yin X, Fan S, Zhang L, Cheng L (2013) Tribological behavior of three-dimensional needled ceramic modified carbon/carbon composites in seawater conditions. Compos Sci Technol 87:50–57

Cai P, Li Z, Wang T, Wang Q (2015) Effect of aspect ratios of aramid fiber on mechanical and tribological behaviors of friction materials. Tribol Int 92:109–116

Chang L, Zhang Z, Ye L, Friedrich K (2007) Tribological properties of high temperature resistant polymer composites with fine particles. Tribol Int 40(7):1170–1178
Chen K, Jia B, Liu X (2017) Comparison and analysis dissimilar joint strength of pultrusion GFRP composites. DEStech Trans Eng Technol Res (icaenm):184–191
Chiellini E, Solaro R (1996) Biodegradable polymeric materials. Adv Mater 8(4):305–313
Cisneros-López EO, González-López ME, Pérez-Fonseca AA, González-Núñez R, Rodrigue D, Robledo-Ortíz JR (2017) Effect of fiber content and surface treatment on the mechanical properties of natural fiber composites produced by rotomolding. Compos Interfaces 24(1):35–53
Ciupan Eet al (2017) Characterization of a thermoforming composite material made from hemp fibers and polypropylene. MATEC Web Conf 137
Dhawan V, Singh S, Singh I (2013) Effect of natural fillers on mechanical properties of GFRP composites. J Compos 2013:1–8
Divya GS, Suresha B (2018) Role of metallic nanofillers on mechanical and tribological behaviour of carbon fabric reinforced epoxy composites. Mater Sci Appl 9(9):740–750
Dorri Moghadam A, Omrani E, Menezes PL, Rohatgi PK (2015) Mechanical and tribological properties of self-lubricating metal matrix nanocomposites reinforced by carbon nanotubes (CNTs) and graphene—a review. Compos Part B Eng 77:402–420
Erdemir A (2005) Review of engineered tribological interfaces for improved boundary lubrication. Tribol Int 38(3):249–256
Gbadeyan OJ, Kanny K, Pandurangan MT (2018) Tribological, mechanical, and microstructural of multiwalled carbon nanotubes/short carbon fiber epoxy composites. J Tribol 140(2)
Graupner N, Ziegmann G, Wilde F, Beckmann F, Müssig J (2016) Procedural influences on compression and injection moulded cellulose fibre-reinforced polylactide (PLA) composites: Influence of fibre loading, fibre length, fibre orientation and voids. Compos Part A Appl Sci Manuf 81(October 2015):158–171
Grinev VG et al (2018) The effect of filler type on the mechanical properties of composite materials based on ultra-high-molecular-weight polyethylene. Polym Sci Ser D 11(2):202–208
Holmberg K, Erdemir A (2017) Influence of tribology on global energy consumption, costs and emissions. Friction 5(3):263–284
Jang H, Ko K, Kim SJ, Basch RH, Fash JW (2004) The effect of metal fibers on the friction performance of automotive brake friction materials. Wear 256(3–4):406–414
Jost HP (2006) Tribology: how a word was coined 40 years ago. Tribol Lubr Technol 62(3):24–28
Kane SN, Mishra A, Dutta AK (2016) Preface: international conference on recent trends in physics (ICRTP 2016). J Phys Conf Ser 755(1)
Kano Y, Akiyama S (1996) Estimation of surface tension and surface segregation of poly(ethyl acrylate)/poly(vinylidene fluoride-co-hexafluoro acetone) blends. Polymer (Guildf) 37(20):4497–4503
Klaas NV, Marcus K, Kellock C (2005) The tribological behaviour of glass filled polytetrafluoroethylene. Tribol Int 38(9, SPEC. ISS.):824–833
Komarov AI, Komarova VI, Shipko AA, Ovchinnikov VV, Kovaleva SA (2013) Effect of phase composition of nanostructured refractory modifier on structure and tribological behavior of AK12M2MgN alloy. J Frict Wear 34(5):329–338
Košíková B, Gregorová A, Osvald A, Krajčovičová J (2007) Role of lignin filler in stabilization of natural rubber–based composites. J Appl Polym Sci 103(2):1226–1231
Kumar M, Bijwe J (2010) Role of different metallic fillers in non-asbestos organic (NAO) friction composites for controlling sensitivity of coefficient of friction to load and speed. Tribol Int 43(5–6):965–974
Kumar P, Srivastava VK (2016) A review on wear and friction performance of carbon-carbon composites at high temperature. Int J Appl Ceram Technol 13(4):702–710
Lai D, Wei Y, Zou L, Xu Y, Lu H (2015) Wet spinning of PVA composite fibers with a large fraction of multi-walled carbon nanotubes. Prog Nat Sci Mater Int 25(5):445–452
Latha PS, Rao MV (2018) Investigation into effect of ceramic fillers on mechanical and tribological properties of bamboo-glass hybrid fiber reinforced polymer composites. Silicon 10(4):1543–1550

Li F, ao Hu K, lin Li J, yuan Zhao B (2001) The friction and wear characteristics of nanometer ZnO filled polytetrafluoroethylene. Wear 249(10–11):877–882
Li Z, Okamoto K, Ohki Y, Tanaka T (2011) The role of nano and micro particles on partial discharge and breakdown strength in epoxy composites. IEEE Trans Dielectr Electr Insul 18(3):675–681
Li Z, Gao Y, Moon KS, Yao Y, Tannenbaum A, Wong CP (2012) Automatic quantification of filler dispersion in polymer composites. Polymer (Guildf) 53(7):1571–1580
Li JF, Chen B, Shi Q, Li C, Chu Y, Li C (2017) Tribological properties of novel Cu/NbSe2 composites reinforced with reduced graphene oxide filler. Chalcogenide Lett 14(11):499–510
Li Q, Imanishi N, Takeda Y, Hirano A, Yamamoto O (2002) PEO-based composite lithium polymer electrolyte, PEO-BaTiO 3-Li (C 2 F 5 SO 2) 2 N. Ionics 8(1–2):79–84
Li Y, Wang S, Wang Q (2017) Enhancement of tribological properties of polymer composites reinforced by functionalized graphene. Compos Part B Eng 120:83–91
Llorente J, Ramírez C, Belmonte M (2019) High graphene fillers content for improving the tribological performance of silicon nitride-based ceramics. Wear 430–431(April):183–190
Matuana LM, Park CB, Balatinecz JJ (1998) Cell morphology and property relationships of microcellular foamed pvc/wood-fiber composites. Polym Eng Sci 38(11):1862–1872
Mohanty A, Srivastava VK, Sastry PU (2014) Investigation of mechanical properties of alumina nanoparticle-loaded hybrid glass/carbon-fiber-reinforced epoxy composites. J Appl Polym Sci 131(1):1–7
Muraliraja R, Arunachalam R, Al-fori I, Al-maharbi M, Piya S (2018) Development of alumina reinforced aluminum metal matrix composite with enhanced compressive strength through squeeze casting process. Proc IMechE Part L J Mater Des Appl 1–8
Musanje L, Ferracane JL, Ferracane LL (2006) Effects of resin formulation and nanofiller surface treatment on in vitro wear of experimental hybrid resin composite. J Biomed Mater Res Part B Appl Biomater 77(1):120–125
Nabinejad O, Liew WYH, Debnath S, Rahman ME, Cao C, Davies IJ (2019) Tribological behavior of unsaturated polyester hybrid composites containing wood flour and carbon nanotubes. SN Appl Sci 1(7):1–9
Naidu PP, Raghavendra G, Ojha S, Paplal B (2019) Effect of g-C3N4 nanofiller as filler on mechanical properties of multidirectional glass fiber epoxy hybrid composites. J Appl Polym Sci 48413(Figure 2):1–9
Nassar MMA, Arunachalam R, Alzebdeh KI (2017) Machinability of natural fiber reinforced composites: a review. Int J Adv Manuf Technol 88(9–12):2985–3004
Nirmal U, Hashim J, Lau STW (2011) Testing methods in tribology: a review. Reg Tribol Conf (RTC), Langkawi, Malaysia 6(3):221–229
Ojha S, Acharya SK, Gujjala R (2014) Characterization and wear behavior of carbon black filled polymer composites. Procedia Mater Sci 6(Icmpc):468–475
Omrani E, Moghadam AD, Menezes PL, Rohatgi PK (2016) Influences of graphite reinforcement on the tribological properties of self-lubricating aluminum matrix composites for green tribology, sustainability, and energy efficiency—a review. Int J Adv Manuf Technol 83(1–4):325–346
Palanikumar K, AshokGandhi R, Raghunath BK, Jayaseelan V (2019) Role of calcium carbonate(CaCO3) in improving wear resistance of polypropylene(PP) components used in automobiles. Mater Today Proc 16:1363–1371
Palutkiewicz P, Trzaskalska M, Bociąga E (2019) The influence of blowing agent addition, talc filler content, and injection velocity on selected properties, surface state, and structure of polypropylene injection molded parts. Cell Polym 35(4):159–192
Pandian A, Jailani S (2019) Development and investigation of jute/linen fibre reinforced polymer composite. SAE Tech Pap (October)
Qiao HB, Guo Q, Tian AG, Pan GL, Xu LB (2007) A study on friction and wear characteristics of nanometer Al_2O_3/PEEK composites under the dry sliding condition. Tribol Int 40(1):105–110
Qu X, Zhang L, Ding H, Liu G (2004) The effect of steel fiber orientation on frictional properties of asbestos-free friction materials. Polym Compos 25(1):94–101

Raja DBP, Retnam BSJ (2019) Effect of short fibre orientation on the mechanical characterization of a composite material-hybrid fibre reinforced polymer matrix. Bull Mater Sci 42(3)

Ravi Kumar BN, Suresha B, Venkataramareddy M (2009) Effect of particulate fillers on mechanical and abrasive wear behaviour of polyamide 66/polypropylene nanocomposites. Mater Des 30(9):3852–3858

Saba N, Tahir PM, Jawaid M (2014) A review on potentiality of nano filler/natural fiber filled polymer hybrid composites. Polymers (Basel) 6(8):2247–2273

Sabeel Ahmed K, Khalid SS, Mallinatha V, Amith Kumar SJ (2012) Dry sliding wear behavior of SiC/Al2O3 filled jute/epoxy composites. Mater Des 36:306–315

Şahin Y (2018) Dry wear and metallographic study of PTFE polymer composites. Mech Compos Mater 54(3):403–414

Sakka MM, Antar Z, Elleuch K, Feller JF (2017) Tribological response of an epoxy matrix filled with graphite and/or carbon nanotubes. Friction 5(2):171–182

Salasinska K, Barczewski M, Górny R, Kloziński A (2018) Evaluation of highly filled epoxy composites modified with walnut shell waste filler. Polym Bull 75(6):2511–2528

Sam-Daliri O, Faller LM, Farahani M, Roshanghias A, Araee A, Baniassadi M, ... Zangl H (2019) Impedance analysis for condition monitoring of single lap CNT-epoxy adhesive joint. Int J Adhes and Adhes 88:59–65

Senbet D (2008) Measuring the impact and international transmission of monetary policy: a factor-augmented vector autoregressive (favar) approach. Eur J Econ Financ Adm Sci 7(13):121–143

Shi Y, Feng X, Wang H, Lu X (2008) The effect of surface modification on the friction and wear behavior of carbon nanofiber-filled PTFE composites. Wear 264(11–12):934–939

Shibata K, Yamaguchi T, Urabe T, Hokkirigawa K (2012) Experimental study on microscopic wear mechanism of copper/carbon/rice bran ceramics composites. Wear 294–295:270–276

Shibata K, Yamaguchi T, Hokkirigawa K (2014) Tribological behavior of polyamide 66/rice bran ceramics and polyamide 66/glass bead composites. Wear 317(1–2):1–7

Sideridis E, Venetis J, Kyriazi E, Kytopoulos V (2017) Influence of moisture absorption on the flexural properties of composites made of epoxy resin reinforced with low-content iron particles. Bull Mater Sci 40(4):805–817

Singh RA, Jayalakshmi S, Gupta M (2016) Indian J Adv Chem Sci Wear Frict Adv Compos (Table 1):165–168

Srirangan AK, Paulraj S (2016) Multi-response optimization of process parameters for TIG welding of Incoloy 800HT by Taguchi grey relational analysis. Eng Sci Technol Int J 19(2):811–817

Sudheer M, Prabhu R, Raju K, Bhat T (2014) Effect of filler content on the performance of epoxy/PTW composites. Adv Mater Sci Eng (2014)

Suresha B, Chandramohan G, Prakash JN, Balusamy V, Sankaranarayanasamy K (2006) The role of fillers on friction and slide wear characteristics in glass-epoxy composite systems. J Miner Mater Charact Eng 5(1):87–101

Suresha B, Ramesh BN, Subbaya KM, Ravi Kumar BN, Chandramohan G (2010) Influence of graphite filler on two-body abrasive wear behaviour of carbon fabric reinforced epoxy composites. Mater Des 31(4):1833–1841

Swain PTR, Biswas S (2017) Effect of moisture absorption on the mechanical properties of ceramic filled Jute/Epoxy hybrid composites. IOP Conference Series: Materials Science and Engineering, vol 178, 012010

Tang H et al (2019) Correlation between failure and local material property in chopped carbon fiber chip-reinforced sheet molding compound composites under tensile load. Polym Compos 40(S2):E962–E974

Tzanakis I, Hadfield M, Thomas B, Noya SM, Henshaw I, Austen S (2012) Future perspectives on sustainable tribology. Renew Sustain Energy Rev 16(6):4126–4140

Urquhart AW (1991) Novel reinforced ceramics and metals: a review of Lanxide's composite technologies. Mater Sci Eng: A 144:75–82

Vasilev AP et al (2019) Mechanical and tribological properties of polytetrafluoroethylene composites with carbon fiber and layered silicate fillers. Molecules 24(2)

Vinayagamoorthy R (2018) Friction and wear characteristics of fibre-reinforced plastic composites. J Thermoplast Compos Mater
Wang Q, Xue Q, Liu H, Shen W, Xu J (1996) The effect of particle size of nanometer ZrO2 on the tribological behaviour of PEEK. Wear 198(1–2):216–219
Wang QH, Xue QJ, Liu WM, Chen JM (2000) Effect of nanometer SiC filler on the tribological behavior of PEEK under distilled water lubrication. J Appl Polym Sci 78(3):609–614
Wilson S, Alpas AT (1996) Effect of temperature on the sliding wear performance of Al alloys and Al matrix composites. Wear 196(1–2):270–278
Wulfsberg J, Herrmann A, Ziegmann G, Lonsdorfer G, Stöß N, Fette M (2014) Combination of carbon fibre sheet moulding compound and prepreg compression moulding in aerospace industry. Procedia Eng 81(October):1601–1607
Wypych G (2016) Fillers in different processing methods. Handb Fill 793–821
Wypych G (2016) Fillers in different products. In: Handbook of fillers, p 851
Xian J, Xiaomei L (2004) Friction and wear characteristics of polymer-matrix friction materials reinforced by brass fibers. J Mater Eng Perform 13(5):642–646
Yang Z, Peng H, Wang W, Liu T (2010) Crystallization behavior of poly(ε-caprolactone)/layered double hydroxide nanocomposites. J Appl Polym Sci 116(5):2658–2667
Yoo HM, Kwon DJ, Park JM, Yum SH, Lee W Il (2017) Mechanical properties of norbornene-based silane treated glass fiber reinforced polydicyclopentadiene composites manufactured by the S-RIM process. E-Polymers 17(2):159–166
Zhang Z, Breidt C, Chang L, Haupert F, Friedrich K (2004) Enhancement of the wear resistance of epoxy: Short carbon fibre, graphite, ptfe and nano-tio2. Compos Part A Appl Sci Manuf 35(12):1385–1392
Zhao G, Wang T, Wang Q (2012) Studies on wettability, mechanical and tribological properties of the polyurethane composites filled with talc. Appl Surf Sci 258(8):3557–3564

Effects of Lubrication on Tribological Properties of Composite MoS_2-TiO_2 Coating Material

Avinash V. Borgaonkar, Ismail Syed, and Shirish H. Sonawane

Abstract Molybdenum disulphide (MoS_2) is widely used in various applications because of its lubricating properties. However, its performance needs to further improve. In the present work Molybdenum disulphide (MoS_2) based composite coating with addition of TiO_2 (having different particle size and weight% addition) were developed and coated on AISI52100 steel substrate. The substrate specimens were pre-treated using phosphating to improve the porosity which helps to enhance the bond strength between the coating and steel substrate. A tribological study of this developed composite coating was carried out at different contact pressures and sliding speeds using the pin-on-disc test rig. It was observed that particle size of TiO_2 and its different wt% significantly affects the tribological properties of the developed composite coating. In case of all considered operating conditions with developed samples of composite MoS_2-TiO_2 coating, the sample C (63.3 nm particle size with 15 wt% addition) of TiO_2 depicts about 31% lower coefficient of friction and 39% lower wear rate compared to pure MoS_2 coating.

Keywords Coating · Composite · MoS_2 · Tribological properties · TiO_2

1 Introduction

As a part of surface modification techniques, bonded solid lubricants exhibits excellent tribological properties. Solid lubricants can be effectively employed in unusual circumstances like space application, where liquid lubricants are prohibited due to

A. V. Borgaonkar · I. Syed (✉)
Department of Mechanical Engineering, National Institute of Technology Warangal, Warangal, Telangana 506004, India
e-mail: syedismail7@nitw.ac.in

A. V. Borgaonkar
e-mail: avi.borgaonkar@student.nitw.ac.in

S. H. Sonawane
Department of Chemical Engineering, National Institute of Technology Warangal, Warangal, Telangana 506004, India
e-mail: shirish@nitw.ac.in

M. T. Hameed Sultan et al. (eds.), *Tribological Applications of Composite Materials*, Composites Science and Technology, https://doi.org/10.1007/978-981-15-9635-3_10

their out gassing under vacuum and high temperature (Erdemir 2000), food and textile industries where products are likely to get contaminated with liquid lubricants and greases (Menczcs et al. 2013), at high load and cryogenic temperatures, in strong radiation fields and in corrosive environments (Sharma and Anand (2016). Most commonly MoS_2, graphite and PTFE have been used as solid lubricants in industrial applications. Among these, MoS_2 solid lubricant has the highest load carrying capacity (Gadow and Scherer 2002) and low coefficient of friction (COF) (Xu et al. 2003; Fridrici et al. 2003).Therefore, MoS_2 has been widely used in different applications to reduce the COF and wear rate (Chen and Jia 2001; Xu et al. 2007).

Luo et al. (2011) and Shen et al. (2017) performed rotational fretting wear test of bonded MoS_2 coating on the medium carbon steel substrate. The counter body material is 52100 steel. The study depicts that in comparison with the uncoated specimen, MoS_2 coated specimen exhibits lower COF and wear rate due to its lubricating property. Hiraoka (2001) performed tribological study of the journal bearings with bonded MoS_2 solid lubricant in air as well as vacuum environment. The shaft and bearing bush made up of steel. Before application of the coating, both the substrate surfaces were pre-treated using sand-blasting method. The MoS_2 coating along with phenolic resin binder was applied on the pre-treated surface by spraying technique. The experimental results show that the MoS_2 coating exhibits high endurance life in vacuum as compared with air. This is due to low COF in vacuum. The COF is directly related to the shear stress at the interface of coating and substrate. Another similar kind of study was carried out by Ye et al. (2009) on bonded composite MoS_2-Sb_2O_3 coating using reciprocating tribometer. The substrate made up of AISI 1045 steel was pre-treated using sand-blasting process to improve the bond strength between substrate and coating. The composite MoS_2-Sb_2O_3 coating along with novolac epoxy resin binder was applied on the pre-treated surface by spraying technique. The bonded MoS_2 coating exhibits excellent tribological properties under fretting condition. With increase in load, the coating material transferred onto the counterbody surface which in turns reduced COF. In addition the composite MoS_2-Sb_2O_3 coating possesses high load carrying capacity.

Zhu et al. (2003) studied the tribological performance of the sputtered MoS_2 and composite MoS_2-Ti coatings under fretting and pin-on-disk test rig using different counter body material such as steel and corundum. The test results show that MoS_2 exhibits better tribological performance in fretting with a corundum counterbody as compared to steel. This is due to the low shear strength of the MoS_2 which results into formation of a transfer film onto a counterbody surface under fretting conditions. The MoS_2-Ti composite coating exhibits better tribological performance with pin-on-disk tests as compared to MoS_2. The addition of Ti improves the wear resistance of the coating without affecting the lubrication property of the MoS_2. Arslan et al. (2004) and Bulbul and Efeoglu (2010) confirms that in comparison with the pure MoS_2, composite MoS_2-Ti coating improves the endurance life. Since the addition of Ti in MoS_2, the hardness of the coating material improved which in turn enhances endurance life. The hardness of composite coating improved due to denser and stronger microstructure of Ti. Shang et al. (2018) developed multilayered composite MoS_2/Pb-Ti coating and studied its tribological and corrosion resistance properties. The obtained results were compared with the pure MoS_2 coating to analyse the performance of the both coatings. The results show that multilayered composite MoS_2/Pb-Ti coating exhibits better tribological performance in ambient as well as in high humid environment, since the doping of Ti improves the hardness at the same time dual doping improves the microstructural properties which reduces the COF. However, pure MoS_2 coating exhibits a high COF and low endurance life in high humid environment or used after long-time storage. The multilayered composite MoS_2/Pb-Ti coating also exhibits better corrosion resistance properties in comparison with pure MoS_2 by inhibiting the permeation of oxygen and other corrosive elements in both ambient and in high humid environment.

Kubart et al. (2005) carried out the tribological analysis of MoS_2 and $MoSe_2$ coatings deposited on steel substrate by magnetron sputtering. The study was performed using ball-on-disc high temperature tribometer considering different humidity. The obtained results reveal that the COF of $MoSe_2$ was not affected by humidity. However, the wear rate of $MoSe_2$ in dry air was significantly higher as compared with MoS_2. Moreover in case of humid air sudden increase in the wear rate of MoS_2 has been observed. Whereas the wear rate of $MoSe_2$ remained unaffected. Shankara et al. (2008) performed tribological analysis of different bonded solid lubricants. The different solid lubricants used for analysis were pure MoS_2, composite MoS_2-ZrO_2 and MoS_2-graphite-ZrO_2. The study was carried out in ambient air at room temperature and at 200 °C temperature. At ambient condition the moisture present in air absorbed into the coating and leads to deteriorate its tribological performance. At high temperature due to evaporation of moisture the tribological performance has been significantly improved. The comparative study shows that MoS_2-ZrO_2 and MoS_2-graphite exhibits better tribological performance than the pure MoS_2. The ZrO_2 helps to improve the endurance life of the coating due to improved hardness whereas the graphite is a soft material which helps to reduce the COF values. Vadiraj et al. (2012) performed the tribological analysis of four different solid lubricants such as MoS_2, boric acid, TiO_2 and graphite. The tests performed at constant load varying the sliding speed. The obtained results compared with pure dry condition. The tests result reveals that MoS_2 and graphite helps to reduce the wear rate by 30–50% compared to dry condition at all sliding speeds because of good lubricity and adherence to the surface. The reduction in COF was observed with increase in sliding speed in case of all solid lubricants compared to dry condition. It may because of with increase in speed the temperature and shear force goes on increasing which results into reorientation of lubricant layers. Out of these solid lubricants boric acid experiences high COF and wear rate because of conversion of boric acid into abrasive boric oxide.

Essa et al. (2017) studied the friction and wear characteristics of composite MoS_2-ZnO coating material. The composites prepared by addition of 10% MoS_2 into steel, 10% ZnO into steel and each MoS_2 as well ZnO 10% into the steel. The tribological analysis of these composites have been carried out and compared with the pure steel specimens. The study has been carried out at different temperatures ranging from room temperature to 800 °C. The test results reveal that due to synergetic effect of composite MoS_2-ZnO exhibits enhanced lubrication performance at wide range of temperatures. In all temperature range the composite steel-MoS_2-ZnO exhibits lowest COF whereas composite steel-MoS_2 exhibits lower wear rate as compared to pure steel and other composite materials. Ramamoorthy et al. (2017) analysed the machining performance employing three different conditions such as wet machining with cutting fluid as a lubricant, dry machining without any lubricant and dry machining with solid lubricant. In this study graphite has been used as a solid lubricant. The test results reveal that surface finish has been significantly improved due to lowered COF at the tool workpiece interface with solid lubricant assisted machining in comparison with other two conditions.

Till now the attempts have been made to enhance the tribological properties of the bonded MoS_2 coating with different composition of metals (MoS_2/Pb-Ti, MoS_2-Ti), selenide (MoS_2-WSe_2) and oxides (MoS_2-ZnO, MoS_2-ZrO_2). However, very few researchers reported the effect of particle size and wt% addition of doping material into the base coating material. The tribological behaviour of such composites is still dramatic in nature, and meticulous examination of the contacting surfaces coated with such composite films (with different wt% addition and particle size of other doping material into MoS_2 base matrix) under various operating conditions is still needed.

In the present study, a tribological analysis of developed composite coating MoS_2-TiO_2 (with in-house synthesized TiO_2 having different particle size and with different wt% addition) was carried out on pin-on-disc friction and wear test rig under different operating conditions like contact pressure, speed. Eventually, the findings of this study will enable us to understand the effect of microstructure of coating and its composition on the tribological properties of the composite coating.

2 Materials and Methods

2.1 Materials

In order to develop pure MoS_2 coating, MoS_2 (with average particle size 170–200 nm) powder was purchased from Sisco Research Laboratories Pvt. Ltd., Mumbai, India. Further to prepare the composite MoS_2-TiO_2 coating, the TiO_2 has been synthesized. For the synthesis of TiO_2, the different chemicals were used such as titanium tetra isopropoxide (TTIP), acetone, methanol and sodium hydroxide (NaOH) were purchased from S. D. Fine Chemicals Ltd., Mumbai, India. The procured chemicals were used as in received condition from the supplier unless stated otherwise. A magnetic stirrer and reactor equipped with ultrasonic horn (M/s Dakshin Ltd., Mumbai, India) with 20 mm tip diameter made up of titanium were used separately to synthesize different sizes of TiO_2.

2.2 *Characterization*

For characterization different instruments have been used, those are elaborated in following sub-section as below.

2.2.1 Particle Size Analyser

The Particle Size (PSA) size of nano-dispersions analysis was done using Malvern Zetasizer (Nano S90 version 7.02). Dilute solution of TiO_2 was characterized using dynamic light scattering method.

2.2.2 Surface Roughness Tester

The surface roughness of the pin (before pre-treatment process) and the counter face disc surface (before wear test) was measured using the Surtronic S-100 (Taylor Hobson) Series Surface Roughness Tester. The required surface roughness was achieved by polishing with different grades of silicon carbide papers.

2.2.3 Scanning Electron Microscopy (SEM)

After application of the coating, the morphological study and coating thickness were examined by scanning electron microscopy (SEM) in a higher resolution field emission gun microscope, Hitachi-4800 (Tokyo, Japan), equipped with an energy dispersive x-ray analysis (EDAX) detector (Bruker, XFlash4100, Billerica, MA, USA).

2.2.4 Vickers Microhardness Tester

The Microhardness of the coated samples is measured using ECONOMET VII-1 MD Microhardness tester supplied by Chennai Metco Pvt. Ltd., (Chennai, India) equipped with testing force range 10–1000 gf and automatic loading–unloading dwell period 50–60 s. The maximum resolution is 0.0625 μm.

3 Development and Application of the Composite Coating

3.1 *Development of Composite MoS_2-TiO_2 Coating*

The development of composite MoS_2-TiO_2 includes different steps such as synthesis of TiO_2 powder, particle size analysis of synthesis of TiO_2 powder, mixing of TiO_2 with MoS_2 (having different particle size and wt% addition).

3.1.1 Synthesis of Titanium Dioxide (TiO_2) Particles

The titanium dioxide exists in different phases out of which the rutile phase is chemically stable. TiO_2 is non-toxic. It does not react with any other material at ambient temperature and not soluble in water. It has good thermal stability. Rutile titanium dioxide has a melting point of 1850 °C. The titanium is having high strength and wear resistance. Due to these advantages, TiO_2 is selected as a reinforcement material to form a composite coating. In order to study the influence of TiO_2 particle size on the tribological performance of the composite MoS_2-TiO_2 coating, the different particle size of TiO_2 powder have been synthesized by different methods.

The typical procedure reported by Hernandez-Perez et al. (2012) for synthesizing TiO_2 nanoparticles is by using ultrasound approach described using a flow chart as shown in Fig. 1.

3.1.2 Particle Size Analysis

The average hydrodynamic diameter and poly-dispersity index (PDI, Malvern definition) were finding out by quasi-elastic light scattering. Before putting the samples to analysis it was diluted to 1/300 in De-ionized (DI) water and bath sonicated for 5–10 min for uniform dispersion of particles and it was used for size analysis. The parameters used for carrying put the analysis mentioned in Table 1.

The particle size distribution for the different synthesized TiO_2 samples obtained from the particle size analyser and depicted in Fig. 2.

From the particle size analysis, the size of the synthesized TiO_2 nanoparticles were estimated and displayed in Table 2.

3.2 *Application of the Composite MoS_2-TiO_2 Coating*

The application of coating involves different steps such as pre-treatment of the substrate surface, bonding of the composite MoS_2-TiO_2 coating using sodium silicate as a binder on the pre-treated surface.

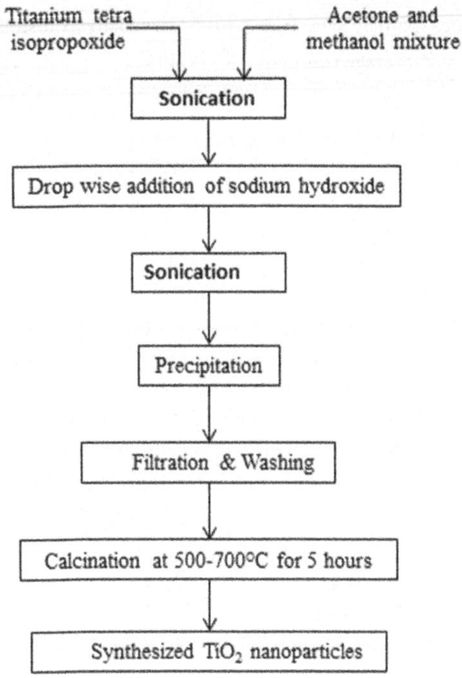

Fig. 1 Synthesis procedure for TiO_2

Table 1 Parameters of particle size analysis

Dispersant name	Water
Dispersant refractive Index	1.33
Viscosity	0.8872
System temperature	25 °C

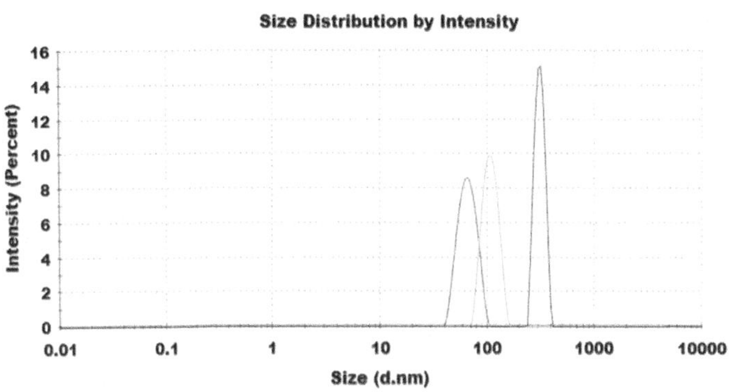

Fig. 2 Particle size analysis of synthesized TiO_2 samples

Table 2 Z-Average values (particle size) from particle size analysis histogram

TiO_2 sample	Z-average (nm)	PDI	% intensity
Sample A	301.5	0.054	95
Sample B	102.6	0.047	91
Sample C	63.3	0.041	93

3.2.1 Pre-treatment Process

The pre-treatment process has been used to enhance the bond strength between coating and substrate. As per the previous findings, different pre-treatment processes such as salt-bath nitriding, sand-blasting, shot peening, micro-arc oxidation, phosphating and abrasive blasting process were used. Among these, more emphasis was given by Duszczyk et al. (2018) on the phosphating as it creates micro-porous which helps to trap the solid lubricant into the interstices between the phosphate crystals which results in enhanced bonding strength as reported by Rajagopal and Vasu (2000). Due to this advantage, phosphating was used as a pre-treatment process in the present analysis. For the phosphating, the required chemicals such as phosphoric acid (H_3PO_4), magnesium carbonate ($MgCO_3$), sodium nitrate ($NaNO_2$), and sodium hydroxide (NaOH) were used. These were purchased from Sigma Aldrich Ltd., Mumbai, India.

3.2.2 Phosphating Process

Sankara Narayanan (2005) elaborated about phosphating process which involves different steps such as degreasing, pickling, rinsing, phosphating, rinsing and drying. The phosphating process was carried out as mentioned by Pokorny et al. (2016).

3.2.3 Bonding of the Composite MoS_2-TiO_2 Coating on the Pre-treated Substrate Surface

In the present study sodium silicate (Na_2SiO_3) is used as a binder to hold together the MoS_2 and TiO_2 powders in order to apply on the substrate surface. While forming the gel type mixture it is found that, less amount of Na_2SiO_3 fails to absorb the powders, whereas excess amount causes poor adhesion. After carrying number of trials, optimum proportions of MoS_2 and Na_2SiO_3 for pure MoS_2 coating have been found out (by wt% 1:2.2).

In case of composite coating to finalise optimum % of TiO_2 to be mixed with MoS_2 which will provide better tribological properties, the % of TiO_2 varied from 5 to 25%. In proportion to that amount of Na_2SiO_3 was added. These prepared coating applied onto the phosphated steel samples by brushing. These coated samples dried and cured at 150 °C for two hours in a furnace.

4 Experimental Analysis

4.1 Pin on Disc Friction and Wear Test Rig

The friction tests were performed using pin-on-disk tribometer (Magnum, India) and referring to ASTM Standard G99-95 (2010). A schematic set up of the pin-on-disk test rig is shown in Fig. 3. The disk material is EN-31 and comprised of 165 mm diameter and thickness of 8 mm. The substrate material for pin is AISI 52100 steel (12 mm diameter and 25 mm length).

While estimating the tribological performance parameters of the coating, such as friction coefficient (COF) and wear rate the sliding distance was kept constant as 3000 m for every test. This was achieved by changing the disc speed and the track radius. The weight loss i.e. (difference between initial weight and final weight) is used to estimate the wear rate of the coating. In order to get the reliable data each experiment was repeated three times.

Before performing each test, the disks surface was cleaned using acetone, and then dried thoroughly. The tests have been performed as per the detailed test parameters and the operating conditions mentioned in Table 3.

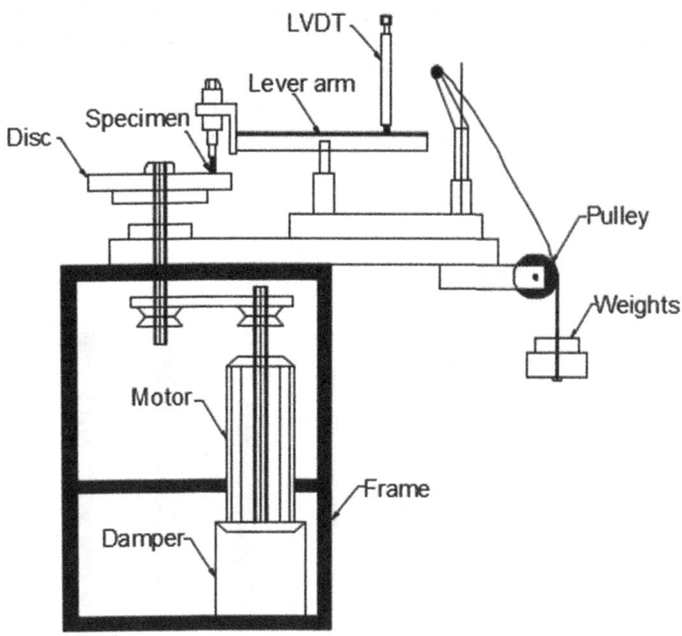

Fig. 3 Pin-on-disc friction and wear test rig schematic set up

Table 3 Test parameters and ranges

S. No.	Parameters	Operating conditions
1	Normal load	176, 442, 707 kPa
2	Sliding speed	1, 2, 3 m/s
3	Track radius	40 mm
4	Sliding distance	3000 m

5 Results and Discussion

The tribological properties such as COF and wear rate of the composite MoS_2-TiO_2 coating have been studied. While analysing the performance, the effect of particle size and different wt% addition of TiO_2 has been considered and elaborated in the following sub-section.

5.1 *Effect of Particle Size and Different Wt% of* TiO_2 *on COF*

The effect of particle size and wt% addition of TiO_2 on the COF at a contact pressure of 176 kPa and a sliding speed of 1 m/s is shown in Fig. 4. Form the test results it has been observed that composite coating exhibits better COF as compared to pure MoS_2 coating. In composite MoS_2-TiO_2 coating, the 15 wt% addition of TiO_2 shows a lower COF. In addition, at 15 wt% of TiO_2, a lower particle size (Sample C) of TiO_2

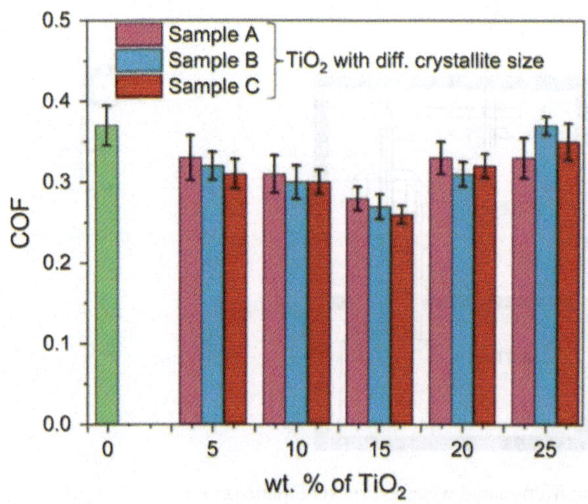

Fig. 4 Effect of addition of different wt% TiO_2 on COF with operating conditions 176 kPa contact pressure and 1 m/s sliding speed

depicts a further reduction in COF among the other considered particle size samples. The Sample C at 15 wt% addition of TiO_2 exhibits near about 31% reduction in COF value as compared to the pure MoS_2 coating. The pure MoS_2 coating shows higher values of COF as compared to other samples due to poor bonding between the substrate and coating, the coating was worn out at a faster rate from the substrate (see Fig. 6a). This is in line with the previous study reported by Shankara et al. (2008) and Shang et al. (2018), where they reported that pure MoS_2 coating exhibits low wear resistance due to which it worn out at a higher rate. The addition of TiO_2 into MoS_2 base matrix resulted into improved tribological properties of composite MoS_2-TiO_2 coating. However, the higher amount of addition reinforcement material leads to improper mixing between the additive and base matrix which leads to poor bonding due to which a reverse trend in the tribological properties was observed as reported by Ding et al. (2010).

5.2 *Effect of Particle Size and Different Wt% of TiO_2 on Wear Rate*

The effect of particle size and wt% addition of TiO_2 on the wear rate at a contact pressure of 176 kPa and a sliding speed of 1 m/s is shown in Fig. 5. Form the test results it has been observed that composite coating exhibits higher wear resistance as compared to pure MoS_2 coating. In composite MoS_2-TiO_2 coating, the 15 wt% addition of TiO_2 shows a lower wear rate. In addition, at 15 wt% of TiO_2, a lower particle size (Sample C) of TiO_2 depicts a further reduction in wear rate among the other considered particle size samples. The Sample C at 15 wt% addition of TiO_2

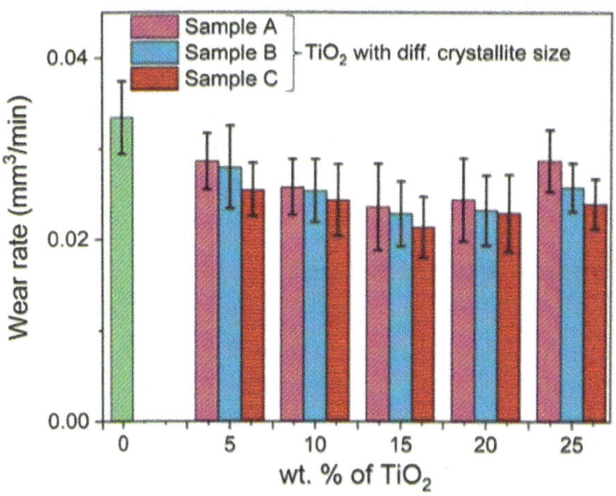

Fig. 5 Effect of addition of different wt% TiO_2 on wear rate with operating conditions 176 kPa contact pressure and 1 m/s sliding speed

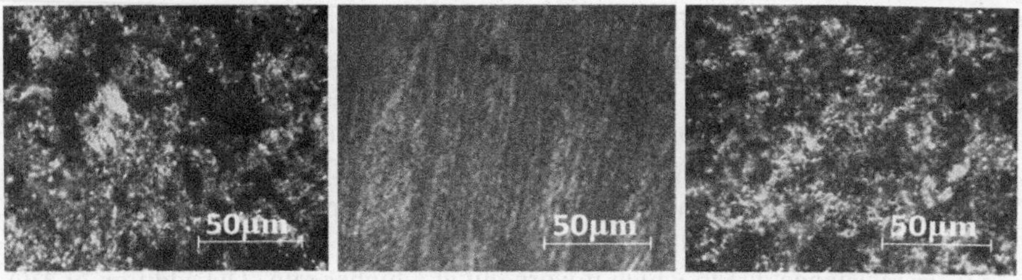

Fig. 6 Pin surface after wear test. **a** Pure MoS_2, composite coating with TiO_2 addition of **b** 15 wt%, **c** 25 wt%

exhibits near about 39% reduction in wear rate value as compared to the pure MoS_2 coating.

It can be depicted from Fig. 6b that in composite coating with the addition of TiO_2 upto 15 wt%, the coating still remains on the pin surface after the test, due to which it exhibits lower COF and wear rate. However, for higher concentration i.e. 25 wt% of TiO_2 the coating has been worn out from the substrate as shown in Fig. 6c. This may be due to improper mixing of TiO_2 particles into MoS_2 matrix which may lead poor bonding between them and lead to an increase in COF as well as wear rate.

In order to confirm this EDAX analysis of the pure as well as composite MoS_2-TiO_2 coating before and after wear test has been carried out with same operating conditions. Figure 7 shows the EDAX spectrum and mapping analysis for different elements present onto the coated surface. The initial spectrum (a) represents different elements which constitute for the composite MoS_2-TiO_2 coating with 15 wt% addition of TiO_2 (i.e. TiO_2 with sample C) before wear test. It shows that different element present such as Mo, Ti, Na, Si, O and C and the spectrum (c) for the same coating after wear test. While the other spectrum (b) represents EDAX spectrum for pure MoS_2 coating after wear test which shows some traces of Fe along with above mentioned elements. The EDAX analysis clearly shows that under same operating conditions pure MoS_2 coating wears out at a faster rate due to poor bonding strength. After worn out of the coating, asperity contact between the contacting surfaces has taken place.

6 Conclusion and Future Scope

Within the scope of this study, a successful attempt has been made for the development of composite MoS_2-TiO_2 coating material with different wt% of TiO_2 and particle size. The tribological properties of the developed composite coating at different contact pressure and sliding speed have been investigated. The test results reveal that, in comparison with the application of pure MoS_2 coating, the composite MoS_2-TiO_2 coating exhibits excellent tribological performance in all considered operating conditions due to the synergistic effect of both MoS_2 and TiO_2. At ambient conditions,

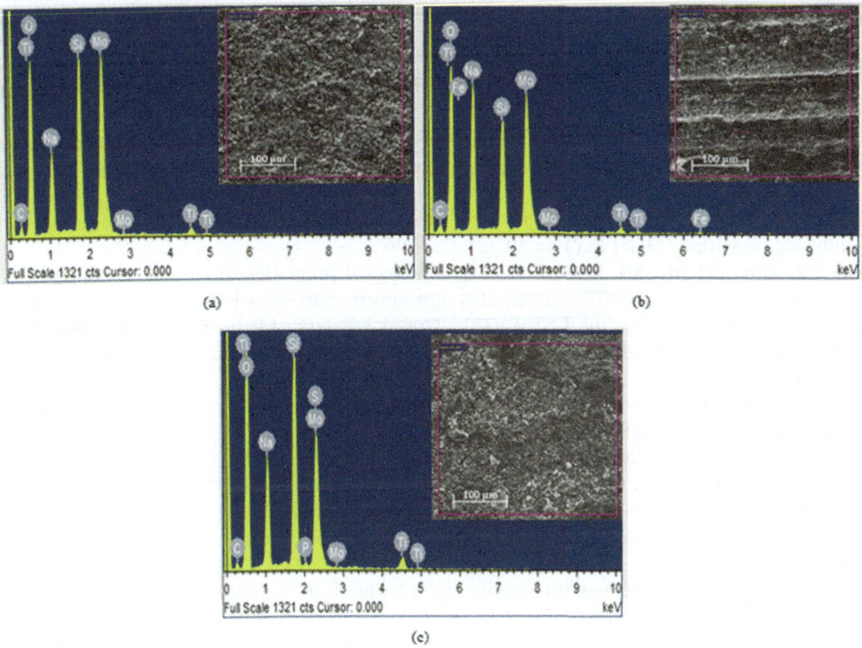

Fig. 7 EDAX spectrum **a** before wear test and **c** after wear test for composite MoS_2-TiO_2 coating with 15 wt% addition of TiO_2 (sample C), **b** pure MoS_2

the introduction of TiO_2 into the MoS_2 matrix helps to improve the bond strength without spoiling the lubricating property of MoS_2 and in turns lead to enhance the endurance life of the coating. In comparison with all considered combinations, the sample C (63.3 nm particle size) with 15 wt% of TiO_2 depicts the lowest COF and wear rate. The enhanced tribological performance of composite MoS_2-TiO_2 coating turns into prominent replacement coating material as compared to pure MoS_2 coating. Further research for the improvement of the performance of the composite MoS_2-TiO_2 coating is needed to explore the followings:

- To study the tribological performance of composite MoS_2-TiO_2 coating with different environmental conditions such as high temperature, vacuum and moisture.
- To study the influence of this composite MoS_2-TiO_2 coating on the tribological performance in combination with surface texturing.

References

Arslan E, Bulbul F, Efeoglu I (2004) The structural and tribological properties of MoS_2-Ti composite solid lubricants. Tribol Trans 47(2):218–226

ASTM AS. G99 (2010) Standard test method for wear testing with a pin-on-disk apparatus, pp 1–5

Bulbul F, Efeoglu I (2010) MoS_2-Ti composite films having (002) orientation and low Ti content. Crystallogr Rep 55(7):1177–1182

Chen JM, Jia JH (2001) Investigation of the tribological behavior of an aluminum alloy with embedded materials. Mater Sci Eng 302(2):222–226

Ding XZ, Zeng XT, He XY, Chen Z (2010) Tribological properties of Cr-and Ti-doped MoS_2 composite coatings under different humidity atmosphere. Surf Coat Technol 205(1):224–231

Duszczyk J, Siuzdak K, Klimczuk T, Strychalska-Nowak J, Zaleska-Medynska A (2018) Manganese phosphatizing coatings: the effects of preparation conditions on surface properties. Materials 11(12):1–22

Erdemir A (2000) Solid lubricants and self-lubricating films. In: Bhushan B (ed) Modern tribology handbook. CRC Press, US, pp 4–39

Essa FA, Zhang Q, Huang X, Ali MKA, Elagouz A, Abdelkareem MA (2017) Effects of ZnO and MoS_2 solid lubricants on mechanical and tribological properties of M50-steel-based composites at high temperatures: experimental and simulation study. Tribol Lett 65(3):1–29

Fridrici V, Fouvry S, Kapsa P, Perruchaut P (2003) Impact of contact size and geometry on the lifetime of a solid lubricant. Wear 255(7–12):875–882

Gadow R, Scherer D (2002) Composite coatings with dry lubrication ability on light metal substrates. Surf Coat Technol 151:471–477

Hernandez-Perez I, Maubert AM, Rendon L, Santiago P, Herrera-Hernandez H, Diaz-Barriga Arceo L, Garibay Febles V, Palacios Gonzalez E, Gonzalez-Reyes L (2012) Ultrasonic synthesis: structural, optical and electrical correlation of TiO_2 nanoparticles. Int J Electrochem Sci 7:8832–8847

Hiraoka N (2001) Wear life mechanism of journal bearings with bonded MoS_2 film lubricants in air and vacuum. Wear 249(10–11):1014–1020

Kubart T, Polcar T, Kopecky L, Novak R, Novakova D (2005) Temperature dependence of tribological properties of MoS2 and MoSe2 coatings. Surf Coat Technol 193(1–3):230–233

Luo J, Zhu MH, Wang YD, Zheng JF, Mo JL (2011) Study on rotational fretting wear of bonded MoS_2 solid lubricant coating prepared on medium carbon steel. Tribol Int 44(11):1565–1570

Menezes PL, Nosonovsky M, Ingole SP, Kailas SV, Lovell MR (eds) (2013) Tribology for scientists and engineers. Springer, New York

Pokorny P, Tej P, Szelag P (2016) Discussion about magnesium phosphating. Metalurgija 55(3):507–510

Rajagopal C, Vasu KI (2000) Properties of phosphate coatings and influencing factors. In: Conversion coatings: a reference for phosphating, chromating, and anodizing processes. Tata McGraw-Hill, New York, pp 62–70

Ramamoorthy R, Venkatesan T, Rajendran R (2017) Solid lubricant assisted machining—an environmental friendly clean technology to improve the surface quality. SAE Tech Pap 28(1):1–4

Sankara Narayanan TSN (2005) Surface pretretament by phosphate conversion coatings—a review. Rev Adv Mater Sci 9:130–177

Shang K, Zheng S, Ren S, Pu J, He D, Liu S (2018) Improving the tribological and corrosive properties of MoS_2-based coatings by dual-doping and multilayer construction. Appl Surf Sci 437:233–244

Shankara A, Menezes PL, Simha KRY, Kailas SV (2008) Study of solid lubrication with MoS_2 coating in the presence of additives using reciprocating ball-on-flat scratch tester. Sadhana 33(3):207–220

Sharma SM, Anand A (2016) Solid lubrication in iron based materials—a review. Tribol Ind 38(3):318–331

Shen MX, Cai ZB, Peng JF, Peng XD, Zhu MH (2017) Antiwear properties of bonded MoS_2 solid lubricant coating under dual-rotary fretting conditions. Tribol Trans 60(2):217–225

Vadiraj A, Kamaraj M, Sreenivasan VS (2012) Effect of solid lubricants on friction and wear behaviour of alloyed gray cast iron. Sadhana 37(5):569–577

Xu J, Zhu MH, Zhou ZR, Kapsa P, Vincent L (2003) An investigation on fretting wear life of bonded MoS_2 solid lubricant coatings in complex conditions. Wear 255(1–6):253–258

Xu J, Zhou ZR, Zhang CH, Zhu MH, Luo JB (2007) An investigation of fretting wear behaviours of bonded solid lubricant coatings. J Mater Process Technol 182(1–3):146–151

Ye Y, Chen J, Zhou H (2009) An investigation of friction and wear performances of bonded molybdenum disulfide solid film lubricants in fretting conditions. Wear 266(7–8):859–864

Zhu X, Lauwerens W, Cosemans P, Van Stappen M, Celis JP, Stals LM, He J (2003) Different tribological behavior of MoS_2 coatings under fretting and pin-on-disk conditions. Surf Coat Technol 163:422–428

Tribology of Composite Materials and Coatings in Manufacturing

M. H. Sulaiman, N. A. Raof, and A. N. Dahnel

Abstract The chapter presents studies regarding the tribological performance of composite materials and multilayer composite coated tools in manufacturing processes carried out by the authors. Two manufacturing processes were investigated—metal forming and metal cutting. In metal forming, the study aimed to explore lubricant-free forming utilizing multilayer DLC composite hard coating as the potential tool coating. The experimental studies on the coating include characterization of the coating, and tribological analysis of the coating using commercially available pin-on-disk, laboratory tribology simulative test and industrial ironing of stainless steel. In order to examine the influence of temperature and contact pressure along the tool/workpiece interface on friction, Finite Element analysis was performed. Meanwhile, in metal cutting, two environmentally benign machining techniques were investigated to determine their potentials in delaying tool wear progression. First, sustainable machining by coupling multilayer ceramic composite coated-tool with cryogenic coolant as the cutting fluid. Second, the machining of Carbon Fibre Composite and Titanium alloys stacks using Ultrasonic Assisted Drilling (UAD) technique. Both techniques include investigations on machining conditions with varied cutting tool speeds. The examinations on the experimental results were focused on temperature, tool wear, surface integrity and metallurgical structure of near-surface region.

Keywords Manufacturing · Lubricant-free forming · Cryogenic machining · Composite coating · Ultrasonic assisted drilling (UAD) · Carbon fibre composite (CFC) · Titanium alloy composite

M. H. Sulaiman (✉) · N. A. Raof · A. N. Dahnel
Department of Manufacturing and Materials Engineering, Kulliyyah of Engineering, International Islamic University Malaysia, P.O. Box 10, 50728 Kuala Lumpur, Malaysia
e-mail: hafissulaiman@iium.edu.my

N. A. Raof
e-mail: natashar@iium.edu.my

A. N. Dahnel
e-mail: aishahnajiah@iium.edu.my

M. T. Hameed Sultan et al. (eds.), *Tribological Applications of Composite Materials*, Composites Science and Technology, https://doi.org/10.1007/978-981-15-9635-3_11

1 Introduction

While lubrication has been the leading edge to improve tribological performance in manufacturing processes, environmental-friendly manufacturing with or without inclusion of lubricants is new and presents extremely demanding solutions for sustainable manufacturing. Driven by environmental concerns and the enforcement of restrictive legislations to ban the use of lubricants, extensive studies on environmental-friendly manufacturing, either by looking at new alternative to lubrication or manufacturing techniques, has attracted great attention from worldwide researchers. Since a large amount of lubricant is necessary for a high volume manufacturing production, this will further increase the risks to health hazards. Up to these days, efficient lubricant containing environmentally hazardous extreme pressure additives is being employed in production as the only reliable solution to impede tool wear. Such lubricants are activated by gradual temperature increase and due to repetitive sliding between the workpiece and the tool surface, thereby a good boundary film is created by the reaction with the material of workpiece, usually involving the combination with coated tools for improved tool wear resistance (Bay 2013; Ceron et al. 2014). Aside from the good boundary lubricants, increasing trends of using vegetable oils (Syahrullail et al. 2011; Zulhanafi and Syahrullail 2019) and tailor-made ionic liquids (Amiril et al. 2017) in manufacturing and industrial engineering has become future prospects of better lubrication with anti-friction and anti-wear properties as opposed to a lot of biolubricants, and mineral oils that are synthetic and petroleum-based.

Manufacturing processes with effective boundary lubricants alone may not necessarily improve tool wear resistance unless other mechanical engineered surface techniques are applied. For instance, anti-seizure tool materials, structured surface topographies, and anti-seizure tool surface treatments (either by coatings or by thermochemical diffusion). To date, some of them are being used in real manufacturing processes, i.e., application of structured workpiece surfaces with the use of big rolls roughened by Shot Blast Texturing (SBT) or Electro Discharge Texturing (EDT) (Kijima and Bay 2007). Apart from that, hard coating is largely used today. Composite coating, for example, is known to decrease tool wear of approximately 2–50 times in abrasive wear conditions while for sliding conditions, beyond four times compared to the reference uncoated steel (Holmberg et al. 2014). For hard coatings in regard with manufacturing applications, the design of coating architecture is important to release residual stress and to avoid the initiation and generation of cracks so that the life span of a tool can be lengthened. In metal stamping, multilayer Diamond-Like Carbon (DLC) composite coating has shown positive influence to promote an improved adhesion strength for DLC and the tool substrate, thus a longer tool lifetime (Sulaiman et al. 2017a). The multilayer coating structure allows the coating in restraining and decelerating crack initiation and propagation rate. Owing to successive microcrystalline coatings with a smoother surface and columnar grain structures, the multilayer DLC composite coating worked satisfactorily so that the coating damage resistance is improved, and even performed well in dry friction condition (Sulaiman et al. 2017b).

A study of WC-Co tools with monolayer, MCD/NCD coating and multilayer diamond/β-SiC composite hard coating using Rockwell and scratch tests has demonstrated that the multilayer composite tool is superior to other hard coatings. The experiment revealed the positive impact of using multilayer diamond/β-SiC composite hard coating which improved the mechanical properties, by means of high hardness, high Young modulus and residual stress that is low, leading to enhanced adhesion and crack propagation resistance (Yuan et al. 2020). Composite hard coating has such mechanical properties that are excellent that it has been supported by optimized thermal expansion coefficient matching amidst the substrate and coating as well as contributed directly by internal mechanical interlock effect of the co-deposition of diamond and β-SiC phase. As a result, this has improved interfacial adhesion of the substrate and MCD/NCD multilayer top coating, in which the diamond functions as high strength structure phase while the bonding phase is represented by β-SiC. A similar result in metal cutting, engineered nano-scale multilayer $TiAlN/Al_2O_3$ composite coating has enabled for an optimum combination between the high adhesion strength and the tool substrate along with the work material adhesion to the tool surface that is minimum (Vereshchaka et al. 2014). The experimental results after the dry machining concluded that the multilayer $TiAlN/Al_2O_3$ composite coating can work effectively to decrease the wear rates, thereby promoting longer operational life of the cutting tool significantly. On the other hand, there is a number of researches on exploring new environmentally benign machining techniques of composites materials. For example, drilling of aerospace composite materials using cryogenic lubrication technique (Barnes 2013; Barnes and Ascroft 2015) and using Ultrasonic Assisted Drilling (UAD) technique (Dahnel et al. 2015, 2016). Both techniques are feasible alternatives to provide low friction and wear rate at controlled temperatures below 300 °C besides enhancing the tool lifetime and drilling hole quality. However, utilizing cryogenic lubrication technique during the machining of the composite materials has no significant influence on tool wear, despite reduced temperature between work material and cutting tool.

The aim of this chapter is to present the discussion of the mechanical and tribological behaviour of composite tool coatings with environmental-friendly lubrication system or without applying lubricants at all, and the tribological performance in environmental-friendly machining techniques of composite materials. Most work are centered on evaluating the tribo-mechanical performances with and without lubricants on how severe the tool wear and product surface quality are. Some studies were carried out at elevated temperatures and hence the discussions are made separately for room and high temperature, categorized by the types of manufacturing processes. The broader impact of the chapter ends with concluding remark of the present tribological composite work in manufacturing processes which are needed for exploring the tribological behavior of new and greatly classes of composite specifically, and for tribology generally.

2 Metal Forming

2.1 Forming of Stainless Steel Sheets with DLC Composite Hard Coating

Studies on eliminating the harmful lubrication in metal forming is growing tremendously. Although adopting anti-seizure hard coatings is imperative to improve tribological performance of tool surface in metal forming, the coated tool lifetime is limited when forming is operated under dry friction condition. This is associated with coating architectures and adhesion bond that is weak which connects top coating layer and tool substrate at high loads (Chuan et al. 2013; Merklein et al. 2015). A solution to this adhesion strength issue is the introduction of interface layer, i.e. TiAlN (Wang et al. 2007; Biksa et al. 2010; Lukaszkowicz 2011; Sulaiman 2017). With the aim to enable lubricant-free forming and to promote environmental-friendly tribosystem, a new multilayer DLC/CrCN/CrN/TiAlN composite coating was developed and deposited onto the tool substrate. The composite coating was tested and evaluated using three different tribological test methodologies. First, tribological pin-on-disk test (Sulaiman et al. 2019a). Second, simulative tribology test (Sulaiman et al. 2017a). Lastly, simulation of industrial ironing of stainless steels (Sulaiman et al. 2019b). All involved approaches adopted experiments at room and elevated temperatures. To investigate the effects on the tool wear following the experimentation, the measurement on the workpiece surface roughness was taken using a tactile roughness profilometer. This was followed by the detection of material transfer to the die surface in the area of contact utilizing a light optical microscope (LOM). Lastly, in order to examine the wear scar elements at the top of coated die surface following the experiment, scanning electron microscopy and energy-dispersive X-ray spectroscopy (SEM–EDX) were used. Before the experimentation, a cleaning process was performed on all worn surfaces by immersing them repetitively in acetone bath. A Finite Element simulation combining analyses on mechanical and thermal were conducted to discover material deformation along with heat generation impact towards the stress conditions of workpiece during forming operation and the friction at the interface of die–workpiece.

2.1.1 Properties of Multilayer DLC Composite Coating

Figure 1 displays the transection perspective of the multilayer DLC/CrCN/CrN/TiAlN composite coating. The coating is formed by DLC film (top layer) and TiAlN film (bottom layer) on top of the substrate (tool steel Vanadis 4, with 62 HRC surface hardness). The die roughness following the coating was equivalent as the previous one: $Ra = 0.02$ µm. The coating was accumulated through the process of Physical Vapour Deposition (PVD) which is related to unbalanced magnetron sputtering (Neergaard 2017). The table next to Fig. 1, presents the coating's mechanical properties and surface characteristics, in which

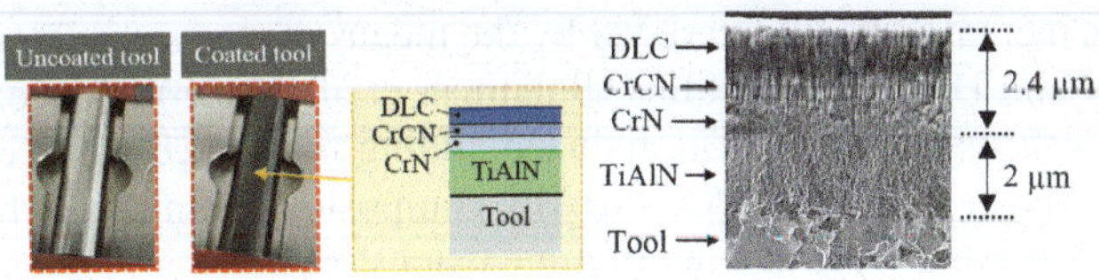

Properties	Value
Hardness	3,000 HV
Roughness *Ra*	0.026 µm

Fig. 1 Design, microstructure, surface characteristics and mechanical properties of the multilayer DLC composite coating

the uncertainty values constitute three to five times of the variations observed in the examined regions. The framework for every DLC/CrCN/CrN/TiAlN composite coating layer disclosed their coating films thickness which were 2.4 and 2 μm, each. Based on the SEM images, TiAlN film (intermediate layer) was found to bond ideally with DLC coating film (top layer) and the Vanadis 4 substrate (bottom layer). It was due to Cr interlayer, as specified by the EDX results in Fig. 2 (top), in which the stress gradient at the interface amidst the DLC and TiAlN coating films was released (Kim et al. 2016). It was also indicated from the EDX results that adhesion strength was enhanced because of the use of the elements, Cr and Ti, in the multilayer DLC/CrCN/CrN/TiAlN composite coating, which then produced low energy, a chemical bond that is stable and strong, and a low stress gradient at the interface (Hong et al. 2001b; Strano et al. 2013; Yasa et al. 2012). A composition that is congregated and rough was exhibited from the first DLC layer, different compared to the framework of the TiAlN layer which was fine and smooth. Good composite coating adhesion is explicable by the composition of CrCN/CrN interlayer under the first DLC layer, that is big and has columnar grain (Kaynak 2014; Dhar et al. 2001). Figure 2 (bottom right) shows the thick microstructure of TiAlN coating that has a steady cylindric architecture with several pores along with inclusions. Meanwhile, Fig. 2 (middle) shows TiAlN coating that contains high level of Al, which causes the multilayer DLC/CrCN/CrN/TiAlN composite coating to enhance its rigidity. It is therefore suggested that the TiAlN interlayer is weak in cutting stress for the sake of enhancing the strength of adhesion amidst the DLC film and the Vanadis 4 substrate,

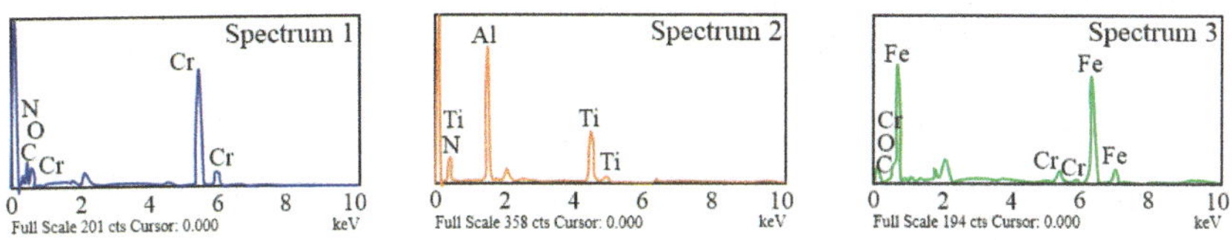

DLC/CrCN/CrN coating

Element	Weight%	Atomic%
Cr	73.54	41.92
N	13.88	29.38
C	8.77	21.63
O	3.81	7.07

TiAlN coating

Element	Weight%	Atomic%
Ti	45.01	24.80
Al	31.33	30.64
N	23.65	44.56

Tool steel, Vanadis 4

Element	Weight%	Atomic%
Fe	87.82	72.08
Cr	5.16	4.55
O	3.57	10.22
C	3.45	13.15

Fig. 2 Elemental compositions of the multilayer DLC/CrCN/CrN/TiAlN composite coating

as the adhesion strength is increased by Cr interlayer by the means of stress gradient reduction between TiAIN and DLC coating films (Sulaiman et al. 2019b). Hence, the combination of these effects could result in reduced sliding-originated surface tensile stresses of the multilayer DLC/CrCN/CrN/TiAlN composite coating, which helps with the prevention of severe wear and die lifetime extension.

2.1.2 Tribological Pin-on-Disk Experiment with Multilayer DLC Composite Coating

Commercially available pin-on-disk test was adopted for studying tribological effects of the composite hard coating. Figure 3 shows the experimental setup for two conditions; lubricated and dry friction for a friction pair comprised a 100Cr6 steel ball along with a Vanadis 4 tool steel flat surface with and without the multilayer DLC/CrCN/CrN/TiAlN composite coating. Table 1 shows the list of test parameters application in the pin-on-disk experiment. In order to examine the wear scar on DLC coated and uncoated tool steel surfaces following the experimentation, a tactile roughness profilometer and a Light Optical Microscope (LOM) were applied.

The remarkable influence of engineered tool surface by using hard coating is apparent. When there is no application of coating to the tool steel at the time of operation, bigger friction coefficient is observable easily in Fig. 4 (left). For specimens with multilayer DLC/CrCN/CrN/TiAlN composite coating, the value

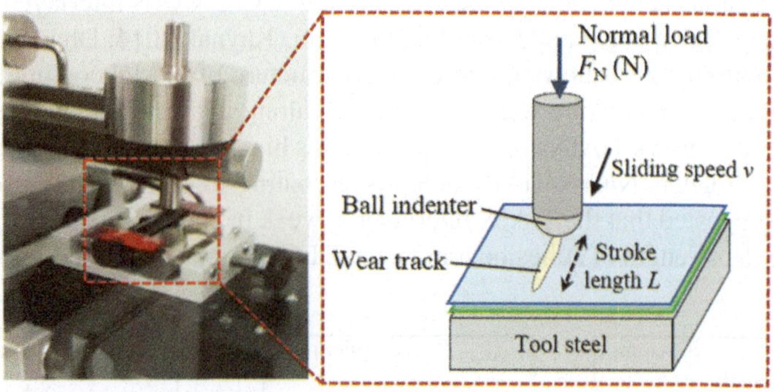

Fig. 3 Schematics of pin-on-disk experimental setup

Table 1 Test parameters

Test parameters	Values
Sliding speed v	100 mm/s
Normal load F_N	10 N
Stroke length L	16 mm
Max. strokes	500 laps
Temperature	32 °C

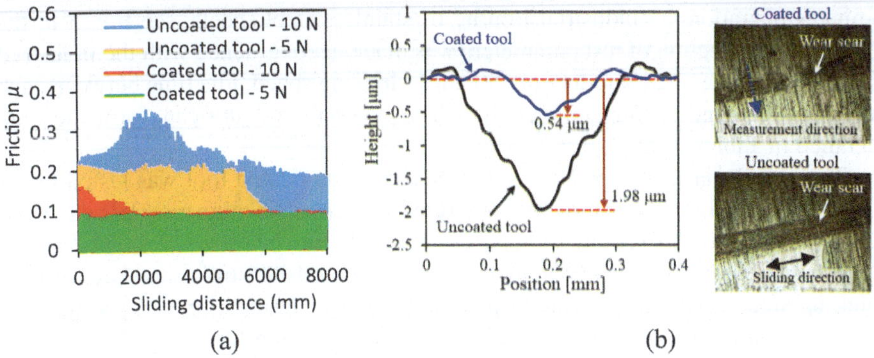

Fig. 4 Friction μ (left) and wear depth profiles after pin-on-disk experiment (right)

of friction coefficient in stable-state condition was observed at its lowest in both conditions of lubricated and dry friction. This explains the capability of multilayer DLC/CrCN/CrN/TiAlN composite coating in frictional effects reduction and also its capability to reach value that is stable after a long experiment. The positive influence of using such composite coating is supported by roughness measurement on the wear scar profiles with an optical profilometer, in which there was a discovery of only minor scratches on the wear track of the multilayer DLC/CrCN/CrN/TiAlN composite coated tool, see Fig. 4 (right). In having good mechanical properties and wear resistance for the multilayer DLC/CrCN/CrN/TiAlN composite coating, the coating characteristics have shown positive improvements in adhesion between the coating and tool substrate to lessen the tool steel wear, hence the tool lifetime is improved.

2.1.3 Tribological Strip Reduction Test (SRT) with Multilayer DLC Composite Coating

In performing the off-line evaluation of the same multilayer DLC/CrCN/CrN/TiAlN composite coated tool as described in Fig. 1, a Strip-Reduction Test (SRT) was chosen, see Fig. 5 to the far right. The SRT was chosen because the simulative test

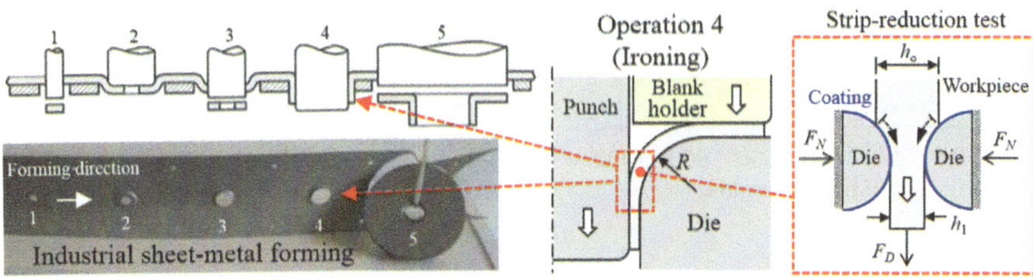

Fig. 5 Experimental strip reduction emulating industrial ironing (Ceron and Bay 2013)

imitates operation 4—industrial ironing of stainless steel sheets, see Fig. 5 to the far left. The selection of test parameters was made in congruence with the industrial production process: 24% reduction, 50 mm/s drawing speed, idle time between each stroke of 1.8 s and 10 mm sliding length. The materials and lubricant for the test are detailed in Tables 2 and 3, respectively.

The multilayer DLC/CrCN/CrN/TiAlN composite-coated tool was tested under dry friction condition and lubrication with environmentally benign mineral oil with no Extreme Pressure (EP) additives, refer to Table 3. Figure 6a shows constant drawing load and stable tool rest temperature despite following 1000 strokes, and none of the pick-up signs on the tool surface was remarked. The verification is made through sheet roughness *Ra* measurement as in Fig. 6b, in which the original workpiece roughness was found to be higher than sheet surfaces performed under conditions of lubricated and dry friction. Hence, the findings have revealed the capability of the multilayer DLC/CrCN/CrN/TiAlN composite coating to ironing the stainless steel sheets with no lubrication, or else will be highly susceptible to galling. Thereby, adopting the multilayer composite coating film by depositing an interlayer metallic coating film like TiAlN, CrCN and CrN in between the DLC film and the tool substrate can therefore improve adhesion strength of the DLC composite coating

Table 2 Test materials for the strip-reduction test

Components	Dimension (mm)	Roughness *Ra* (μm)
Tool (Vanadis 4)	Ø15 × 34	0.02
Workpiece (EN1.4307)	*W*30 × *t*1.0	0.14

Table 3 Properties of the test lubricant

Oil type	Product name	Kinematic viscosity η (cSt @ 40 °C)
Mineral oil	CR5 Houghton Plunger	660

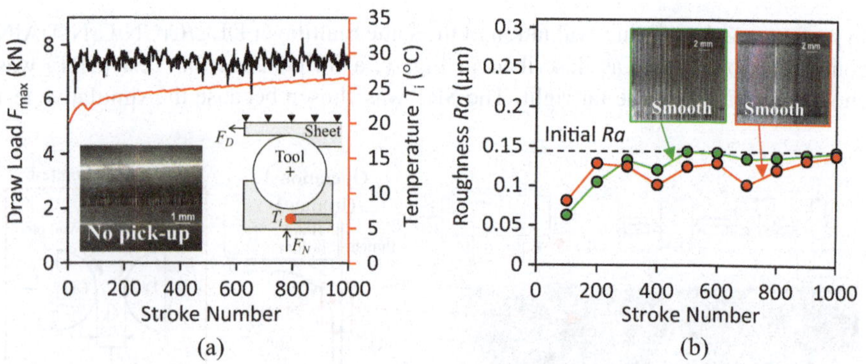

Fig. 6 **a** Forming load F_D and **b** workpiece roughness *Ra* in ironing with dry friction condition

and even perform well under the extreme test conditions in ironing of stainless steel. A coupled mechanical and thermal simulation analysis utilizing Finite Element (FE) software LS-Dyna shown in Fig. 7a has supported the experimental findings. As seen in Fig. 7b, a minimum quantity of the hazard free lubricant is sufficient to minimize the friction, and no significant difference of the normal pressure along the contact region, reaching 400 and 1000 MPa in dry and lubricated conditions, see Fig. 8a. However, it is anticipated that the temperature would be higher at the tool/workpiece assemblage by the Finite Element analysis using strip reduction test in dry condition as opposed to that of lubricated, see Fig. 8b. This indicates the temperature change ΔT alongside the assemblage of the tool/workpiece has been reduced due to a smaller friction coefficient and thereby, no lubricant film breakdown. This further suggests that the composite coating improves wear resistance for the DLC coated tool surface for the sake of protecting the DLC coated tool for a longer tool lifetime.

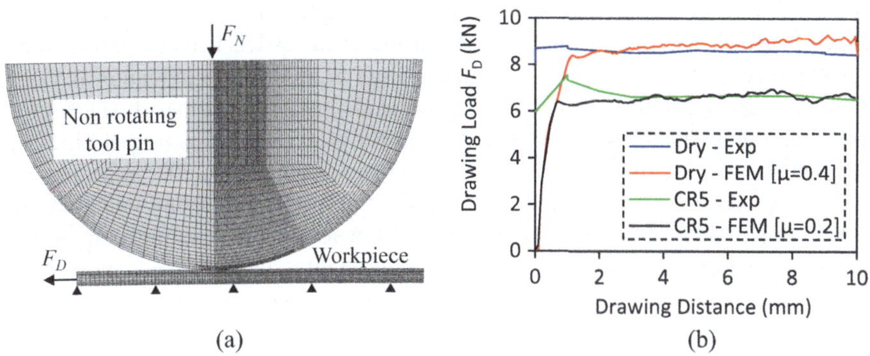

Fig. 7 **a** Numerical simulation of ironing, and **b** friction coefficient

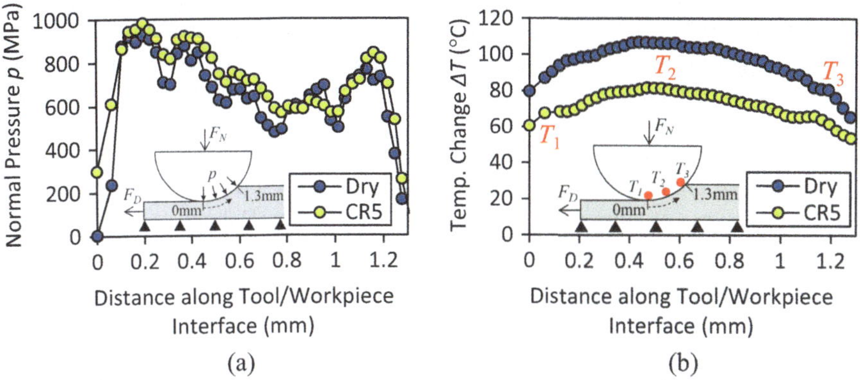

Fig. 8 Distributions of **a** contact pressure and **b** temperature along the tool/workpiece interface

3 Machining

Machining is important in almost all manufacturing processes. However, drawbacks of poor machinability are becoming prevalent in industry. This is especially pronounced when machining aerospace and automotive composite materials. The primary problems concerning machining in regard with such work materials are mostly associated with tool wear and also surface quality. Contributed by the generation of heat which occurs near the tool/workpiece sliding surfaces, many machining techniques have been developed so that the amount of heat generated can be controlled, for instance, the application of cutting fluid. Recent trends of environmental-friendly machining with the use of cryogenic cutting fluid are becoming popular. The cryogenic machining employed varied cryogen types like helium (LHe), liquid nitrogen (LN), ethane, argon, methane, and oxygen in reducing the cutting temperature (Shokrani et al. 2012). Due to this, researches have been conducted concerning the impacts of cryogenic application in various kinds of work materials, namely Inconel 718 (Courbon et al. 2013; Kaynak 2014; Wang and Rajurkar 2000), Ti-6Al-4V (Wang and Rajurkar 2000; Bermingham et al. 2011; Hong et al. 2001a, b; Strano et al. 2013; Yasa et al. 2012), various steels (Dhar et al. 2001; Paul and Chattopadhyay 1996; Venugopal et al. 2007; Pušavec et al. 2011; Rotella et al. 2012), shape memory alloys (Kaynak et al. 2011, 2013a, b), magnesium alloys (Pu 2012) and the rest of engineering materials (Wang and Rajurkar 1997), which showed positive improvements towards the machining outputs like the cutting temperature, cutting forces, surface quality, and tool wear. In drilling, two elements that are acknowledged in common to be hard in materials machining are Carbon Fibre Composite (CFC) and titanium (Ti) alloys. Therefore, it is highly desirable for techniques that can substitute conventional drilling as alternative in which drilling performance of such materials can be improved. A new alternative to replace the conventional way of drilling of CFC/Ti stacks with Ultrasonic Assisted Drilling (UAD) has demonstrated that the life of Tungsten Carbide (WC) drills has improved by 300%. Owing to titanium adhesion and flank wear reduction, reducing cutting temperatures is achievable as cutting tool vibration during drilling resulted in improved evacuation of the hot titanium chips from the cutting zone and leading to the cooling of cutting tool (Dahnel et al. 2015; Pecat and Brinksmeier 2014). In regard with burr formation upon drilling, the report observation stated that there was improved hole quality with less burr on aluminum alloy 1100 and Inconel 738-LC holes with the use of Ultrasonic Assisted Drilling (UAD) (Dahnel et al. 2015). This, however, requires ultrasonic amplitude of 4–10 μm and the frequency of 20–21 kHz (Chang and Bone 2005; Azarhoushang and Akbari 2007). While for CFC drilling, surface roughness and circularity of the drilled holes experienced 50% improvement with the aid of ultrasonic (Makhdum et al. 2014). The UAD techniques have shown positive influence to enhance tool wear resistance with remarkable cutback of thrust force and torque. This section reported studies on tribological composite in machining processes—cryogenic turning with composite coated tool (Raof et al. 2019) along with Ultrasonic Assisted Drilling (UAD) of CFC/Ti stacks (Dahnel et al. 2015, 2016;

Dahnel 2017). Comparisons were made between the investigated machining techniques and the conventional ones. The assessment of machinability was conducted for its thrust forces, surface integrity, tool wear, burr and delamination. Dominant types of tool wear were also studied.

3.1 *Cryogenic Turning*

AISI 4340 alloy steel that had gone through quenching and tempering, with the diameter of 100 mm and 317 HB hardness, was the material for the test. Figure 9 shows a typical lath-martensitic structure microstructure of the work material was examined before the turning experiment. A CNC lathe machine with a CVD TiCN/Al_2O_3 composite ceramic carbide insert was used to turn on the work material. Table 4 shows the listing for cutting parameters. For cryogenic flushing, a connection was made between a flexible hose and Liquid Nitrogen (LN) tank, and a copper pipe functioned as the nozzle pointing to the clearance face of the insert.

3.1.1 Effects of Cryogenic Turning Condition on Temperature, Tool Wear and Surface Integrity

The measured cutting temperatures occurring during turning in dry and cryogenic flushing are shown in Fig. 10. Cryogenic LN application during turning has managed to cut down the cutting temperature as much as 35–55% in comparison to dry turning, particularly when this was done at cutting speeds that were higher. Cryogenic

Fig. 9 Microstructure of the work materials before the turning experiment

Table 4 Test parameters in turning experiment

Parameters	Description
Cutting speed (m/min)	160, 200, 240
Feed rate (mm/rev)	0.3
Depth of cut (mm)	1.0
Coolant	Dry and cryogenic (LN)

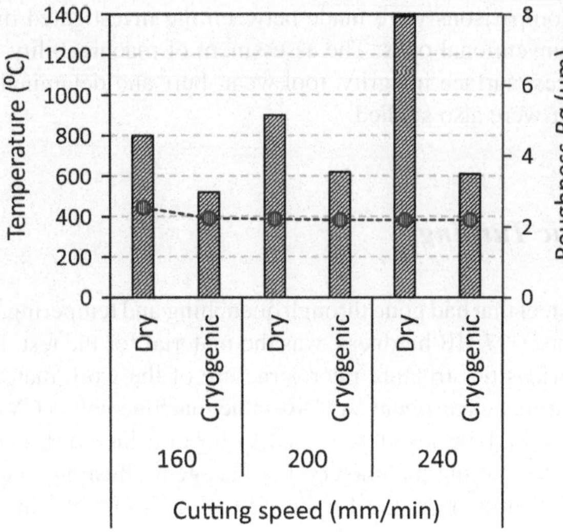

Fig. 10 Cutting temperature and Roughness *Ra* of the work materials (Kijima and Bay 2007; Holmberg et al. 2014)

machining has lower temperatures which is beneficial as machined surface amendments due to thermal activity can be reduced. Nevertheless, in terms of *Ra* values of the work material at distinct cutting speeds, there was not much apparent and significant difference observed in cryogenic turning. It was discovered that when using higher speeds, effective result was generated in producing cryogenic cooling effect that controlled the generation of heat caused by cutting speeds. This phenomenon is related to the formation of chip morphology that occurs during the cutting. When higher speeds are used, the chip will become curlier and thinner. Sudden cooling by cryogenic will harden the chips and improve their fragility. Thus, this will ease the LN to penetrate into the chip-tool interface, hence the reduction in cutting temperature.

In this research, the impact of cutting tool wear towards the roughness of machined surface was also examined. Figure 11 illustrates the readings of average surface roughness as a function of cutting tool condition. The measurement of *Ra* values occurred at flank wear, $V_B = 0$ µm (new tool), $V_B \geq 0.15$ µm (medium wear), and $V_B \geq 0.3$ µm (wear). Based on the observation, a few patterns of the graph relation were recorded between the roughness of surface and tool wear. Various conditions of machining resulted in better surface roughness when a worn tool was used in cutting. This might be explained by the efficient flattening of the tool nose due to increased worn flat on the tool flank (More et al. 2006), which in turn increasing the tool nose radius as in Fig. 12, hence a better quality of machined surface is produced. Using a worn tool at cutting speed of 240 m/min in cryogenic condition generated critical reduction in roughness value which might be due to the combination of higher speed and cryogenic application that improves surface roughness as in previous discussion.

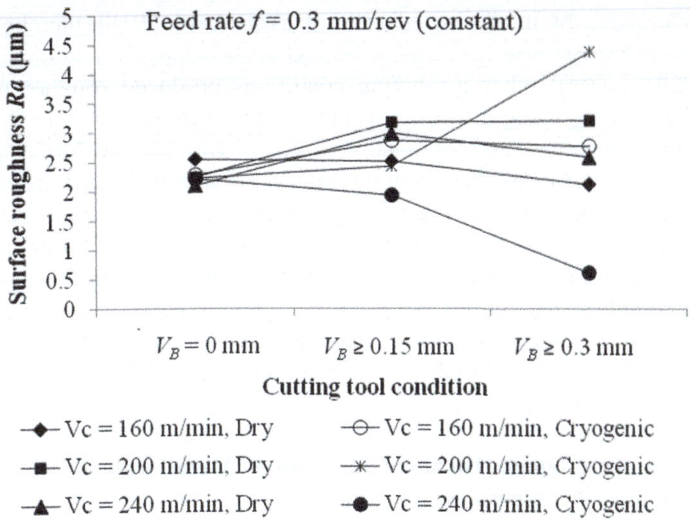

Fig. 11 Roughness *Ra* of AISI 4340 alloy steel surface with different cutting speeds under dry and cryogenic environment (f = 0.3 mm/rev)

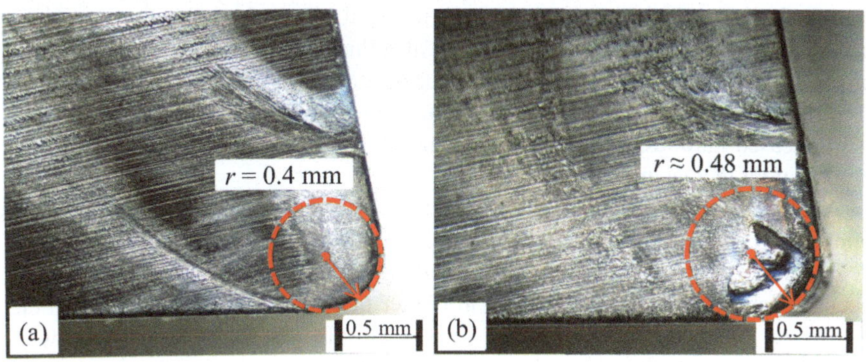

Fig. 12 Measurements of tool nose radius and surface condition of CVD $TiCN/Al_2O_3$ composite ceramic carbide insert **a** before and **b** after machining

The improved wear resistance of the CVD $TiCN/Al_2O_3$ composite ceramic as protective coating for the carbide insert is explicable by the examination of microstructure of the CVD $TiCN/Al_2O_3$ composite hard coating. Figure 12c shows the cross-section of $TiCN/Al_2O_3$ composite coating (Gassner et al. 2018). TiCN coating coupled with Alumina Oxide Al_2O_3 top layer were accumulated on cemented carbide insert. Due to its hardness and firmness which are high, the use of the $TiCN/Al_2O_3$ composite coated carbide insert exhibited less flank wear for the cryogenic turning. In applications involving elevated temperatures and shear stresses as in turning of steels, the Alumina Oxide Al_2O_3 top layer coupled with TiCN coating could enhance the cutting performance by increasing adhesion (Gassner et al. 2018), thereby improve tool life.

Meanwhile, several other machining conditions produced rougher machined surface when cutting using a worn tool. Increasing machining time leads to deterioration of tool sharpness and degradation of surface roughness (More et al. 2006). Apart from that, one factor that may contribute towards bad finishing surface is the work material adhesion between the tool edge and the tool flank face. When cutting was performed using a worn tool at the cutting speed of 200 m/min in cryogenic condition, it was found that the roughness value increased significantly. In this machining condition, the cutting tool used was found to have fracture wear on the tool flank face when observed closely. This might result in more deterioration of the machined surface as opposed to the rest of machining conditions.

3.1.2 Effect of Cryogenic Turning Condition on Metallurgical Structure of Near-Surface Region

In this study, the machining introduced two surface layers that have different characteristics: the refined grain layer (RL) and also the transition layer (TL), see Fig. 13. There are distinct attributes between the two layers; the mechanical features and chemical structure in comparison to bulk material. Many factors have been identified to influence the layers' properties and also thickness which are the temperature produced and the rate of heating–cooling during machining, original size of grain, and the original mechanical features that belong to the bulk material. As seen in Fig. 14, the increased hardness of the machined surface is greater, while lower hardness is marked with increasing profile depth.

A greater hardness value was observed for cryogenically machined surfaces as opposed to dry machined surfaces in all variations of speeds. The reason is the degradation of thermal softening effect and the strain hardening of the work which occur at low temperatures, resulting in higher density of refined carbide particles in cryogenically machined surfaces. This is closely associated to ultrafine white

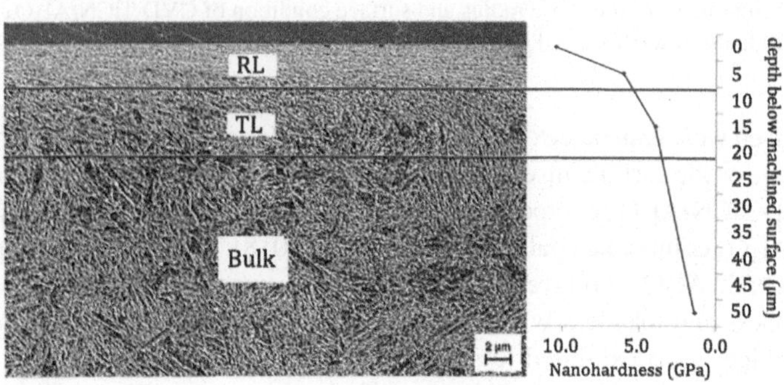

Fig. 13 Hardness distribution profile of the machined surface

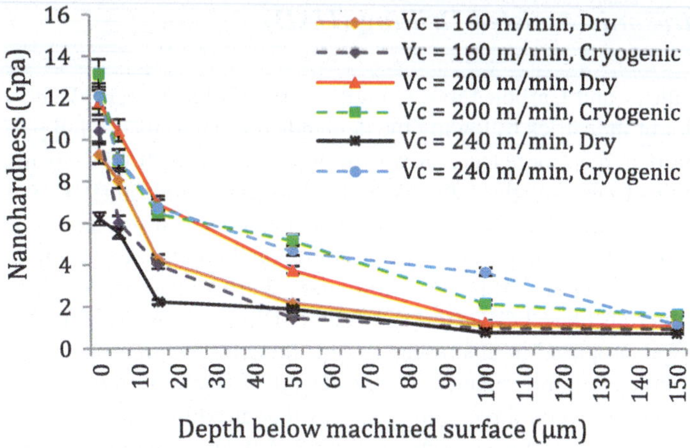

Fig. 14 Hardness distribution of machined surface (Holmberg et al. 2014)

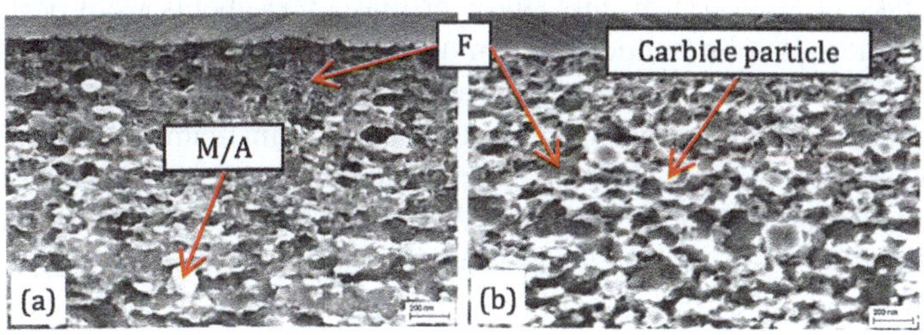

Fig. 15 Ultrafine particles of the machined surface in **a** dry, and **b** cryogenic cutting (160 m/min)

globular particles with a diameter smaller than 200 nm as noted in dry and cryogenic machining, see Fig. 15 for 50k× magnification of a Fe-SEM microstructure of machined cross-section. Significant heat builds up in dry turning has led to a high hardness characteristics of M/A island particles, Fig. 15a, and this has been an influential contributing factor in initiating cracks and brittle fractures. Alternatively, cryogenic turning is advantageous since it has shown positive impact in controlling the heat production below the austenitizing temperature. This, in turn, leads to enhancement towards carbide particles precipitation into martensite matrix, and encourages carbide particles purification with a more identical ultrafine white globular carbide particle distributions within the martensite matrix, see Fig. 15b, that has increased the hardness (Shokrani et al. 2013; Jawahir et al. 2016). This investigation has revealed that excessive heat caused by high cutting speed is undesirable. As a result, there will be thermal damage occurring at the machined surface and subsurface which affects the surface quality, increases tensile residual stresses, and reduces the dimensional accuracy that belongs to the work material.

3.2 Ultrasonic Assisted Drilling (UAD)

Titanium alloys and Carbon Fibre Composite (CFC) have been placed in a limelight in a lot of industries following the demands for materials of high performance, lightweight that keep increasing. It was in preference to drill CFCs using tungsten carbide cutting tools, with the cutting speed of 100–200 m/min and 0.01–0.05 mm/rev feed rates (Liu et al. 2012). Nevertheless, such materials drilling can possibly lead to stratification (CFC), formation of burr (titanium) and accelerated tool failure (Wang et al. 2014; Shyha et al. 2011). Therefore, new alternative to drill such difficult-to-machine materials like CFC/Ti stacks is desired. Ultrasonic Assisted Drilling (UAD) technique was proposed in this research. This includes a comprehensive evaluation on UAD tribological performance on CFC/Ti6Al4V stacks with regard to tool wear as well as tool life when three cutting speeds that are different and consistent feed rate are used instead of conventional drilling. UAD can be understood as a process of composite machining of which its cutting motion is such a superimpose of that of conventional drilling, having high frequency ultrasonic vibration in centre direction (Babitsky et al. 2007). In this study, drilling experiments were performed on aerospace materials—4 mm thick CFC (multidirectional (0°, 45°, 90°, 135°) carbon fibres with Bismaleimide (BMI) resin) and 4 mm thick titanium alloy Ti6Al4V. DMG Ultrasonic 65 Monoblock machine tool and 6.1 mm diameter tungsten carbide 2-flutes twist drills were used in the experimentation to compare between conventional drilling and UAD of CFC/Ti6Al4V stacks. The empirical setup for conventional drilling of CFC/Ti6Al4V stacks (with cutting fluid) and with ultrasonic using 0.05 mm/rev feed rate is shown in Fig. 16. Table 5 illustrates the cutting parameters.

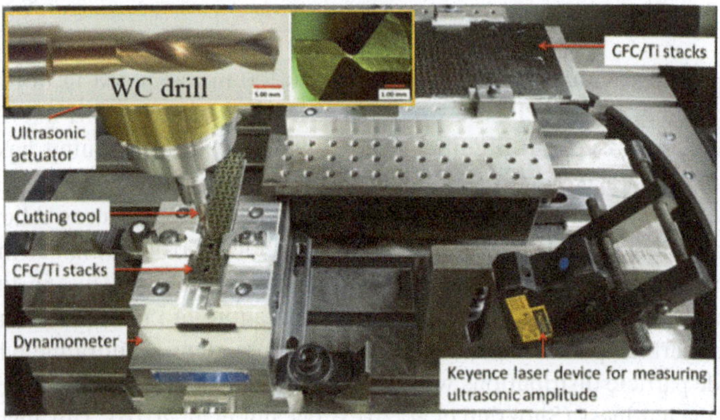

Fig. 16 Experimental setup for ultrasonic assisted drilling (UAD)

Table 5 Drilling parameters

Drilling experiment		Cutting speed (m/min)	Total no. of holes
1	Conventional	25	40
	UAD		
2	Conventional	50	80
	UAD		
3	Conventional	75	80
	UAD		

3.2.1 Effects of UAD Drilling Speed on Tool Wear and Surface Integrity

Figure 17 shows the comparison between the flank wear rate among the drills utilized in conventional drilling and of UAD with 25, 50 and 75 m/min cutting speeds. Based on the graph, the result of using UAD lowered the rate of tool wear and extended the tool life at all cutting speeds that were used. According to ISO 3685 standard, once the flank wear reaches 300 μm, that is when the tool life ends (Tool-life testing with single-point turning tools 1993). The observation shows that when CFC/Ti6Al4V stacks drilling was conducted, the characteristics of tool wear were the adhesion of titanium, edge chipping, along with dull cutting edges. Using 75 m/min cutting speed, conventional drilling showed failure of the cutting tool after 28 holes were being drilled, while UAD showed failure after 34 holes as the edge chipping and wear reached 300 μm. Meanwhile, when lower cutting speed of 50 m/min was used, it extended the tool life in which conventional drilling lasted with 62 holes drilled, and UAD with 80 holes. Therefore, in regard with tool life, the optimum cutting speed

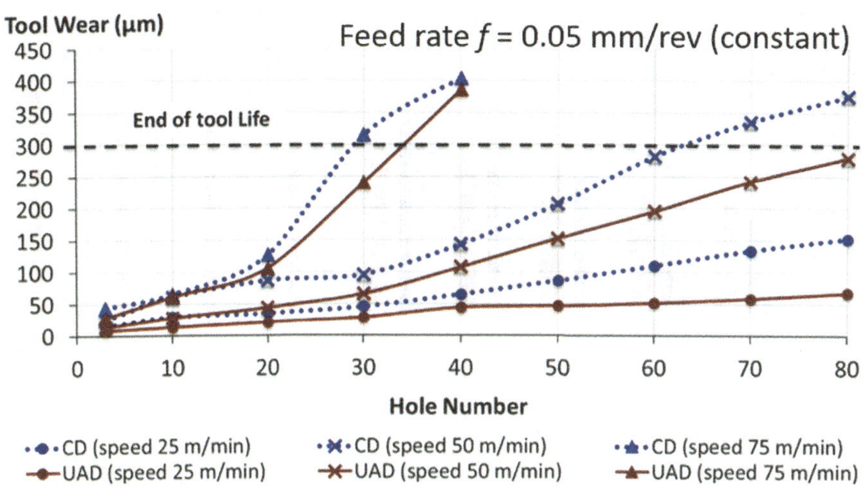

Fig. 17 Tool wear at different cutting speeds of 25, 50 and 75 m/min for UAD and conventional drilling of CFC/Ti6Al4V stacks

is 25 m/min as it showed the lowest tool wear rate, and UAD is proven to provide tool life which is longer compared to conventional drilling, see Fig. 17.

As opposed to drilling conventionally, UAD resulted in lower tool wear which reduced the thrust forces (for both CFC and titanium) that is desirable to improve the hole quality. Figures 18 and 19 show the result in UAD using 25 m/min cutting speed which led to 10–42 N lower thrust forces for CFC, while 36–78 N lower thrust forces for titanium alloy, as opposed to conventional drilling. Meanwhile, when cutting

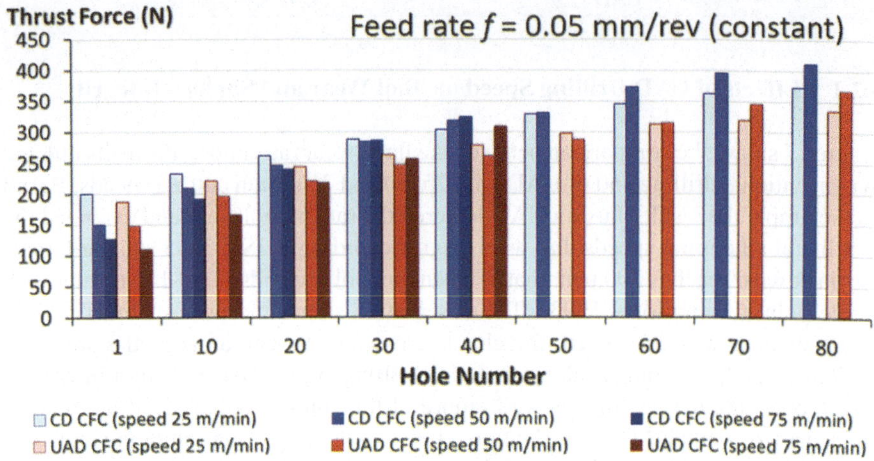

Fig. 18 Thrust forces at different cutting speeds of 25, 50 and 75 m/min for conventional drilling (CD) and UAD of CFC stacks

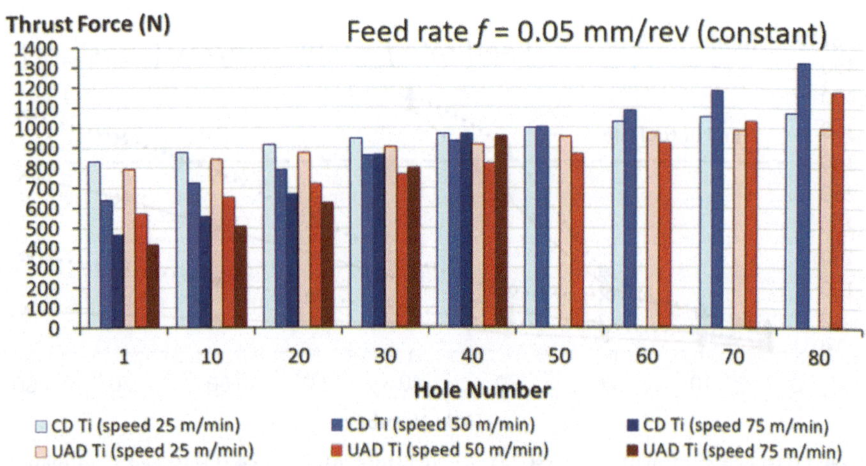

Fig. 19 Thrust forces at different cutting speeds of 25, 50 and 75 m/min for conventional drilling (CD) and UAD of Ti6Al4V stacks

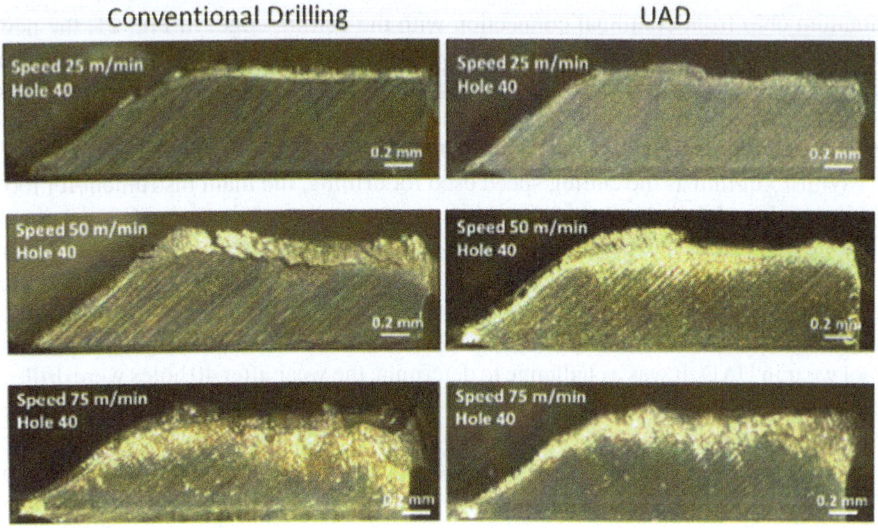

Fig. 20 Tool wear at different cutting speed of 25, 50 and 75 m/min for UAD and conventional drilling of CFC/Ti6Al4V stacks (images were taken after drilling 40 holes)

speed was changed to 50 m/min in UAD, lower thrust force was generated by 3–58 N for CFC, and 70–164 N for titanium alloy compared to conventional drilling. When the cutting speed was raised to 75 m/min, the outcome was similar in which the thrust forces were lower in UAD (by 15–40 N for CFC; by 13–66 N for titanium) in comparison with conventional drilling.

Figure 20 exhibits the cutting edges state after 40 holes were drilled in CFC/Ti6Al4V stacks with 25, 50 and 75 m/min cutting speeds. As observed, using higher cutting speeds led to more adhesion of titanium, which occurred most probably as a result of higher cutting temperatures being used. Despite there was no measurement of cutting temperatures in this study, Li and Shih's work (Li and Shih 2007) in regard with drilling of titanium-only had established that increasing the cutting speed from 24 to 73 m/min raised the cutting temperature from 480 to 1060 °C.

Titanium is a material that is reactive and the stimulation of chemical reaction between titanium and the material of cutting tool will occur when the temperature during drilling increases (Hosseini and Kishawy 2014). During the drilling process, instead of sliding along the cutting edges, there were separation and adhesion on the cutting edges for part of the titanium chip that was in contact with the tool. When there was removal of titanium, chipping of the cutting edges happened and part of the tool material was also taken away. In addition, in regard with hole quality, it was not desired for the adhered material upon the cutting edges as it is unstable and has uneven surface which causes the dimension to be inaccurate and the machined part to have poor surface finishing (Dahnel et al. 2015). Based on the observation as in Fig. 20, during UAD, there was fewer titanium adhesion on the cutting edges in comparison with conventional drilling; caused by the vibration of tool that partly prevented the titanium chip from continual connection with the cutting edge. In Fig. 21, the new cutting edge state is shown, while the state of chipped cutting edge is shown in Fig. 22, in which irregular and coarse tungsten carbide grains were disclosed. As a consequence, such conditions caused the cutting edges to be more prone towards bigger tool fracture during drilling, and resulted in rapid tool failure.

With 75 m/min as the cutting speed used for drilling, the main instrument for tool wear is edge chipping resulting from removal of adhered titanium. Conventional drilling had more edge chipping as opposed to UAD. However, drilling with lower cutting speed of 25 m/min resulted in no edge chipping at all until 80 holes for both conventional drilling and with ultrasonic; the drills were also wearing out by galling instrument. Recurrent cutting is the reason that correlates to slower progression of tool wear in UAD. It was a challenge to determine the wear after 40 holes were drilled in the stacks with the cutting speeds of 50 and 75 m/min resulting from compelling titanium amount that covered the majority of the cutting edges. Figure 20 shows the images taken of the cutting edges following the drills exit through titanium layer in the stacks. Therefore, after 40 holes were drilled, for clear cutting edges and graph

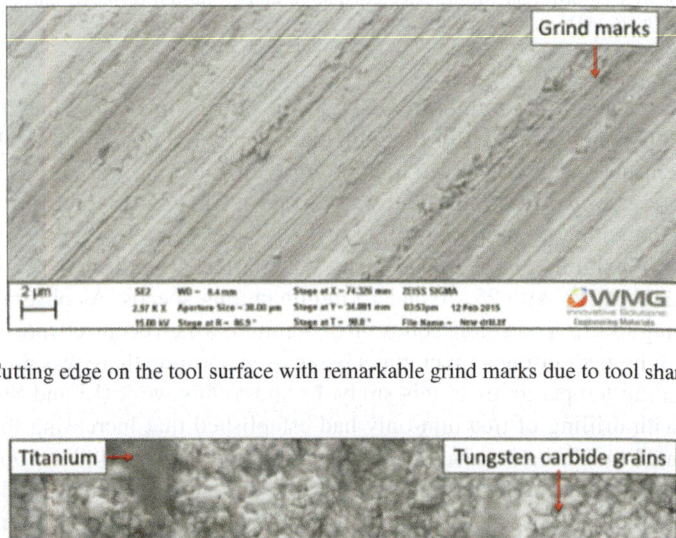

Fig. 21 Cutting edge on the tool surface with remarkable grind marks due to tool sharpening

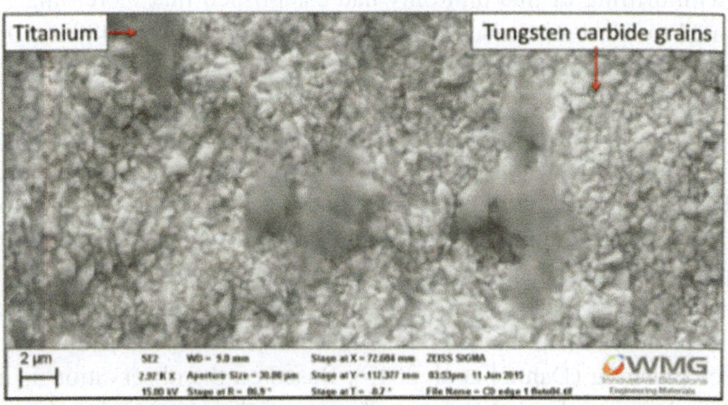

Fig. 22 Chipped cutting edge of tungsten carbide tool

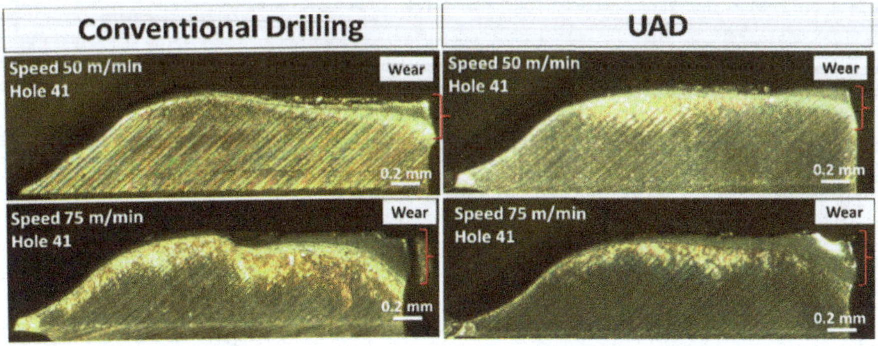

Fig. 23 Tool wear images just right after CFC drilling and before drilling titanium

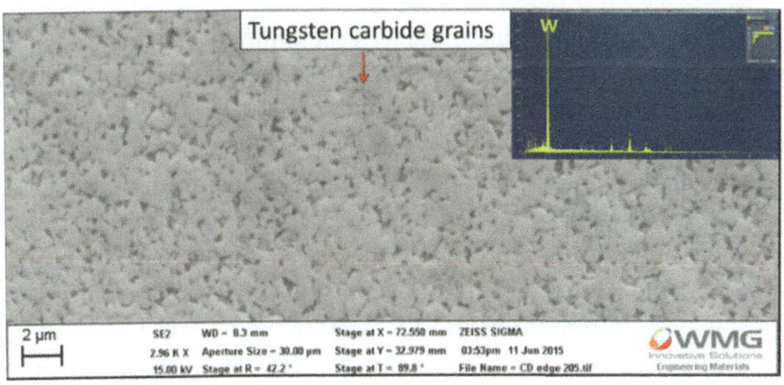

Fig. 24 Abrasive wear on the tungsten carbide tool surface

plotting of the tool wear rate as in Fig. 17, the measurement of the wear was taken after CFC plate was drilled and before titanium plate was drilled as in Fig. 23. The tungsten carbide grains underwent grounding and smoothening by rough carbon fibres as shown in Fig. 24. The observation was done on the edge rounding, in which abrasive tool wear characteristic was evident, as a result of abrasive carbon fibres that rub against the cutting edges.

3.2.2 Effects of UAD Machining Condition on CFC Delamination and Ti-Alloy Exit Burr

An analysis was conducted on delamination factor that is a ratio of CFC delamination extent at the hole entrance to the hole diameter. Figure 25a, b demonstrate the comparison between conventional drilling and UAD, in terms of CFC delamination at the hole entrance after 40 holes were drilled. The result showed that for UAD, the delamination length at the hole entry for the 40th hole was 3 mm, while for conventional drilling was 1.5 µm; hence UAD produced further length of hole

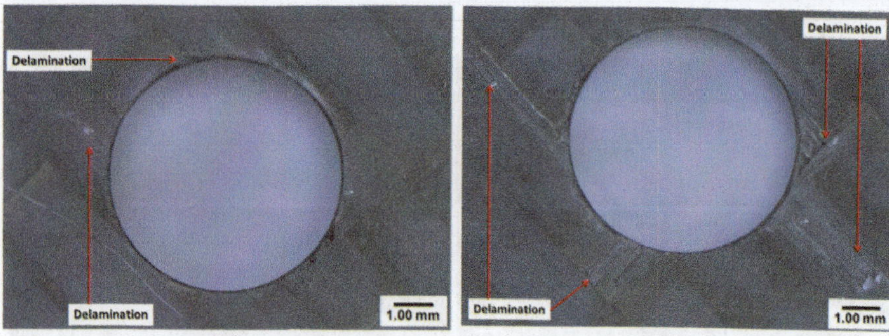

(a) (b)

Fig. 25 CFC entry delamination at hole 40 produced by **a** conventional drilling, and **b** UAD

delamination. Moreover, there were more CFC delamination during UAD because of the cutting tool which experienced more vibration energy effects. When the drilling penetrated Ti6Al4V, the back and forth oscillation of the drill flutes against CFC surface, combining with tool wear and high thrust forces, exerted more pressure on the CFC layers which then pulled the layers, hence causing the CFC to separate or delaminate.

Figure 26a, b illustrate the burr size comparison between conventional drilling and UAD, around the 40th hole. It was found that the burr produced at the 40th hole after drilled conventionally had 111 μm of thickness and 162 μm height, each. Meanwhile, the burr of the 40th hole drilled by UAD had 68 μm of thickness and 128 μm height which meant, the burr thickness was smaller by 39% and its height was smaller by 21%. Minimization or elimination of the burr formation is highly desired in industry especially in securing assembly of parts. UAD application in

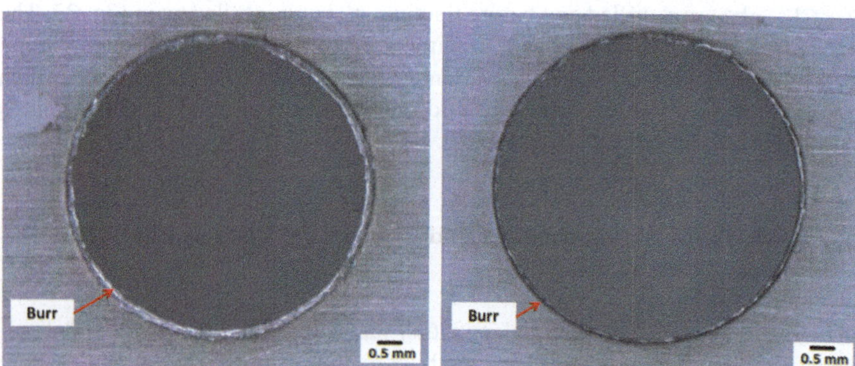

(a) (b)

Fig. 26 Titanium exit burr at hole 40 produced by **a** conventional drilling, and **b** UAD this study has shown its potentials for hole production with fewer or smaller burr as compared to conventional drilling.

4 Concluding Remarks and Future Perspective

The chapter presents a comprehensive investigation on tribological behaviors of tools that were involved in manufacturing processes of various engineering materials. In the field of metal stamping, the use of multilayer DLC composite coating on a tool has shown to be effective to prolong the tool life. This could relate with high hardness, high young modulus and also low residual stress of DLC composite, which lead to improved adhesion strength amidst the DLC and tool substrate. For machining application, alternative techniques which involve cryogenic cutting fluid and ultrasonic assisted cutting tool are recommended in order to enhance the tool life and machined surface quality. The application of Liquid Nitrogen (LN) as cryogenic cutting fluid to turn AISI 4340 alloy steel has demonstrated improvement in reducing cutting temperature by 35–55%. This could lead to a reduction in both tool wear and the alteration of machined surface. Whereas, the utilization of ultrasonic assisted cutting tool for drilling Carbon Fibre Composite and Titanium stacks has prolonged the tool life and enhanced the hole quality as a result of reduction in titanium adhesion, tool wear and improved titanium chip evacuation. The alternative techniques proposed in this chapter are seen as feasible solutions to environmental-friendly manufacturing operations. On another note, more scientific research are needed to develop advanced composite hard coatings and environmentally benign manufacturing techniques in order to realize the requirements of industry to achieve sustainable manufacturing environments, higher productivity and lower production costs. For metal forming, the development of advanced composite hard coatings, ie. nano-multilayer composite coatings, with super hardness, good thermal stability, high oxidation resistance and high wear resistance are a necessity for lubricant-free formation. In machining operations, the cryogenic cooling coupled with ultrasonic assisted machining technique could be a feasible solution in overcoming severe tribological loads at the tool/workpiece interface, and is a future challenge in making manufacturing attractive.

Acknowledgements The authors are grateful for the support from Tribology Innovations for Manufacturing (TriboMAN) IIUM research group. M.H. Sulaiman would like to thank Taiho Kogyo Tribology Research Foundation (TTRF) for the financial support, and CemeCon Scandinavia A/S for the coatings.

References

Amiril SAS, Rahim EA, Syahrullail S (2017) A review on ionic liquids as sustainable lubricants in manufacturing and engineering: recent research, performance, and applications. J Clean Prod 168:1571–1589. https://doi.org/10.1016/j.jclepro.2017.03.197

Azarhoushang B, Akbari J (2007) Ultrasonic-assisted drilling of Inconel 738-LC. Int J Mach Tools Manuf 47:1027–1033

Babitsky V, Astashev V, Meadows A (2007) Vibration excitation and energy transfer during ultrasonically assisted drilling. J Sound Vib 308:805–814

Barnes S (2013) Drilling performance of carbon fiber reinforced epoxy composite when machined dry, with conventional cutting fluid and with a cryogenically cooled tool. In: ASME 2013 international mechanical engineering congress and exposition, pp 1–10

Barnes S, Ascroft H (2015) Study of cutting forces and surface roughness in milling of carbon fibre composite (CFC) with conventional and pressurized CO_2 cutting fluids. In: Proceedings of the ASME 2015 international mechanical engineering congress and exposition, pp 1–8

Bay N (2013) New tribo-systems for cold forming of steel, stainless steel and aluminium alloys. In: Proceedings of 46th International Cold Forging Group (ICFG) plenary meeting, pp 7–04

Bermingham MJ, Kirsch J, Sun S, Palanisamy S, Dargusch MS (2011) New observations on tool life, cutting forces and chip morphology in cryogenic machining Ti-6Al-4V. Int J Mach Tools Manuf 51:500–511. https://doi.org/10.1016/j.ijmachtools.2011.02.009

Biksa A, Yamamoto K, Dosbaeva G, Veldhuis SC, Fox-Rabinovich GS, Elfizy A, Wagg T, Shuster LS (2010) Wear behavior of adaptive nano-multilayered AlTiN/MexN PVD coatings during machining of aerospace alloys. Tribol Int 43:1491–1499. https://doi.org/10.1016/j.triboint.2010.02.008

Ceron E, Bay N (2013) A methodology for off-line evaluation of new environmentally friendly tribo-systems for sheet metal forming. CIRP Ann Manuf Technol 62:231–234. https://doi.org/10.1016/j.cirp.2013.03.062

Ceron E, Olsson M, Bay N (2014) Lubricant film breakdown and material pick-up in sheet forming of advanced high strength steels and stainless steels when using environmental friendly lubricants. Adv Mater Res 967:219–227. https://doi.org/10.4028/www.scientific.net/AMR.966-967.219

Chang S, Bone G (2005) Burr size reduction in drilling by ultrasonic assistance. Robot Comput Integr Manuf 21:442–450

Chuan P, Ghani JA, Jameelah M, Ria T (2013) Characterization of TiCN and TiCN/ZrN coatings for cutting tool application. Ceram Int 39:1293–1298

Courbon C, Pusavec F, Dumont F, Rech J, Kopac J (2013) Tribological behaviour of Ti6Al4V and Inconel718 under dry and cryogenic conditions—application to the context of machining with carbide tools. Tribol Int 66:72–82. https://doi.org/10.1016/j.triboint.2013.04.010

Dahnel AN (2017) Conventional and ultrasonic assisted drilling of carbon fibre reinforced polymer/titanium alloy stacks (Doctoral dissertation, University of Warwick)

Dahnel AN, Ascroft H, Barnes S, Gloger M (2015) Analysis of tool wear and hole quality during ultrasonic assisted drilling (UAD) of carbon fibre composite (CFC)/titanium alloy (Ti6Al4V) stacks. In: ASME 2015 international mechanical engineering congress and exposition, pp 1–10

Dahnel AN, Ascroft H, Barnes S (2016) The effect of varying cutting speeds on tool wear during conventional and ultrasonic assisted drilling (UAD) of carbon fibre composite (CFC) and titanium alloy stacks. Procedia CIRP 46:420–423. https://doi.org/10.1016/j.procir.2016.04.044

de Neergaard A (2017) Personal communication with M.H. Sulaiman

Dhar NR, Paul S, Chattopadhyay AB (2001) The influence of cryogenic cooling on tool wear, dimensional accuracy and surface finish in turning AISI 1040 and E4340C steels. Wear 249:932–942. https://doi.org/10.1016/S0043-1648(01)00825-0

Gassner M, Schalk N, Tkadletz M, Pohler M, Czettl C, Mitterer C (2018) Influence of cutting speed and workpiece material on the wear mechanisms of CVD TiCN/α-Al2O3 coated cutting inserts during turning. Wear 398–399:90–98

Holmberg K, Laukkanen A, Ghabchi A, Rombouts M, Turunen E, Waudby R, Suhonen T, Valtonen K, Sarlin E (2014) Computational modelling based wear resistance analysis of thick composite coatings. Tribiology Int 72:13–30. https://doi.org/10.1016/j.triboint.2013.12.001
Hong SY, Ding Y, Jeong W (2001a) Friction and cutting forces in cryogenic machining of Ti–6Al–4V. Int J Mach Tools Manuf 41:2271–2285. https://doi.org/10.1016/S0890-6955(01)00029-3
Hong SY, Markus I, Jeong W (2001b) New cooling approach and tool life improvement in cryogenic machining of titanium alloy Ti-6Al-4V. Int J Mach Tools Manuf 41:2245–2260. https://doi.org/10.1016/S0890-6955(01)00041-4
Hosseini A, Kishawy H (2014) Cutting tool materials and tool wear. In: Davim J (ed) Machining of titanium alloys. Springer, Berlin
Jawahir IS, Attia H, Biermann D, Duflou J, Klocke F, Meyer D, Newman ST, Pusavec F, Putz M, Rech J, Schulze V, Umbrello D (2016) Cryogenic manufacturing processes. CIRP Ann Manuf Technol 65:713–736. https://doi.org/10.1016/j.cirp.2016.06.007
Kaynak Y (2014) Evaluation of machining performance in cryogenic machining of Inconel 718 and comparison with dry and MQL machining. Int J Adv Manuf Technol 72:919–933. https://doi.org/10.1007/s00170-014-5683-0
Kaynak Y, Karaca HE, Jawahir IS (2011) Cryogenic machining of NiTi shape memory alloys. In: 6th international conference and exhibition on design and production of machines and dies/molds, pp 123–128
Kaynak Y, Karaca HE, Noebe RD, Jawahir IS (2013a) Analysis of tool-wear and cutting force components in dry, preheated, and cryogenic machining of NiTi shape memory alloys. Procedia CIRP 8:498–503. https://doi.org/10.1016/j.procir.2013.06.140
Kaynak Y, Karaca HE, Noebe RD, Jawahir IS (2013b) Tool-wear analysis in cryogenic machining of NiTi shape memory alloys: a comparison of tool-wear performance with dry and MQL machining. Wear 306:51–63. https://doi.org/10.1016/j.wear.2013.05.011
Kijima H, Bay N (2007) Contact conditions in skin-pass rolling. CIRP Ann Manuf Technol 56:301–306. https://doi.org/10.1016/j.cirp.2007.05.070
Kim KS, Kim HK, La JH, Lee SY (2016) Effects of interlayer thickness and the substrate material on the adhesion properties of CrZrN coatings. Jpn J Appl Phys 55:01AA02. https://doi.org/10.7567/JJAP.55.01AA02
Li R, Shih A (2007) Tool temperature in titanium drilling. J Manuf Sci Eng 129:740–749
Liu D, Tang Y, Cong W (2012) A review of mechanical drilling for composite laminates. Compos Struct 94:1265–1279
Lukaszkowicz K (2011) Review of nanocomposite thin films and coatings deposited by PVD and CVD technology. Nanomaterials 145–162. https://doi.org/10.5772/25799
Makhdum F, Phadnis VA, Roy A, Silberschmidt VV (2014) Effect of ultrasonically assisted drilling on carbon fibre reinforced plastics. J Sound Vib 333:5939–5952
Merklein M, Schmidt M, Wartzack S, Tremmel S, Andreas K, Häfner T, Zhao R, Steiner J (2015) Development and evaluation of tool sided surface modifications for dry deep drawing of steel and aluminum alloys. Dry Met Form Open Access J 1:121–133
More AS, Jiang W, Brown WD, Malshe AP (2006) Tool wear and machining performance of cBN–TiN coated carbide inserts and PCBN compact inserts in turning AISI 4340 hardened steel. J Mater Process Technol 180:253–262. https://doi.org/10.1016/j.jmatprotec.2006.06.013
Paul S, Chattopadhyay AB (1996) The effect of cryogenic cooling on grinding forces. Int J Mach Tools Manuf 36:63–72. https://doi.org/10.1016/0890-6955(95)92629-D
Pecat O, Brinksmeier E (2014) Tool wear analyses in low frequency vibration assisted drilling of CFC/Ti6Al4V stack material. Procedia CIRP 14:142–147
Pu Z (2012) Cryogenic machining and burnishing of AZ31B magnesium alloy for enhanced surface integrity and functional performance
Pušavec F, Govekar E, Kopač J, Jawahir IS (2011) The influence of cryogenic cooling on process stability in turning operations. CIRP Ann Manuf Technol 60:101–104. https://doi.org/10.1016/j.cirp.2011.03.096

Raof NA, Ghani JA, Haron CHC (2019) Machining-induced grain refinement of AISI 4340 alloy steel under dry and cryogenic conditions. J Mater Res Technol 8:4347–4353

Rotella G, Umbrello D, Dillon OW, Jawahir IS (2012) Evaluation of process performance for sustainable hard machining. J Adv Mech Des Syst Manuf 6:989–998

Shokrani A, Dhokia V, Newman ST (2012) Environmentally conscious machining of difficult-to-machine materials with regard to cutting fluids. Int J Mach Tools Manuf 57:83–101. https://doi.org/10.1016/j.ijmachtools.2012.02.002

Shokrani A, Dhokia V, Muñoz-Escalona P, Newman ST (2013) State-of-the-art cryogenic machining and processing. Int J Comput Integr Manuf 26:616–648. https://doi.org/10.1080/0951192X.2012.749531

Shyha I, Soo S, Aspinwall D, Bradley S, Perry R, Harden P, Dawson S (2011) Hole quality assessment following drilling of metallic-composite stacks. Int J Mach Tools Manuf 51:569–578

Strano M, Chiappini E, Tirelli S, Albertelli P, Monno M (2013) Comparison of Ti6Al4V machining forces and tool life for cryogenic versus conventional cooling. Proc Inst Mech Eng Part B J Eng Manuf 227:1403–1408. https://doi.org/10.1177/0954405413486635

Sulaiman MH (2017) Development and testing of tailored tool surfaces for sheet metal forming

Sulaiman MH, Christiansen P, Bay N (2017a) A study of DLC coatings for ironing of stainless steel. In: 36th International Deep Drawing Research Group (IDDRG), pp 1–6

Sulaiman MH, Christiansen P, Bay N (2017b) A study of anti-seizure tool coatings for ironing of stainless steel. J Tribol Spec Issue WTC2017 17:1–11

Sulaiman MH, Farahana RN, Mustaffa MN, Bienk K (2019a) Tribological properties of DLC coating under lubricated and dry friction condition. IOP Conf Ser Mater Sci Eng 670:012052

Sulaiman MH, Farahana RN, Bienk K, Nielsen CV, Bay N (2019b) Effects of DLC/TiAlN-coated die on friction and wear in sheet-metal forming under dry and oil-lubricated conditions: experimental and numerical studies. Wear 438–439:203040

Syahrullail S, Zubil BM, Azwadi CSN, Ridzuan MJM (2011) Experimental evaluation of palm oil as lubricant in cold forward extrusion process. Int J Mech Sci 53:549–555. https://doi.org/10.1016/j.ijmecsci.2011.05.002

Tool-life testing with single-point turning tools. STN ISO 3685 (1993)

Venugopal KA, Paul S, Chattopadhyay AB (2007) Growth of tool wear in turning of Ti-6Al-4V alloy under cryogenic cooling. Wear 262:1071–1078. https://doi.org/10.1016/j.wear.2006.11.010

Vereshchaka AA, Vereshchaka AS, Mgaloblishvili O, Morgan MN, Batako AD (2014) Nano-scale multilayered-composite coatings for the cutting tools. Int J Adv Manuf Technol 72:303–317. https://doi.org/10.1007/s00170-014-5673-2

Wang Z, Rajurkar K (1997) Wear of CBN tool in turning of silicon nitride with cryogenic cooling. Int J Mach Tools Manuf 37:319–326

Wang ZY, Rajurkar KP (2000) Cryogenic machining of hard-to-cut materials. Wear 239:168–175. https://doi.org/10.1016/S0043-1648(99)00361-0

Wang MX, Zhang JJ, Yang J, Wang LQ, Li DJ (2007) Influence of Ar/N2 flow ratio on structure and properties of nanoscale ZrN/WN multilayered coatings. Surf Coatings Technol 201:5472–5476. https://doi.org/10.1016/j.surfcoat.2006.07.015

Wang X, Kwon P, Sturtevant C, Kim D, Lantrip J (2014) Comparative tool wear study based on drilling experiments on CFRP/Ti stack and its individual layers. Wear 317:265–276

Yasa E, Pilatin S, Çolak O (2012) Overview of cryogenic cooling in machining of Ti alloys and a case study. J Prod Eng 2:1–9

Yuan Z, Guo Y, Li C, Liu L, Yang B, Song H, Zhai Z, Lu Z, Li H, Staedler T, Huang N, Jiang X (2020) New multilayered diamond/β-SiC composite architectures for high-performance hard coating. Mater Des 186:108207

Zulhanafi P, Syahrullail S (2019) The tribological performances of super olein as fluid lubricant using four-ball tribotester. Tribol Int 130:85–93. https://doi.org/10.1016/j.triboint.2018.09.013

Tribo-analysis of Polymer Composite in Spur Gear

Hemalata Jena and Jitendra Kumar Katiyar

Abstract In the recent years, several studies have been carried out the feasibility of the application of polymer composite spur gears in tribological applications. Due to the heterogeneous nature of composite materials, its wear property behaves differently during meshing of gears. Hence, the present chapter studied different types of polymers and polymer composites and their wear behaviours and wear test in spur gear application. It also provides the importance of fillers in polymer in wear related application. This is an attempt to present a review on tribological performance of polymer and its composites. Tribological analysis is an important role to decide the friction and wear behaviour of polymer spur gear during the meshing of gears. The process parameters like temperature, applied torque, stress and speed affect the friction and wear property of polymer spur gear. The failure of polymer spur gear is widely affected by these parameters generated on tooth. The optimum value of all these parameters reduces the wear and improves the life of gear. Further, the addition of proper functional fillers on polymer are also able to improve the wear property. It is a novel technique to add fillers in neat polymer which is not only a cost effective but also a method to modify the various property of the polymers.

Keywords Spur gear · Reinforcement · Polymer composite · Friction and wear · Fatigue

H. Jena
School of Mechanical Engineering, KIIT Deemed To Be University Bhubaneswar, Odisha 751024, India
e-mail: hemalata.jenafme@kiit.ac.in

J. K. Katiyar (✉)
Department of Mechanical Engineering, SRM Institute of Science and Technology, Tamil Nadu 603203, India
e-mail: jitendrv@srmist.edu.in

M. T. Hameed Sultan et al. (eds.), *Tribological Applications of Composite Materials*, Composites Science and Technology, https://doi.org/10.1007/978-981-15-9635-3_12

1 Introduction

Nowadays, polymer spur gears are widely use to replace metallic gears mainly in low load conditions. Polymer spur gears are most important type of power transmission system which have straight teeth, and are mounted on parallel shafts. Tribo-performance of polymer composite in spur gears get much attention due to its suitability in low load condition and attractive alternatives to traditional metal gears. The other properties like light weight, ability to create low noise, shock and impact and high flexibility in gear geometry are also suitable to make polymer spur gear. But the applications of polymer spur gear are still limited due to its low strength and their teeth are worn away easily. The polymer spur gears are usually failed due to thermal stress which causes wear, pitting, root and pitch crack which do not occur in the metal gear (Yousef 1973). Similarly, the other significant problems of polymer spur gears are:

- Low load-carrying capacity.
- High initial cost of moulding.
- Poor mechanical properties at high and low temperature.
- Poor dimensional stabilities due to poor moisture absorption resistance and high coefficient of thermal expansion.
- Adverse effect when in contact with certain chemicals or lubricants.
- Costs of plastics vary closely with the pricing variation of the petrochemical base material, and thus its cost is more unstable in comparison to the cost of metals.
- Lower accuracy.

These limitations of polymer spur gear as compared to metallic gear restrict its application in a specified domain. But increase in research and development in polymer and its composite, gears are also used in heavy duty application like agricultural, healthcare and automotive engineering. It is observed that 70% decrease in mass, 80% decrease in inertia and up to 9% cut in fuel consumption have been occurred in automotive engineering when polymer spur gears are used instead of metal (Mao et al. 2015). Different polymers are used to manufacture the spur gear which details are given below. Tribological analysis of polymer and polymer composite is a complex phenomenon. For this, mechanical, thermal, wear property etc. are important factor for making an effective gear. Mechanical property like high flexural strength, shear strength to resist the fracture of teeth, compressive strength, stiffness are preferred for the gear. It is required to study the wear behaviour under different loading condition, torque, speed etc. during the meshing of polymer composite gears.

2 Polymers in Spur Gear

Different polymers are used in making the spur gear. These polymers are used for specific application as its properties are varying from polymer to polymer. Polymer spur gears are made of thermoplastic of biodegradable or non-biodegradable polymer as shown in the Table 1. Mostly, thermoplastic polymers are easier to process, and have better flexibility to make a gear than thermosetting polymers. These gears are commonly used in low to medium power transmission like gears used in fax, printers, facsimiles, copy machines and paper mill gears. The high abrasion resistance and self-lubrication property of the polymer are advantageous to use in the gear. Polymer spur gears are usually low load gears, but due to advancement on polymer, they are also used in high load application. Table 2 shows the important mechanical property of polymer which decides a standard gear for power transmission. A brief review of polymers used in gear are given below.

2.1 Acrylonitrile Butadiene Styrene (ABS)

It is a general thermoplastic triblock copolymer derived from petrochemical origin. It is mainly contained acrylonitrile, butadiene and styrene which indicate its name itself. Acrylonitrile is basically a monomer from propylene and ammonia, where butadiene is a petroleum hydrocarbon generated from the C4 fraction of steam cracking and styrene is a dehydrogenation of ethyl (Mukherjee and Saravanan 2019). ABS has good impact strength, rigidity, chemical inertness, adhesive, and resistance to solvent cracking. It is widely used to automotive body parts.

Table 1 Different polymers for gear making (Singh et al. 2018; Adams 1986)

Non-biodegradable Polymer	Biodegradable Polymer
High Density Polyethylene (HDPE)	Polylactide (PLA)
Acetal (Polyoxymethylene-POM)	Polybutylene Succinate (PBS)
Nylon (Polyamide) (PA)	
Polyetheretherketone (PEEK)	
Polybutylene-terephthalate (PBT)	
Polycarbonate (PC)	
Polyphenylene sulphide (PPS)	
Acrylonitrile Butadiene Styrene (ABS)	

Table 2 Mechanical and wear property of different polymer used in gear (Ku et al. 2011; Mitrovic et al. 2018; Xiong et al. 2018; Li et al. 2011; Ramanjaneyulu et al. 2017; Siva Prasad et al. 2012)

Property	Nylon 6	Nylon 66	Acetal	Polyester	ABS	HDPE	PLA	PC	PEEK
Density (gm/cm^3)	1.12–1.14	1.13–1.15	1.41	1.2–1.5	1.0–1.4	0.94	1.30	1.1e-6	1.32
Water absorption 24 h (%)	1.30–1.80	1.0–1.60	0.20	0.1–0.3		0.01–0.2		–	–
Tg (°C)	48	80	–	–	105	−133–100	60	147	–
Tm (°C)	215	250–269	168	–	–	129–140	–	–	343
Coefficient of thermal expansion (mm/mm/°C × 10^5)	8.0–8.86	7.2–9.0	–	–	7.38	12.0–13.0	–	–	–
Tensile strength (MPa)	43.0–79.0	12.4–94.0	70	40–90	27	14.5–38.0	37	55–70	150
Elastic Modulus (GPa)	2.9	2.5–3.9	2.7	–	2.1–7.6	0.4–1.5	4	275	3.5
Elongation (mm)	20–150	35–300	> 15	2.0	3.5–50	2.0–130	6	–	–
Izod impact strength (J/m)	42.7–160	16–654	80	0.15–3.2	–	26.7–1068	–	–	–
Coefficient of friction	–	0.26	0.21	–	–	–	–	–	0.21

2.2 High Density Polyethylene (HDPE)

Polyethylenes (PE) are largely prepared olefin polymers of all the thermoplastic materials. Different families of PE are well established in the polymer market, each having a different structure, behaviour, properties and applications (Feldman and Barbalata 1996). High density polyethylene (HDPE) is one of the families of PE which has good tribological property, cost effective, low temperature, and high moisture resistance. Design and preparation of HDPE gear by compression moulding technique is quite easy, simple and well performed.

2.3 Acetal (Poly-Oxymethylene)

The acetal polymer is strongest, stiffest among all thermoplastics. It is a crystalline thermoplastic polymer that is used either in its pure state or slightly altered state. It has good abrasion resistance, malleability, resilience, fatigue life, electrical properties, and sharp melting point, high strength, stiffness and toughness. Low moisture sensitivity, dimensional stability, low coefficient of friction, eases of machinability, and resistance to solvents and chemicals make it to a special polymer for gear (Ramanjaneyulu et al. 2017). The mechanical and chemical properties can also sustain in a variety of environmental conditions (Zahran 1998). Due to this, it is a major structural constituent of pump and fan propellers, bearing liners, rings and etc. The higher crystallinity level of Polyoxymethylene (POM) (approximately up to 70%) distinguishes it from other structural polymers. Due to this, higher content of crystallinity, local macromolecular arrangements form into lamella-like aggregates (fibrilles). The group of these fibrilles (form of grains) are termed as spherolites. As a result of these special structures, POM exhibits excellent mechanical (Dziadur 2001).

2.4 Polybutylene Succinate (PBS)

Polybutylene Succinate (PBS) is a biodegradable polymer derived from petrochemical resources. It is an aliphatic polyester of 1, 4-butandiol and succinic acid. It is generally used in the packaging field like disposable dishware, flexible packaging, and coated paper etc. (Xu and Guo 2010). It is also mainly used to make eco-friendly fishing gear (Deroine et al. 2010).

2.5 *Nylon (Polyamides)*

Polyamide gears offer low friction, weight, cost along with noise free operation and oil-less condition (Senthilvelan and Gnanamoorthy 2006a). It is a commonly used material for fishing gears due to its persistence in the marine environment for several hundred years. It is also used in the automotive industry. Due to its low load-carrying capacity, small service life, and low heat resistance, it is avoided in particularly intense load, speed, or high ambient temperature conditions application (Lin and Kuang 2008). Nylon of different types (Nylon6, Nylon 6–6, and Nylon 11 etc.) are widely used for making gears.

2.6 *Polyetheretherketone (PEEK)*

Polyetheretherketone (PEEK) shows good resistance to high temperature generated during friction. It is a special kind of polymer in which the glass transition and melting range can be customized by varying the ketone and ether groups in the polymer chain (Friedrich 1995).

2.7 *Polylactide (PLA)*

Polylactide (PLA) is a biomass which is derived from renewable resources like corn starch or sugar cane. It is a non-petrolium polymer unlike most plastics. It is biodegradable polymer which has similar features of polypropylene (PP), polyethylene (PE), or polystyrene (PS). Hence production of PLA gears from already existing manufacturing equipment used for petrochemical industry plastics like PP, PE, PS are cost effective. It is found that 3D printed spur gears manufactured from PLA plastic have better operational characteristics than the ones manufacture from ABS plastic (Aleksandar et al. 2018).

2.8 *Polybutylene-Terephthalate (PBT)*

Polybutylene-Terephthalate (PBT) is a thermoplastic semi-crystalline polymer of polyester group. Some of the remarkable properties of PBT are: high dielectric strength, good electrical properties, high chemical and heat resistance, good strength and modulus at elevated temperatures. Its property like dimensional stability, easy to manufacture due to fast crystallization and fast cooling is made to possible a standard gear (Haruhara et al. 2009).

2.9 Polycarbonate (PC)

Polycarbonate (PC) polymer is having carbonate groups (–O–(C=O)–O–). This amorphous polycarbonate polymer has good thermal, electrical, optical, mechanical properties and UV resistance. Its transparency property is similar to that of the acrylic (Siva Prasad et al. 2012). It is mainly used without any reinforcement phase condition.

2.10 Polyurethane (PUR)

Polyurethane (PUR) is a special kind of polymer which may be a thermosetting or thermoplastic polymer. Two types of polyurethane are common: polyester based and polyether based. It has good abrasion resistance, but it has low resistance to organic solvent and poor stability in direct sunlight (Patel and Purohit 2017).

3 Polymer Composite Gear

Investigations on plastic gear performance have been widely increased due to their several advantages over the metal as discussed above. But in order to improve the performance of the neat polymer in the field of mechanical, wear, thermal etc., it is advisable to add reinforcement as second phase in the neat polymer like fibre or filler. This is known as polymer composite. Researchers have incorporated the fibres like glass, carbon, natural fibre etc. and fillers like graphene, carbon nanotube, MoS_2, glass bead, etc. in the polymer to achieve the better mechanical and wear characteristics in polymer. Jena et al. (2013), have added the cenosphere in bamboo-epoxy composite to improve its mechanical property. Addition of filler/fibre not only reduces the cost of polymer and consumption of petroleum based non-degradable polymer but also improves the property of neat polymer. Short-fibre reinforced polymer composite materials are used as reinforcement in the thermosetting polymer. Table 3 shows the gear geometry of different polymer and polymer composite to gather the knowledge of polymer composite used for gear application. Manufacturing of polymer spur gears are same with the machining process as metal gears, usually milling or hobbing from a blank. It can also be fabricated either by injection moulding or machined from a rod (Singh et al. 2018) and additive manufacturing.

Kurokawa et al. (2000a, b), have used carbon fibre (CF) with PEEK, Polyamideimide, in the ratio of 15 vol.% and 30 wt.%, respectively and polyphenylenesulphide (PPS) containing glass fibre (GF) of 30 wt.%. It is observed that CF-PEEK has high performance load bearing capability and better performance as compared to the other two polymer composite gear. Similarly, glass fibre of 28 wt. % reinforced in POM (Polyoxymethylene) shows 50% better load capacity, compared to neat POM gear pair (Mao et al. 2019). It is observed that polyoxymethylene (POM)

Table 3 Gear geometry of different polymer and polymer composite (Mao et al. 2015; Ramanjaneyulu et al. 2017; Li et al. 2011; Mao et al. 2019; Pandya and Parey 2013; Kurokawa et al. 2003; Singh and Singh 2018; Sukumaran et al. 2012; Sudhagar et al. 2019)

Composite	Manufacturing process	Tooth thickness (mm)	Module (mm)	Standard pressure angle (degree)	No. of teeth	Face width (mm)	Pitch diameter (mm)	Contact ratio
Acetal	Machine cut	3.14	2	20	30	15	–	1.67
Acetal	Injection mould	3.14	2	20	30	17	–	1.67
Acetal	Modelling	25.4	–	20	20	8	127	–
Acetal, Nylon 66, PEEK	Injection mould	3.14	2	20	30	17	–	1.65
POM + 28 wt.% Glass fibre	Injection mould	3.14	2	20	30	17	–	1.67
PC	Modelling	–	10	20	15	6	64.52	1.43
Nylon 12 + 15 wt.% CF, Nylon 6 + 15 wt.% CF, Nylon 46 + 15 wt.% CF, Nylon 66 + 30 wt.% CF, Nylon 46 + 30 wt.% CF	Injection mould	–	1	20	30	8	30	–

(continued)

Table 3 (continued)

Composite	Manufacturing process	Tooth thickness (mm)	Module (mm)	Standard pressure angle (degree)	No. of teeth	Face width (mm)	Pitch diameter (mm)	Contact ratio
ABS, POM, HDPE	Injection mould	–	2	20	20	8	40	1.57
Nylon	Modelling	–	10	20	12	5	Addendum diameter (mm) = 139.47	1
PS, PS+ (5–20 wt. %) Bauhinia Racemosa Fibre, PS+ (5–20 wt. %) Madar Fiber	Gear Hobbing	4.5	2	20	30	18	56.38	–

has minimum wear rate as compared to acrylonitrile butadiene styrene (ABS), high density polyethylene (HDPE) (Singh and Singh 2018). Kurokawa et al. (2000a, b), have studied the wear behaviour of POM composites containing silicon carbide (SiC) and/or calcium salt of octacosanoic acid (Ca-OCA) and polytetrafluoroethylene (PTFE). It is observed that combine effect of SiC and Ca-OCA is decreased the coefficient of friction and wear rate due to the lubrication effect Ca-OCA. Addition of 30 wt. % of glass fibre addition in nylon 66 improve the wear resistance of the composite (Singh 2017). Similarly 28 wt. % of glass fibre in POM gives minimum wear rate (Mao et al. 2019). Kurokawa et al. (2003), have investigated the effect of carbon fibre in nylon 6, nylon 12, nylon 46, and nylon 66 in related to wear rate and it is noted to be minimum at nylon 12. Senthilvelan and Gnanamoorthy (2006b), have prepared the nylon 66 spur gear with addition of short carbon fibre. Due to the influence of fibre/filler in neat polymer, several authors have attempted to improve the wear property of the polymer which is a new emerging technique in the composite research (Jena et al. 2016; Patnaik et al. 2010). Table 4 also shows that addition of fibre/filler in polymer are capable to modify the coefficient of friction and wear resistance of the neat polymer.

Table 4 Coefficient of dry friction of polymers at friction with steel (Starzhinsky 2013; Kurokawa et al. 2000a, b)

Materials	Static coefficient of friction	Dynamic coefficient of friction	Wear rate (10^{-5} m^2/N)
POM	0.20–0.35	0.35–0.45	35.0
PA6	0.20–0.40	0.20–0.30	1.0
PA6.6	0.30–0.60	0.25–0.50	5.0
PEEK	0.45–0.55	0.30–0.50	10.0
PPS	0.25–0.40	0.25–0.40	2.0
PI(polyimide)	0.35–0.50	0.30–0.40	2.0
PBT	0.41–0.46	–	–
PC	0.35	–	–
PAI(polyamide imide)	0.15–0.30	–	–
Graphite filled Nylon 6	0.20–0.40	–	–
Glass fibre filled Nylon 6	0.28–0.32	–	–
Glass fibre filled PBT	0.19–0.24	–	–
Neat POM SiC filled POM SiC and Ca-OCA filled POM PTFE filled POM	0.45 0.70 0.30 0.30	– – – –	80.0 0.2 3.0 0.03

4 Friction and Wear of Polymer Spur Gear

Wear property like wear resistance and low coefficient of friction and thermal property like high heat resistance are preferred for making an effective gear. When plastic gears are mesh with same polymer spur gear or metal gear, the heat generation in the friction of these pairs is more prominent than for metal–metal pairs. The wears are mostly generated due to sliding, corrosion, and erosion (Jena 2019). The gears are encountered mostly with sliding wear where removals of materials are due to the mechanical, chemical, or other external interaction. Wear in gears is never acceptable as it reduces the kinematics accuracy and causes the tooth fracture. Estimation of wear can be done by using factors like: wear rate, coefficient of friction. Coefficient of friction of polymer plays an important role in heat liberation during meshing of gear tooth. Because the objective is to decrease the heat generation in the friction unit of polymer spur gear. Thermoplastics gear are used for meshing pairs without lubrication can be employed only at low and intermediate loads and operating temperatures up to 100°. Table 4 shows the coefficient of friction of different polymers and polymer composites which are measured at 23° at dry friction against steel. From the table it is observed that addition of filler/fibre in neat polymer, the coefficient of friction has been modified.

The wear rate of polymers is also important factor for determination of effective gear. Table 4 shows the wear rate of polymer which is also measured at 23 °C at dry friction against steel. Efficiency (ρ) of designing kinematic power transmissions is also an important factor for spur gears which can be determined from the following expressions:

$$\rho = 1 - C\mu\pi\left(\frac{1}{z_1} - \frac{1}{z_2}\right)$$

where

$$C = \frac{(F_t + 2.920)}{(F_t + 0.174)}$$

μ = Coefficient of friction.
Z_1 = Number of teeth in driver gear.
Z_2 = Number of teeth in driven gear.

Efficiency values for a cylindrical gear pair, bevel gear pair are 0.96–0.98 and 0.95–0.97 respectively (Starzhinsky 2013).

Yousef et al. (2015) have developed an in-house universal test rig which can be able to measure the wear characteristics of different types of polymer spur gears such as spur, helical, bevel and worm. The test rig has three different units. First unit is used to investigate the bevel gears, second unit used to investigate the spur and helical gear and third unit is used to investigate the worm gear. The developed universal test rig is shown in Fig. 1. It is observed that only two types of wear i.e. rolling and sliding are found in the gear. Moreover, the wear rate in helical, bevel and

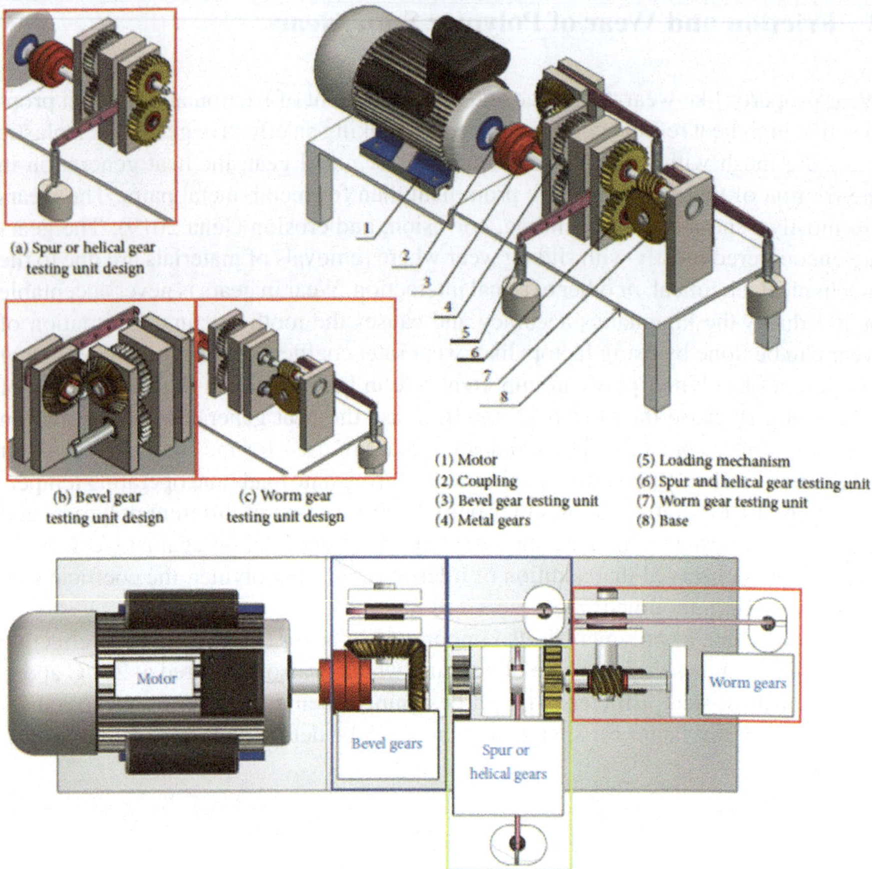

Fig. 1 Developed universal test rig (Reproduced with permission of Yousef et al. 2015)

worm gear is more as compare to spur gear. This is because of the presence of axial load, bending stress and contact stresses. The different phases present in different types of gears are presented in Table 5.

5 Process Influence the Wear Behaviour of Polymer Spur Gear

Every polymer material has different elastic modulus and thermal conductivity. So that polymer spur gears behaves differently as from metallic gears. Tooth deflection, melting of tooth, pitting effect, surface and subsurface cracks, fracture on tooth, etc. are responsible for failure of polymer spur gears in various investigations. Due to these failure modes, Jain et al. (2019) have categorised the failure criteria into three.

Table 5 Wear phase in different types of polymer spur gears (Reproduced with permission of Yousef et al. 2015)

	Running-in phase	Linear phase	Rapid-wear phase
Acetal pair of spur gears	Duration from 0 to 15 $\times 10^3$ Cycles weight loss 2.3 mg	Duration from 15 $\times 10^3$ to 200 $\times 10^3$ cycles weight loss 11 mg	Not found
Acetal pair of helical gears	Duration from 0 to 16 $\times 10^3$ Cycles weight loss 4 mg	Duration from 16 $\times 10^3$ to 85 $\times 10^3$ cycles weight loss 14 mg	Duration from 85 $\times 10^3$ to 200 $\times 10^3$ cycles weight loss 26 mg
Acetal pair of bevel gears	Duration from 0 to 45 $\times 10^3$ Cycles weight loss 7 mg	Duration from 53 $\times 10^3$ to 145 $\times 10^3$ cycles weight loss 16 mg	Duration from 53 $\times 10^3$ to 200 $\times 10^3$ cycles weight loss 29 mg
Acetal pair of worn gears	Duration from 0 to 12 $\times 10^3$ Cycles weight loss 4.4 mg	Duration from 12 $\times 10^3$ to 90 $\times 10^3$ cycles weight loss 19 mg	Duration from 90 $\times 10^3$ to 200 $\times 10^3$ cycles weight loss 35 mg

These are failure due to wear, failure due to temperature and failure due to cyclic loading. The material properties of polymer composites are affected by all three types of failure. Nozawa et al. (2009) have studied the polymer metal hybrid gears. They have used nylon66/poly (phenylene ether) polymer sheet over steel gear teeth and performed the test at 1000 RPM speed and high torque against steel gear for evaluation of friction coefficient, surface morphology under molybdenum grease. It is observed that a periodical torque change when speed is varied. Further, noise is abruptly reduced as comparison to steel/steel pair but when polymer sheet delaminated then it is increased further. The delamination is occurred due to poor adhesive strength. Yakut et al. (2009), have studied the load carrying capacity and failure of PC/ABS spur gear. They have applied three different loads with two different speeds on FZG test rig. For calculation of specific wear rate, they have used the following formula

$$W_v = \frac{V}{2zmbN}$$

where W_v is the wear volume in mm^3, z is the number of teeth in pinion gear, m is the module of gear in mm, b is the tooth width in mm and N is the total number of revolution. It is reported that at lower load and speed, the accumulated heat on steel gear spreads out easily but at higher load, thermal failure is occurred due to the melting of tooth.

Further, Hoskins et al. (2011), have discussed about the noise generation due to friction and wear between two mating gears even though polymer spur gears are considered as a low noise component because of their lower modulus of elasticity and self-lubrication property. They observe the noise level between two mating gears using acoustic emission. For study, they have taken different polymer materials and tested under different load and speed. It is observed that the sound level is proportional to load but inversely proportional to speed in case of POM but for other materials it is proportional to both speed and load which is shown in Fig. 2. Moreover, surface roughness of materials also affects the noise level and wear is directly proportional to the surface roughness.

Mostly three process parameters are reported by researchers which adversely affected the wear behaviour of polymer spur gears. These are cyclic loading, temperature, mechanical properties of polymers (Yakut et al. 2009; Chernets et al. 2018; Myshkin and Kovalev 2018). Some researchers also reported the failure of spur gears due to contact stresses and bending stresses (Lu et al. 2019; Miler et al. 2019). At lower load and speed, the temperature generated on gear tooth is very well spread out but when speed and load increases, the temperature increased drastically and crossed the thermal equilibrium because of the more frictional forces at the interface. Due to which the failure occurs on gear tooth is known as failure due to thermal damage

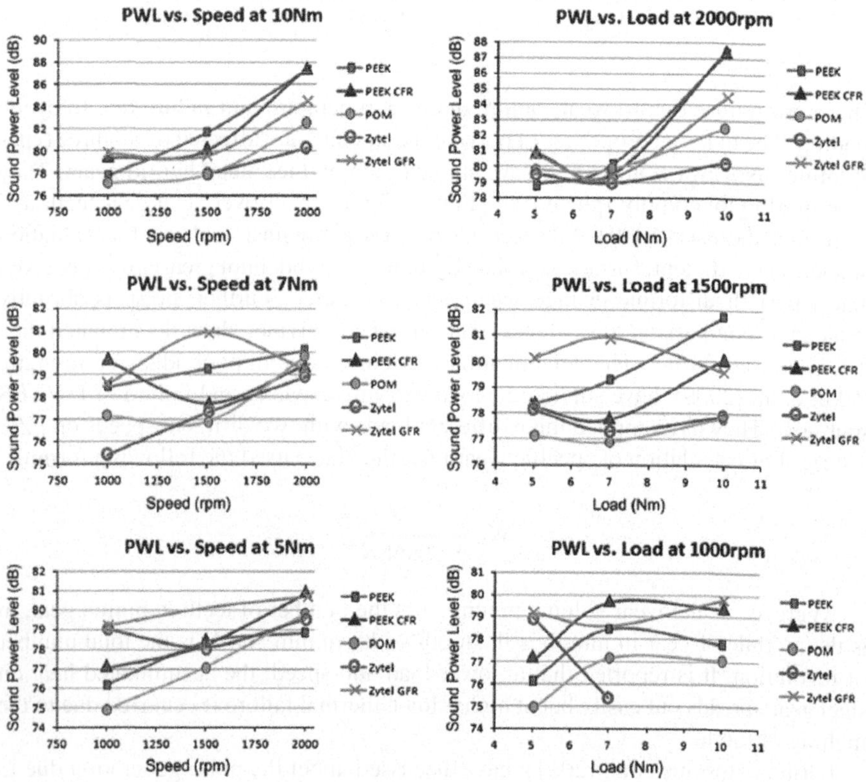

Fig. 2 Variation on sound level under different load and speed for different polymer materials (Reproduced with permission of Hoskins et al. 2011)

which makes the materials brittle and materials structure shows the glassing properties. To reduce the wear of gear tooth, the process parameters such as temperature, applied torque, speed are very important to optimize so that higher life of gear tooth can be achieved.

Yousef et al. (2013), have blended the carbon nanotube (CNT) in polymer composite and fabricated a spur gear. The test is performed at 1420 RPM speed, 13 and 16 Nm torque for 200×10^3 cycles. It is observed that addition of 15 wt% of CNT, the wear resistant is increased. This is because of the change in mechanical properties such as young's modulus, stiffness etc. Afifi et al. (2018) have added the multilayer graphene nanoplatelets (MLNGPs) in varying concentration (0.1–0.5 wt%) on polyamine 6 for fabrication of spur gear using melt mixing method followed by injection moulding process. The fabricated gear is having the capacity to derive the system i.e. 1.5 hp. The addition of fillers make the gear of chemically resistance, thermally stable and moisture resistance. The obtained results reveal that 0.3 wt % MLGNPs gives the 40% higher Young's modulus, 25% higher microhardness, 37% higher storage modulus, and 14% higher glass transition temperature. Further wear is decreased by 35% at 16 Nm torque and 54% at 13 Nm torque. Kalin and Kupec (2017) have studied the failure due to fatigue on polymer spur gears.

They have reported that temperature from 30° to 70° influences the fatigue failure of gears. It is observed that increasing in temperature reduces the fatigue life of polymer spur gear. Furthermore, the increment in temperature decreases the efficiency of gear.

Singh and Siddhartha (2017,2018) have developed the in-situ method for fabrication of functionally graded material-based polymer composite gears. Durability analysis of polymer-based gears such as ABS, HDPE and POM under various torque levels and different rotational speed are conducted by using power absorbing type gear test rig (CM-9108) of DUCOM instruments, India. The gear is manufactured by using injection moulding process. They have reported that the failure in ABS gear occurred because of the excessive wear of the gear teeth whereas HDPE fails because of cracking on the root of the teeth. POM gear shows the better life from ABS and HDPE even after 2 million cycles. They have also reported that the increment in torque increases surface temperature. The following relation shows the relationship among rotational speed, contact period and strain rate of gear tooth. It indicates that contact period decreases at higher speed, strain rate and increases with reduced contact period at a constant torque.

$$\text{Contact of a tooth} = \frac{1}{\text{Number of tooth} \times \text{Speed}}$$

$$\text{Stain rate or rate of loading} = \frac{\text{Torque acting on the tooth}}{\text{Contact Period} \times \text{Contact ratio}}$$

It is also observed that loading rate on gears is inverse relationship with contact period and contact ratio. Moreover, Zorko et al. (2019), have conducted durability test on steel/PEEK gear pair under various torque as well as dry and lubricated conditions

Fig. 3 Test rig and area of the temperature measurements at 1.2 Nm and 1600 rpm and failed after 5.94×10^6 cycles (Reproduced with permission of Zorko et al. 2019)

at room temperature. In all test steel gear act as a driver and polymer spur gear act as a driven. The combination is shown in Fig. 3. From test, they reported that the life of polymer spur gear increased by 1.23 times at greasy environment. Further, some additional treatment by trovalisation (it is a surface treatment process in a vibratory machine using abrasive materials) extend the life by 2.54 time.

After test, they have also observed the mode of failure of polymer spur gear and reported that in both cases flanks of the polymer spur gears are heavily worn and also the cracks in Dedendum flank area which is shown in Fig. 4.

Sarita and Senthilvelan (2019), have prepared the gear through injection moulded and performed durability test on power absorption test rig under lubricated condition. It is observed that the combination of rolling and sliding between gear tooth causes the hysteresis heating and frictional heating. Further fatigue performance of the gear have shown cracks, local softening, and plastic deformation on gear tooth. They have reported the summary of failure modes which is shown in Table 6.

Miler et al. (2019), have predicted the friction coefficient of polyoxymethylene spur gear paper under dry film lubricant using full factorial method. It is reported that friction coefficient is widely influenced by sliding velocity, relative curvature and applied load. From optimization, it is observed that radius of relative curvature above 5 mm have no influence in friction coefficient. Mao et al. (2019), have tested the polymer composite gears with or without glass fibre reinforcement. They have manufactured gears by injection molding process and performed the friction test polymer against polymer spur gear. It is noticed that around 28% and 50% increment between with and without reinforcement polymer composite gear in performance and

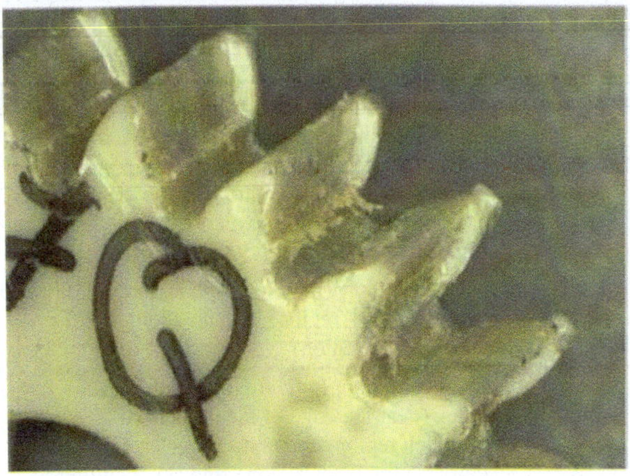

Fig. 4 PEEK-gear wear during operation with no added lubrication at torque 1.0 Nm, rotational speed 1600 rpm, wear after 9.13×10^6 cycles (Reproduced with permission of Zorko et al. 2019)

loading capacity, respectively. It is also reported that below the critical value, specific wear rate is very low. Furthermore, Lu et al. (2019), have investigated the failure mode in PEEK and steel gear pair under lubricated condition. They have performed the test in FZG test rig with C type gear pair and observed the pitting induced tooth breaking under lower and moderate loading condition while under heavy load condition, it is observed a tooth root breakage because of the insufficiency of the bending strength. It is observed a micro pits near the pitch line which is shown in Fig. 5.

6 Conclusions

Present study reveals that polymer is widely and effectively used to prepare the spur gear which play an important role in power transmission system. Thermoplastic polymer is more convenient as compared to thermosetting polymer for fabrication of polymer spur gear. Moreover, the tribological study of polymer spur gears are very important because it is widely dependent upon the properties of polymers and it decides the life of the polymer spur gear. The failure of polymer spur gear is widely affected by temperature, applied torque, speed and stresses generated on tooth. The optimum value of all these parameters are very important to increase the life of gear. These failures can be overcome by adding the fillers on polymer but the selection of appropriate fibre/filler content also very important according to the application of polymer spur gear. Fillers like graphene, carbon nanotube etc. and fibre like glass, carbon are added in the neat polymer to improve its wear property.

Table 6 Summary of failure mode at different torques (Reproduced with permission of Sarita and Senthilvelan 2019)

Test conditions	Loads (Nm)	Failure mode of test gear	Life (10^5 cycles)
Dry	1.8	Pitting (pitch region), subsurface crack (near edge of the face and front face of the pitch region), plastic flow (middle of the face region) and mild sliding marks (lower flank region)	8.6
Lubricated	1.8	No plastic flow (middle of the face region), scuffing (flank region), and no sliding marks (lower flank region)	14.4
Dry	3.5	Local softening (pitch region), subsurface crack (near the edge of face and front face of the pitch region) and abnormal wear/plastic deformation (root region), and sever sliding marks (lower flank region)	2.8
Lubricated	3.5	Plastic flow (face region), and scuffing (flank region), and very sliding marks (lower flank region)	5.7
Dry	4	All modes of failure similar to the 3.5 Nm dry condition failure but comparatively severe in nature, subsurface crack (front face of the region), and accumulated sliding material (pitch region)	1.4
Lubricated	4	Flake off of fragment (middle of the face region), scuffing and ripple (flank region), and sever sliding marks (lower flank region)	4.3
Dry	4.5	Severe plastic flow (middle of the face region), subsurface cracks (near the edge of face and front face of pitch region), severe thermal damage (pitch region) and abnormal wear/plastic deformation (root region)	1.29
Lubricated	4.5	No subsurface cracks (near the edge of the face region), and sever scuffing (root region)	2.8

7 Future Scope

- In depth study is required of influence of fibre/filler in polymer for preparation of spur gear
- Bevel gears, step gears, herring bone gears, worm gears, and others should be studied in the wear related application.
- Study the effect of manufacturing process of spur gear on wear behaviour.

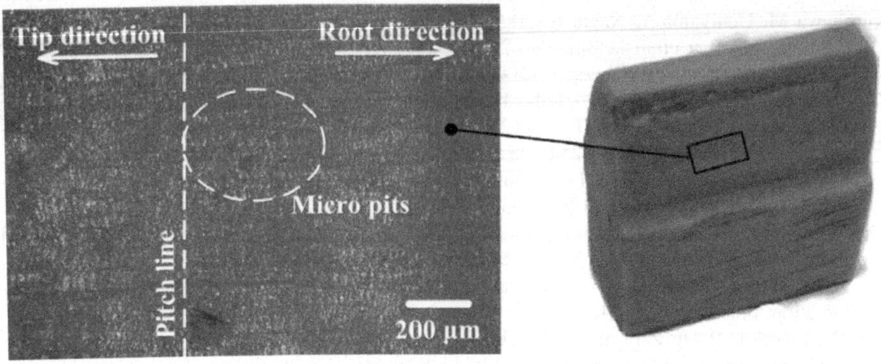

Fig. 5 The micro-pits on the PEEK gear under the case of 15 Nm (Reproduced with permission of Lu et al. 2019)

References

Adams CE (1986) Plastic gearing: selection and application. M Dekker, New York, pp 39–65

Afifi EM, Elshalakny AB, Osman TA et al (2018) Investigation of gear performance of MLNGPs as an additive on polyamide 6 spur gear, Fullerenes. Nanotubes Carbon Nanostruct 26(6):351–359

Aleksandar D, Zarko M, Radivoje M et al (2018) The influence of material on the operational characteristics of spur gears manufactured by the 3d printing technology. J Mech Eng Strojnícky Časopis 68(3):261–270

Chernets MV, Shilko SV, Pashechko MI et al (2018) Wear resistance of glass- and carbon-filled polyamide composites for metal-polymer gears. J Fric Wear 39(5):361–364

Deroine M, Pillin I, Le Maguer G et al (2010) Development of new generation fishing gear: a resistant and biodegradable monofilament. Polym Test 74:163–169

Dziadur WC (2001) The effect of some elastomers on the structure and mechanical properties of polyoxymethylene. Mater Charact 46:131–135

Feldman D, Barbalata A (1996) Synthetic polymers: technology properties applications. Springer, New York

Friedrich K, Lu Z, Hager AM (1995) Recent advances in polymer composites tribology. Wear 190:139–144

Haruhara J, Hirakawa T, Hanabusa K et al (2009) Flame retardant polybutylene terephthalate resin composition—US Patent App. 11/991, 380

Hoskins TJ, Dearn KD, Kukureka SN et al (2011) Acoustic noise from polymer gears—a tribological investigation. Mater Des 32:3509–3515

Jain M, Patil S, Ghosh SS (2019) A review on failure characteristics of polymeric gears. AIP Conf Procs. https://doi.org/10.1063/1.5123979

Jena H, Pandit MK, Pradhan AK (2013) Effect of cenosphere on mechanical properties of bamboo–epoxy composites. J Reinf Plast Compos 32(11):794–801

Jena H, Pandit MK, Pradhan AK (2016) Study of solid particle erosion wear behaviour of bamboo fiber reinforced polymer composite with cenosphere filler. Adv Polym Technol. https://doi.org/10.1002/adv.21718

Jena H (2019) Study of tribo-performance and application of polymer composite. In: Automotive tribology, 1st edn. Springer, Berlin, pp 65–99

Kalin M, Kupec A (2017) The dominant effect of temperature on the fatigue behaviour of polymer gears. Wear 376–377:1339–1346

Ku H, Wang H, Pattarachaiyakoo N et al (2011) A review on the tensile properties of natural fibre reinforced polymer composites. Compos Part B 42:856–873

Kurokawa M, Uchiyama Y, Nagai S (2000a) Performance of plastic gear made of carbon fiber reinforced polyether-ether-ketone: Part 2. Tribo Int 33:715–721
Kurokawa M, Uchiyama Y, Nagai S (2000b) Tribological properties and gear performance of polyoxymethylene composites. J Tribo 122:809–814
Kurokawa M, Uchiyama Y, Iwai T et al (2003) Performance of plastic gear made of carbon fiber reinforced polyamide 12. Wear 254:468–473
Li W, Wood A, Weidig R et al (2011) An investigation on the wear behaviour of dissimilar polymer gear engagements. Wear 271:2176–2183
Lin AD, Kuang JH (2008) Dynamic interaction between contact loads and tooth wear of engaged polyamide gear pairs. Int J Mechan Sci 50:205–213
Lu Z, Liu H, Zhu C et al (2019) Identification of failure modes of a PEEK-steel gear pair under lubrication. Int J Fatig 125:342–348
Mao K, Langlois P, Hu Z et al (2015) The wear and thermal mechanical contact behaviour of machine cut polymer gears. Wear 332–333:822–826
Mao K, Greenwood D, Ramakrishnan R et al (2019) The wear resistance improvement of fibre reinforced polymer composite gears. Wear 426–427:1033–1039
Miler D, Hoi M, Domitran Z et al (2019) Prediction of friction coefficient in dry-lubricated polyoxymethylene spur gear pairs. Mech Mach Theor 138:205–222
Mitrovic R, Miskovic Z, Ristivojevic M et al (2018) Determination of optimal parameters for rapid prototyping of the involute gears. IOP Conf Ser Mater Sci Eng. https://doi.org/10.1088/1757-899X/393/1/012105
Mukherjee P, Rani A, Saravanan P (2019) Polymeric materials for 3D bioprinting, In: Ahmad N, Gopinath P, Dutta R (eds) 3D printing technology in nanomedicine, 1st edn. Elsevier (Chapter 4)
Nozawa J, Komoto T, Kawai T, Kumehara H et al (2009) Tribological properties of polymer-sheet-adhered metal hybrid gear. Wear 266:893–897
Pandya Y, Parey A (2013) Experimental investigation of spur gear tooth mesh stiffness in the presence of crack using photo elasticity technique. Eng Fail Anal 34:488–500
Patil A, Patel A, Purohit R (2017) An overview of polymeric materials for automotive applications. Mater Today Proc 4:3807–3815
Patnaik A, Satapathy A, Chand N et al (2010) Solid particle erosion wear characteristics of fibre and particulate filled polymer composites: a review. Wear 268:249–263
Ramanjaneyulu S, Suman KNS, Phani Kumar S et al (2017) Design and development of graphene reinforced acetal copolymer plastic gears and its performance evaluation. Mater Today Proc 4(8):8678–8687
Sarita B, Senthilvelan S (2019) Effects of lubricant on the surface durability of an injection molded polyamide 66 spur gear paired with a steel gear. Tribo Int 137:193–211
Senthilvelan S, Gnanamoorthy R (2006a) Effect of gear tooth fillet radius on the performance of injection molded nylon 6/6 gears. Mater Des 27:632–639
Senthilvelan S, Gnanamoorthy R (2006b) Carbon fibre reinforced nylon 66 spur gears: development and performance. Appl Compos Mater 13(1):43–56
Siddhartha AK, Sidhartha SPK (2018) Noise emission from ABS, POM and HDPE spur gears—a comparative study. Mater Today Proc 5:18038–18044
Singh AK, Siddhartha (2017) Thermal and wear behaviour of glass fibre-filled functionally graded material-based polyamide 66 spur gears manufactured by a novel technique. J Tribo 140(2):1–16
Singh PK, Sidharth SAK (2018) An investigation on the thermal and wear behaviour of polymer-based spur gears. Tribo Int 118:264–272
Siva Prasad V, Altaf Hussain S, Pandurangadu V et al (2012) Modelling and analysis of spur gear for sugarcane juice machine under static load condition by using FEA. Int J Mod Eng Res 2(4):2862–2866
Sudhagar S, Mugesh Raja V, Sathees Kumar S, Jonathan Samuel A et al (2019) The wear behavior and service life of Madar and Bauhinia Racemosa reinforced polyester hybrid composites for gear applications. Mater Today Proc. https://doi.org/10.1016/j.matpr.2019.07.738

Sukumaran J, Ando M, DeBaets P et al (2012) Modelling gear contact with twin-disc setup. Tribo Int 49:1–7

Starzhinsky VE (2013) Polymer gears. Springer, Berlin, pp 2592–2602

Xiong X, Shen SZ, Alam N et al (2018) Mechanical and abrasive wear performance of wovenflax fabric/polyoxymethylene composites. Wear 414–415:9–20

Xu J, Guo BH (2010) Microbial succinic acid, its polymer poly (butylene succinate), and applications. In: Plastics from bacteria. Springer, Berlin. https://doi.org/10.1007/978-3-642-03287-5_14

Yakut R, Düzcükoğlu H, Demirci MT (2009) The load capacity of PC/ABS spur gears and investigation of gear damage. Arch Mater Sci Eng 40(1):41–46

Yousef S, Khattab A, Zaki M et al (2013) Wear characterization of carbon nanotubes reinforced polymer gears. IEEE Trans Nano Tech 12(4):616–620

Yousef S, Osman TA, Khattab M et al (2015) A new design of the universal test rig to measure the wear characterizations of polymer acetal gears (spur, helical, bevel, and worm). Adv Tribol. 10.1155/2015/926918

Yousef SS, Burns DJ, McKinlay W (1973) Techniques for assessing the running temperature and fatigue strength of thermoplastic gears. Mech Mach Theor 8:175–185

Zahran RR (1998) Effect of g-irradiation on the ultrasonic and structural properties of polyoxymethylene. Mater Let 37:83–89

Zorko D, Kulovec S, Duhovnik J et al (2019) Durability and design parameters of a Steel/PEEK gear pair. Mech Mach Theor 140:825–846

Tribological Evaluation of Solid Lubricant Enriched in Modified Jatropha-Based Oil as Minimum Quantity Lubrication (MQL) Oil for Composite Material

N. Talib, R. M. Nasir, E. A. Rahim, W. K. Lee, H. Abdullah, and A. Saleh

Abstract The use of lubricant during the machining process plays an important role to reduce friction and wear. Mineral-based oil is the most widely used lubricant that provided high-quality lubrication properties. However, mineral-based oil has poor biodegradability and causes long-term pollution to the environment and harmful to human. Implementation of environmental-friendly lubricant was encouraged to achieve sustainable manufacturing practices. The inherent biodegradability of vegetable-based oil with solid particle offers greater benefit to the environment and lubrication performance. The study aims to evaluate the influence of green solid particle (hexagonal boron nitride, hBN) enriched in the modified jatropha oil (MJO) through tribology testing using four-ball tribotester machine. hBN particle was added in MJO at various concentration ratio; 0.05 wt% and 0.5 wt%. The MJO samples were compared with the crude jatropha oil and commercial synthetic ester. The tribology testing was conducted according to ASTM D4712. The value of coefficient of friction, wear scar diameter, worn surface analysis and surface roughness were evaluated. The lowest concentration of hBN particles in MJO (MJO + 0.05 wt% hBN) has reduced the coefficient of friction with smaller wear scar diameter and better surface roughness quality. The worn surface analysis from the ball lubricate by MJO + 0.05 wt% hBN had light and shallow grooves. The study proved that MJO + 0.05 wt% hBN exhibits better lubrication ability and suitable as an alternative for the environmental-friendly lubricant especially for minimum quantity lubrication (MQL) oil.

N. Talib (✉) · E. A. Rahim · W. K. Lee · H. Abdullah
Faculty of Mechanical and Manufacturing Engineering, Universiti Tun Hussein Onn Malaysia, 86400 Batu Pahat, Johor, Malaysia
e-mail: fazillah@uthm.edu.my

R. M. Nasir
School of Mechanical Engineering, Universiti Sains Malaysia (Engineering Campus), Seri Ampangan, 14300 Nibong Tebal, Seberang Perai Selatan, Pulau Pinang, Malaysia

A. Saleh
Faculty of Engineering Technology, Universiti Tun Hussein Onn Malaysia, Pagoh Higher Education Hub, 84600 Pagoh, Muar, Johor, Malaysia

M. T. Hameed Sultan et al. (eds.), *Tribological Applications of Composite Materials*, Composites Science and Technology, https://doi.org/10.1007/978-981-15-9635-3_13

Keywords Tribology · Solid lubricant · Vegetable oil · Minimum quantity lubrication

1 Introduction

Lubricant is substance formulated using various types of additive to be applied in several applications such as engine oil, grease, hydraulic fluids, metalworking fluid and air compressor lubricant. As reported by Lukoil (2013), the global consumption of mineral-based oil production will gradually increase and by 2025, the quantity would reach 105 million barrels of oil per day. In response to the demand for sustainable product, environment-friendly lubricant was recently explored to reduce the use of hazardous mineral-based oil that negatively affect the environment and human (Zainal et al. 2018). The use of edible and non-edible vegetable-based oil as lubricant promising good lubricating properties and excellent friction as well as the wear behaviour. Vegetable-based oil has a high viscosity index (VI) and high flash point compared to the mineral-based oil which showed excellent lubrication film behaviour (Mobarak et al. 2014). Ullah and Dhar (2018) experimentally investigated the machining of kevlar composite material using vegetable-based oil cutting fluid. They conducted drilling process at three different machining conditions; dry, conventional-based cutting fluid (VG-68) and the olive oil. It was found that the olive oil perform well than the dry and conventional-based cutting fluid conditions. The olive oil tends to lower the cutting temperature and displayed better surface roughness quality. This is because of the olive oil has good lubrication properties and successfully deposited at the sliding surfaces thus forms a lubrication layer.

1.1 The Potential Used of Solid Particles as Lubricant Oil

Lubricant promotes a thin film layer that acts between moving parts to reduce friction and wear. Additionally, thin film of solid lubricant which form at two sliding solid bodies is favourable to reduce the amount of interaction due to various operating speed and loads. Gunda and Narala (2016) indicated that an optimum particle size and concentration of particle added in the lubricant oil prolong the life time of the surfaces. The experiment used molybdenum disulphide (MoS_2) particles at three different sizes: 10 μm, 30 μm, and 50 μm and the concentration of MoS_2 were varied from 0.1 to 0.5 wt%. MoS_2 particles was mixed with SAE 40 mineral-based oil. Tribological effect through pin-on-disc wear and friction testing was investigated. The result found that 0.2 wt% of MoS_2 particle demonstrated low friction reduction and the smallest size of MoS_2 (10 μm) acts as an effective solid lubricant film.

Reeves et al. (2013) experimentally investigated the tribological effect of bio-based lubricant with hexagonal boron nitride (hBN) particle using pin-on-disk tribometer. hBN particles blended in canola oil at various hBN particle sizes: 70 nm,

0.5, 1.5, and 5.0 μm. The result showed that smaller particles size (70 nm) have the ability to establish a thin film layer between the contact surfaces. The green solid lubricant hBN consisting of lamellar crystal structures produces protective film layer that attached to the contact surfaces thus enhanced the anti-wear ability of sliding surfaces. Çelik et al. (2013) analysed the friction and wear behaviour of hBN particles in SAE10W mineral-based oil. Three types of hBN added in oil sample at different viscosity values were prepared and compared with SAE10W. The experiment was carried out through ball-on-disc tribometer. The result concluded that the high viscosity of hBN sample form a sufficient lubrication layer that completely cover asperities at the sliding surfaces. The mending effect occurred and exhibited the lowest friction and wear rate. Furthermore, Abdullah et al. (2014) experimentally investigated the effect of additive particles on the tribological performance of engine oil using a four-ball tester. The 70 nm sized hBN and alumina (Al_2O_3) were mixed in SAE 15W40 diesel engine oil. They found that the presence of 0.05 vol% of hBN particles in SAE 15W40 had a ball bearing effect which altered the sliding friction to a rolling friction between the contact surfaces. The hBN particle provide the anti-wear effect by lowering the coefficient of friction value, reduce wear rate with smoother worn surface. Nguyen et al. (2012) carried out ball-milling experiment through minimum quantity lubrication (MQL) method. In the experiment, 0.1 and 0.5 wt% of hBN particles and 0.1 wt% nano-graphite platelets were mixed with vegetable oil (Unist-Coolube 2210). They concluded that the mixture of 0.5 wt% of hBN particle (particles diameter < 15 μm) substantially improved the wear behaviour by reducing flank and central wear of the ball mill tool.

Mosleh et al. (2019) analysed the influence of MQL with nanofluids when drilling titanium. This study used MoS_2 and hBN particles at 70–100 nm in size added in the commercial MQL lubricant, Boelube 70104. The result show that both MQL oil with MoS2 and hBN had fewer frictional torque variation at lower peaks. They noted that the nanoparticles exhibited anti-wear behaviour thus enhanced the drilling performance. Furthermore, Sen et al. (2019) indicated that the application of nanofluid as MQL oil significantly enhanced the machining performance. They summarized that the addition of nanoparticle such as Al_2O_3, MoS_2, nano-diamond, silicon dioxide (SiO_2), nanoplatelet and carbon nanotube (CNT) in MQL-based oil providing better lubrication thin film layer at the tool-chip contact zone. The nanoparticles added in the MQL-oil was able to withstand the high contact pressure at the sliding surfaces causes lower values of surface roughness, reduction in cutting force and cutting temperature and less tool wear. This phenomenon occurred due to the several lubrication mechanism of nanoparticle at the sliding surfaces such as rolling effect, surface protective film, mending effect and polishing effect.

1.2 Implementation of MQL Method During Machining of the Composite Material

MQL method was implemented through the machining process to reduce lubricant oil usage. This method represents a green machining technology when a small quantity of lubricant is mixed with compressed air to form an aerosol. The lubricant was projected to the cutting zone using a nozzle. Sharma et al. (2016) had reviewed that MQL method was used in many machining operations such as turning, milling and grinding processes. The finding shows that MQL method efficiently reduces the friction coefficient thereby reducing the cutting force. Besides, the workpiece surface roughness had been improved by reduction of adhesion occur at the cutting tool, thus prolonging the tool life.

In recent years, the MQL method was employed in the machining of composite materials. Teti (2002) stated that composite material is difficult to cut thus increase the wear rate of the cutting tool. However, the widespread use of composite material in automotive, aerospace and building material industries challenging the researchers to investigate the machining criteria for composite material. Previous study by Adibi et al. (2017) which used MQL method during grinding of carbon fiber-reinforced SiC matrix composites (CMCs). The machining parameter were varies in terms of cutting speed, feed rate, depth of cut and lubricant method (dry, MQL and flood) which using the corn oil as the MQL oil. The results indicated that the usage of MQL method obtained the lowest value of grinding energy subsequently reduced the grinding force. The projected oil mist from the MQL method formed excellent lubrication layer in the contact surfaces thus generated better workpiece surface quality. Senthilkumar et al. (2018) carried out the drilling process of carbon fibre reinforce plastics (CFRP) using MQL method using LRT 30 oil as the MQL oil. They initiated that MQL efficiently supplied the lubricant at the cutting zone during the machining process, thus reducing force and produce better hole quality. Furthermore, James and Annamalai (2018) evaluated the turning performance of aluminium metal matrix composite (MMC) by MQL method. The MMC used in this study was AA6061 while the reinforcement material was ZrO_2. They found that the practise of MQL method had decreased the workpiece surface roughness and tool wear. This method also prevent the built-up-edge formation thus enhanced the surface roughness quality.

Helmy et al. (2018) analysed the effect of different machining condition on ultrasonic assisted edge trimming of multidirectional CFRP composites. The experiment was carried out using 5-Axis DMG Ultrasonic 20-Linear milling machine. They used flood and MQL methods of coolant. The result shows that MQL method increased the cutting force due to poor heat transfer at the cutting zone and shortage of coolant to flush chips. However, MQL method still revealed comparable result with flood coolant and promoted positive impact by reducing the machine coolant contamination. Furthermore Wang et al. (2017), studied the secondary cutting edge corners of one-shot drill bit when drilling CFRP at varies cooling position. They had performed drilling process of aerospace grade T800 CFRP using GONA 5 axis machine centre with dry and MQL coolant conditions. MQL was internally supplied through coolant holes of the drill bits. The coolant was projected out at the secondary cutting edge from the supply outlets. The result showed that the delivering of MQL coolant to the secondary cutting edge had reduced the tool wear and lower the smaller maximum drilling torque value.

Moreover Schneider and Beckenlechner (2016), conducted a drilling process of CFRP using minimum quantity dry lubrication (MQDL)-process. They used compressed air with graphite powder as a lubrication medium during the drilling process. The result showed that minimal amounts of graphite powder in MQDL process tended to reduce the tool wear. In order to applied sustainable manufacturing through environmentally-friendly machining process, Nandakumar and Rajmohan (2018) reviewed several studies investigating the potential of vegetable oil as MQL lubricant for MMC during the grinding process. They stated that the usage of nanofluid from vegetable oil as MQL oil will has a great impact to the grinding performance in terms of surface roughness quality and grinding force. MQL offers good machining performance from high cooling and lubrication ability. It was found that lack of studies reporting on the influence of nanofluid from vegetable oil as the MQL-based oil for machining of the composite material. The combination use of nanofluid and MQL method are one step towards achieving sustainable machining which are environmentally-friendly, clean and safe method.

Therefore, in this study the influence of solid lubricant (hexagonal boron nitride) enriched in vegetable based-oil from modified jatropha oil was evaluated through tribology testing. The finding from this study was essential in the development of the environmentally-friendly MQL oil.

2 Methodology

2.1 Preparation of the Lubricant Sample

Earliest, the jatropha methyl ester (JME) was formulated through a two-step acid–base catalysed transesterification using crude jatropha oil (CJO). Next, TMP ester was produced from the reaction between JME with trimethylolpropane (TMP) and indicated as modified jatropha-based oil (MJO). The formulation of TMP ester in the MJO increasing the polar functionality which increased the absorption ability on the metal surface (Talib and Rahim 2018). The MJO was blended with hBN particle with the size of 2–5 μm. The experiment was performed at operating temperature of 60 °C for one hour with the speed of 700 rpm using a magnetic stirrer Nguyen et al. (2012). has concluded that the hBN particle that less than 15 μm in size provided a better anti-wear behaviour. The concentration ratios of hBN were blended at 0.05 wt% and 0.5 wt% based on the lubricant weight.

Table 13.1 displayed the physicochemical properties of hBN particle. Figure 13.1a, b presented the powder and the morphological structure of hBN particle

Table 13.1 Physicochemical properties of hBN (Talib and Rahim 2018)

Properties	Values
Density (g/cm^3)	2.3
Young's modulus (MPa)	20–102
Thermal expansion coefficient (10^{-6}/°C)	1
Thermal conductivity (cal/cm s K) at 293 K, directional average	0.08

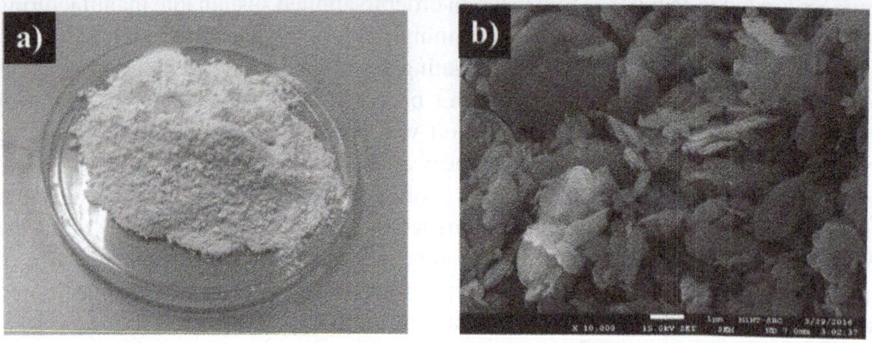

Fig. 13.1 **a** hBN powder and **b** FESEM micrograph of hBN particles

which captured using a field emission scanning microscope (FESEM) at magnification 10000×. hBN is a green solid lubricant that formed extremely stable compounds and produced a lubrication film that is safe to handle (Reeves et al. 2013). Figure 13.2 shows the MJO samples and were compared to the commercial synthetic ester (SE, Unicut Jinen MQL) and the CJO. Table 13.2 shows the properties of the lubricant samples.

Fig. 13.2 Lubricant samples; **a** synthetic ester, **b** crude jatropha oil, **c** MJO + 0.05 wt% hBN and (d) MJO + 0.5 wt% hBN

Table 13.2 Physical properties of lubricant samples (Talib et al. 2017)

Sample	Density at 15 °C (g/cm^3)	Kinematic viscosity, v at 40 °C (mm^2/s)	Kinematic viscosity, v at 100 °C (mm^2/s)	Viscosity index, VI	Flash point (°C)
SE	0.9500	19.05	4.33	137	250.0
CJO	0.9143	30.66	6.67	183	240.0
MJO + 0.05 wt% hBN	0.9221	17.13	4.66	211	222.0
MJO + 0.5 wt% hBN	0.9397	17.56	4.90	228	223.7

2.2 *Tribology Test (ASTMD4712)*

The tribology test was performed according to ASTM D4712 using four ball tribotester equipment as presented in Fig. 13.3a. Every sample was tested with four

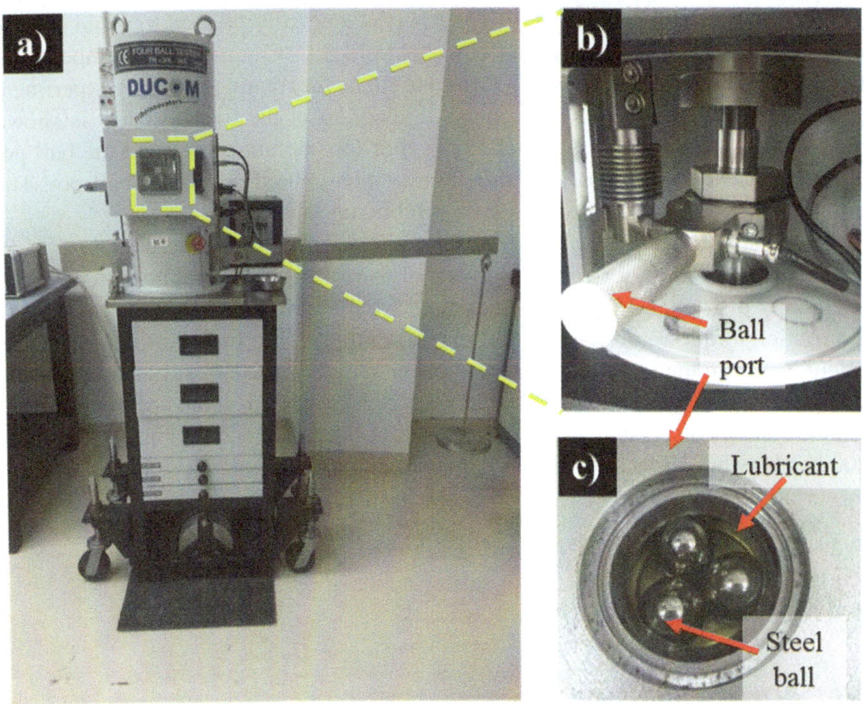

Fig. 13.3 **a** Tribotester machine, DUCOM TR-30L-IAS, **b** ball pot assembly and **c** stationary balls in the ball pot with lubricant sample

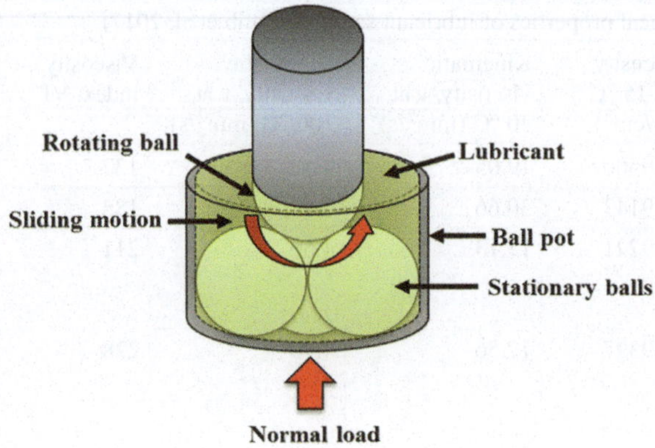

Fig. 13.4 Tribology test diagram (Talib et al. 2017)

steel (AISI 52100) balls with the hardness from 64 to 66 HRC and the diameter of 12.7 mm. The stationary balls were fixed in the ball pot assembly (Fig. 13.3b) and were tightened using a torque wrench. As shown in Fig. 13.3c, 10 ml of sample was placed into the ball pot assembly. After that, the ball pot was installed in the tribotester and a normal load of 392 ± 2 N was gradually pressed to avoid intense stress. Gradually, the sample has been heated to 75 ± 2 °C at a constant temperature. The speed was set at 0.461 m/s (1200 rpm) to rotate the rotating ball. The experiment was conducted for 60 min. The schematic diagram of the tribology test is shown in Fig. 13.4. The heater was switched off after the operating time and the ball pot assembly was removed out. The sample was then poured out of the ball pot. The tribology test was repeated twice for each lubricant sample.

The wear scar diameters (WSD) of the stationary balls were determined based on the average lengths of horizontal and vertical scars using an optical microscope. The coefficient of friction (COF) was measured from the average tangential load from the friction torque values taken at 30 readings/minute. The COF was automatically calculated from the friction torque data using Winducom 2010 software. The worn surface morphology was observed via scanning electron microscopy (SEM) equipped with the energy dispersive x-ray spectroscopy (EDX) at magnifications of 100× and 200×. The worn surface roughness (R_a) was examined according to ISO4288:1996 by surface roughness tester.

3 Results and Discussion

3.1 Friction and Wear

Figure 13.5 displayed the COF and WSD of lubricant samples. SE had the highest

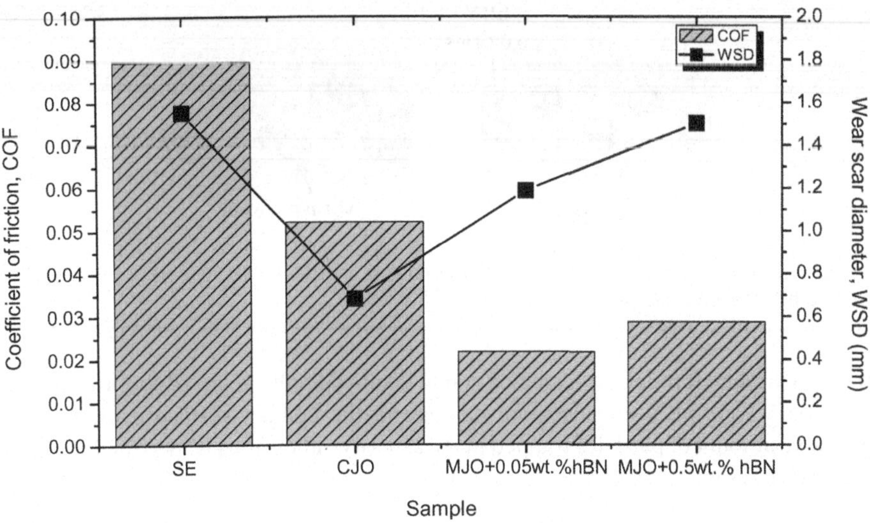

Fig. 13.5 Coefficient of friction and wear scar diameter of lubricant samples

COF and WSD which were 0.0895 and 1.5529 mm, respectively. Meanwhile, the usage of CJO sample had reduced the COF and WSD by 42% and 57% when compared to SE. This occurrence was due to the longer molecular chain of CJO ranging between 16 and 18 carbon atoms per molecule, compared to SE (8 and 10 carbon atoms per molecule). As reviewed by Zainal et al. (2018), the tribological properties was influenced by the length of the carbon chain. However, unnecessary amounts of long-chain fatty acids in vegetable oil lowers the low-temperature behaviour, thus cause oxidation to occur. Therefore, the modification of the crude vegetable oil by chemical modification process and added reformulation of additives was suggested by Shashidhara and Jayaram (2010).

It can be observed that both MJO samples (MJO + 0.05 wt% hBN and MJO + 0.5 wt% hBN) had the lowest COF (0.0217 and 0.0287) compared to CJO and SE. MJO-based oil consists of long molecular and branches chains, while CJO only has long molecular chains. Therefore, MJO offers better low-temperature behaviour and formed a good lubrication film. In term of wear, MJO + 0.05 wt% hBN had reduced 23% of WSD than SE. The addition of 0.05 wt% hBN in MJO yielded superior lubrication properties. This is because of the rolling effect of the particles between the two contact surfaces, which significantly reduced the COF and WSD. The results were in agreement with Rahmati et al. (2014) who noted that the addition of nanoparticle enhanced the lubrication performance. Meanwhile, MJO + 0.5 wt% hBN sample had reduced only 3% of WSD compared to SE. This is because the excessive amount of particles lead to agglomeration, thus increased friction and wear (Padmini et al. 2016). Based on the finding of Talib et al. (2017), the phenomenon of MJO + 0.05 wt% hBN and MJO + 0.5 wt% hBN can be explained in Fig. 13.6. A thin lubrication film layer was developed at the lowest amount of additive particle

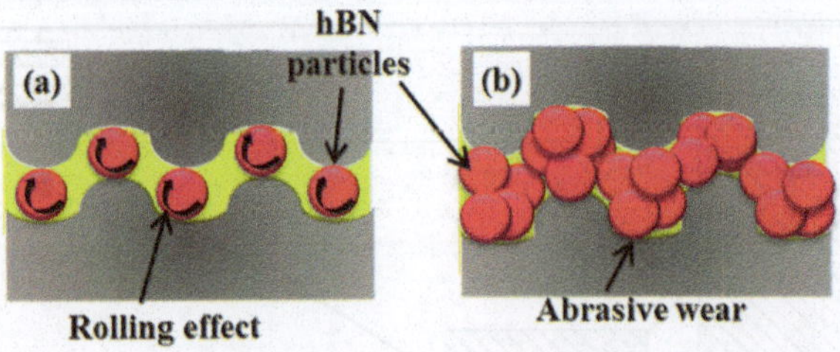

Fig. 13.6 Lubrication film mechanism in MJO samples; **a** MJO + 0.05 wt% hBN and **b** MJO + 0.5 wt% hBN (Talib et al. 2017)

(0.05 wt%) which reduce the area of contact, substantially reduced the friction and wear. 0.05 wt% hBN creates more damage in the contact surface caused of the excessive amount of particles trapped in the asperity valley, thus lead to the formation of abrasive wear.

3.2 *Analysis of Worn Surface*

Figure 13.7 shows the worn surfaces morphology for lubricant samples at the magnifications of 100× and 200×. From Fig. 13.7a, the worn surface of SE was inconsistent rough and had deep grooves. The observation was proved by the result of the surface roughness value of SE which also had the highest of R_a at 0.73 μm. The finding was paralleled with the COF value of SE. The high value of COF (1.5529) showed that SE had poor lubrication film thus increased the wear behaviour at the contact surfaces. The abrasive wear occurred at the metal surfaces. The lubrication film of SE unable to support the applied load and only little part of the metal surfaces separated (Zulkifli et al. 2013). The worn surface of SE also had the highest R_a value which was 0.73 μm. The EDX spectrum in Fig. 13.8a showed that C, Fe, Cr, Mn, Si, S and P were the elements found on the worn surface from SE. Based from Fig. 13.7b, the worn surface of CJO was smooth compared to all samples and had the lowest R_a value (0.13 μm). This results showed that the CJO had good anti-wear behavior. However, a darker region on the worn surface of CJO is caused by the oxidation process due to the contamination of inorganic compound in CJO (Haseeb et al. 2010). The oxidation process was proven from the EDX spectrum (Fig. 13.8b) which showed the addition of O element found on the worn surface from CJO. Besides, there was material transfer on the worn surface of CJO caused by metallic contact.

Figure 13.7c showed the worn surface from MJO + 0.05 wt% hBN which had thin and shallow grooves. However, the worn surface from MJO + 0.5 wt% hBN had deep and wider grooves as displayed in Fig. 13.7d. The roughness of worn surface from MJO + 0.05 wt% hBN recorded the lowest value of 0.32 μm compared with MJO + 0.5 wt% hBN which was 0.54 μm. This phenomenon was in agreement with the finding of Padmini et al. (2016) which revealed the highest concentration

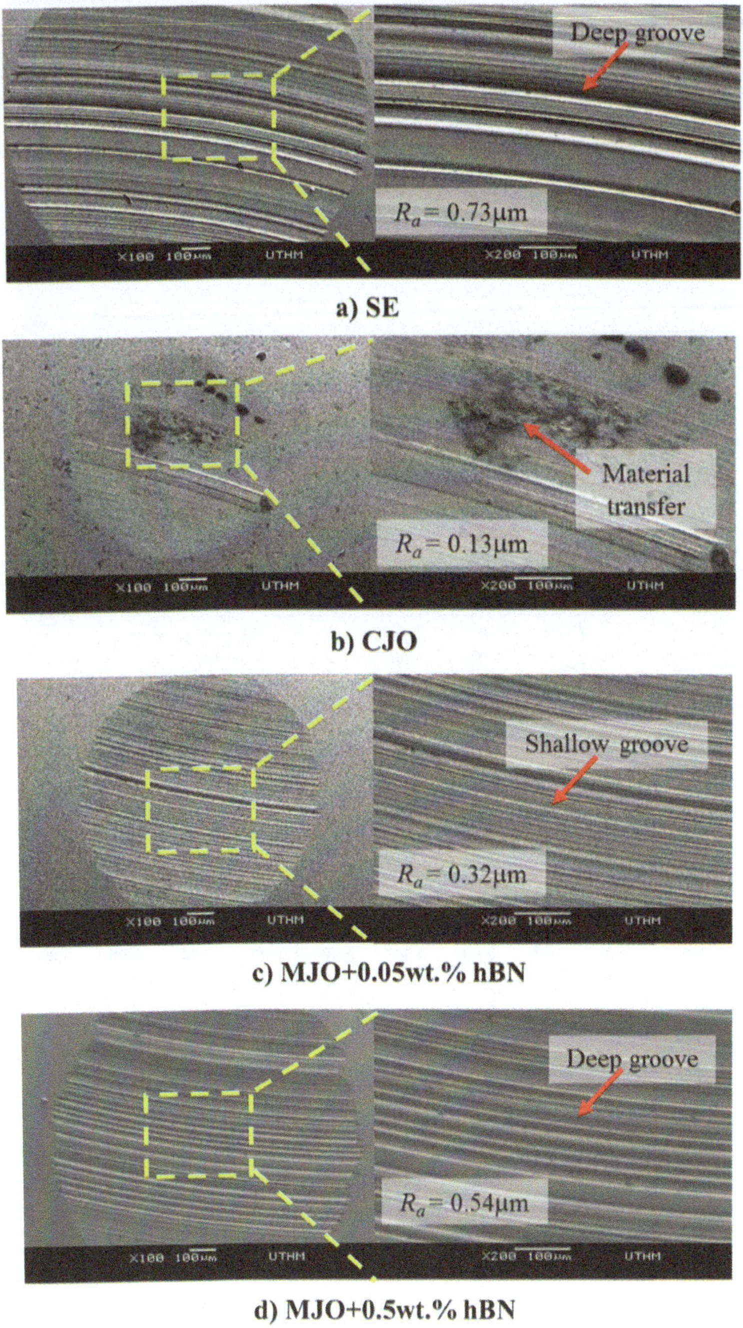

Fig. 13.7 Worn surface micrograph on the stationary balls using SEM

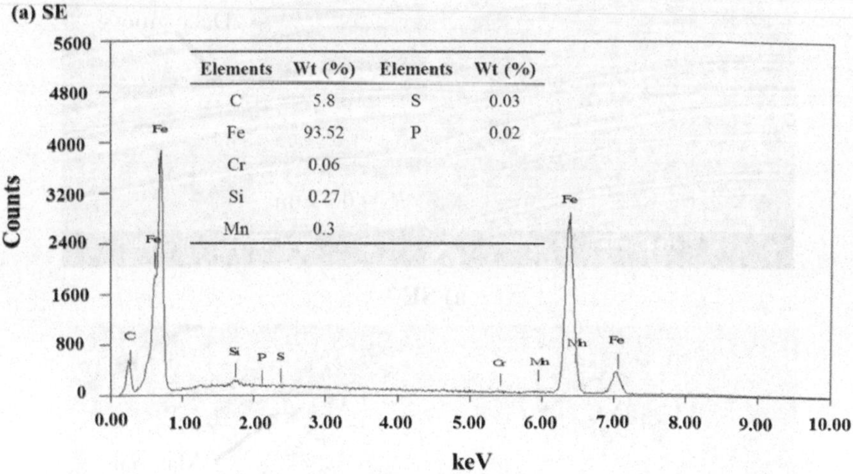

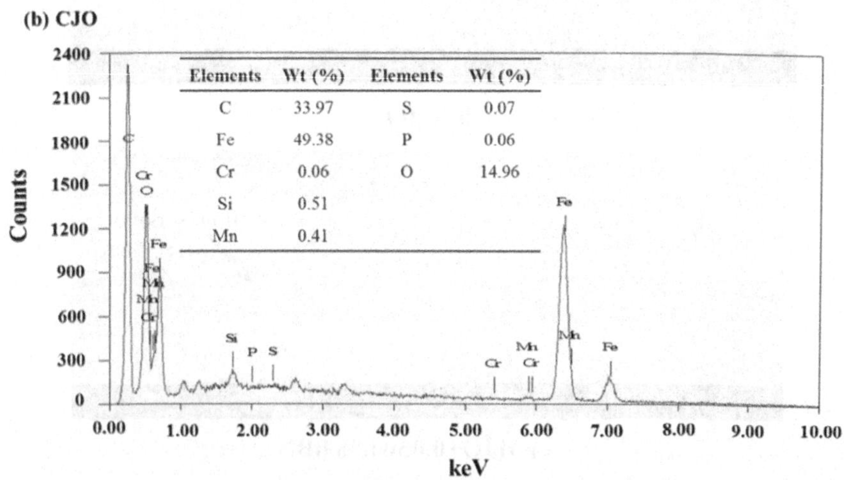

Fig. 13.8 Energy Dispersive X-ray (EDX) spectra at worn surface lubricate by **a** SE, **b** CJO and **c** MJO + 0.05w t% hBN (Talib et al. 2017)

of particles caused non-uniform dispersion in the lubricant, thus broke the micro-joints and creates grooves. The lowest concentration of particles (0.05 wt%) formed a lubrication film that tends to fill the valleys between asperities. The hBN particle deposition on the worn surface is shown in the EDX spectrum displayed in Fig. 13.8c.

(c) MJO + 0.05wt.%

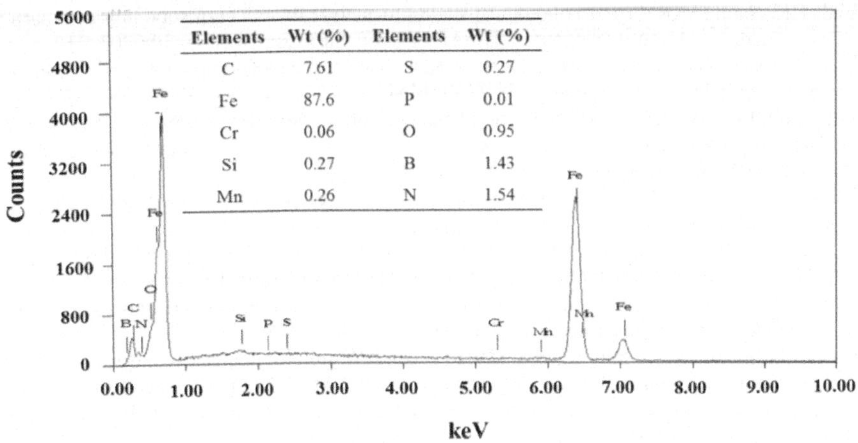

Fig. 13.8 (continued)

4 Conclusion

From this study, the influence of solid lubricant (hBN) in MJO was successfully evaluated through tribology testing. The use of low concentration of hBN particle in MJO resulted in lower COF, WSD and R_a values. hBN particles had formed a thin lubrication layer that provided anti-wear behaviour. The worn surface of ball lubricate by MJO + 0.05 wt% hBN had light and shallow grooves. The results showed that MJO + 0.05 wt% hBN had better lubrication ability and suitable as the alternative for MQL oil. This finding can be used as the fundamental approach of the implementation of solid particle in MQL oil during machining the composite material.

Acknowledgements The authors gratefully acknowledge with thanks to Ministry of Higher Education (MOHE) and Universiti Tun Hussein Onn Malaysia (UTHM) through the Fundamental Research Grant Scheme Vot No. FRGS/1/2018/TK03/UTHM/03/10 and TIER 1 Research Grant (H176).

References

Abdullah MIHC, Abdollah MF, Amiruddin H et al (2014) Effect of hBN/Al_2O_3 nanoparticle additives on the tribological performance of engine oil. J Teknol 3:1–6

Adibi H, Esmaeili H, Rezaei SM (2017) Study on minimum quantity lubrication (MQL) in grinding of carbon fiber-reinforced SiC matrix composites (CMCs). Int J Adv Manuf Technol

Çelik ON, Ay N, Göncü Y (2013) Effect of nano hexagonal boron nitride lubricant additives on the friction and wear properties of AISI 4140 Steel. Part Sci Technol 31:501–506

Gunda RK, Narala SKR (2016) Tribological studies to analyze the effect of solid lubricant particle size on friction and wear behaviour of Ti-6Al-4V alloy. Surf Coat Technol 308:203–212

Haseeb ASMA, Sia SY, Fazal MA, Masjuki HH (2010) Effect of temperature on tribological properties of palm biodiesel. Energy 35:1460–1464

Helmy MO, El-hofy MH, El-hofy H (2018) Effect of cutting fluid delivery method on ultrasonic assisted edge trimming of multidirectional CFRP composites at different machining conditions. Procedia CIRP 68:450–455

James SJ, Annamalai AR (2018) Machinability study of developed composite AA6061-ZrO_2 and analysis of influence of MQL. Metals (Basel) 8

Lukoil (2013) Global trends in oil & gas markets to 2025. In: Lukoil. https://www.lukoil.com/materials/doc/documents/global_trends_to_2025.pdf. Accessed 11 Aug 2016

Mobarak HM, Mohamad EN, Masjuki HH et al (2014) The prospects of biolubricants as alternatives in automotive applications. Renew Sustain Energy Rev 33:34–43

Mosleh M, Shirvani KA, Smith ST et al (2019) A Study of minimum quantity lubrication (MQL) by nanofluids in orbital drilling and tribological testing. J Manuf Mater Process 3:1–12

Nandakumar A, Rajmohan T (2018) Grinding of MMC using MQL based vegetable oil—review. In: IOP conference series: materials science and engineering paper

Nguyen TK, Do I, Kwon P (2012) A tribological study of vegetable oil enhanced by nano-platelets and implication in MQL machining. Int J Precis Eng Manuf 13:1077–1083

Padmini R, Vamsi Krishna P, Krishna Mohana Rao G (2016) Effectiveness of vegetable oil based nanofluids as potential cutting fluids in turning AISI 1040 steel. Tribol Int 94:490–501

Rahmati B, Sarhan AAD, Sayuti M (2014) Morphology of surface generated by end milling AL6061-T6 using molybdenum disulfide (MoS_2) nanolubrication in end milling machining. J Clean Prod 66:685–691

Reeves CJ, Menezes PL, Lovell MR, Jen TC (2013) The size effect of boron nitride particles on the tribological performance of biolubricants for energy conservation and sustainability. Tribol Lett 51:437–452

Schneider M, Beckenlechner R (2016) Minimum quantity dry lubrication in machining of CFRP. Adv Mater Res 1140:296–303

Sen B, Mia M, Krolczyk GM et al (2019) Eco-friendly cutting fluids in minimum quantity lubrication assisted machining: a review on the perception of sustainable manufacturing. Int J Precis Eng Manuf-Green Tech

Senthilkumar M, Prabukarthi A, Krishnaraj V (2018) Machining of CFRP/Ti_6Al_4V stacks under minimal quantity lubricating condition. J Mech Sci Technol 32:3787–3796

Sharma AK, Tiwari AK, Dixit AR (2016) Effects of minimum quantity lubrication (MQL) in machining processes using conventional and nanofluid based cutting fluids: a review. J Clean Prod 127:1–18

Shashidhara YM, Jayaram SR (2010) Vegetable oils as a potential cutting fluid—an evolution. Tribol Int 43:1073–1081

Talib N, Nasir RM, Rahim EA (2017) Tribological behaviour of modified jatropha oil by mixing hexagonal boron nitride nanoparticles as a bio-based lubricant for machining processes. J Clean Prod 147:360–378

Talib N, Rahim EA (2018) Experimental evaluation of physicochemical properties and tapping torque of hexagonal boron nitride in modified jatropha oils-based as sustainable metalworking fluids. J Clean Prod 171:743–755

Teti R (2002) Machining of composite materials. CIRP Ann 51:611–634

Ullah S, Dhar NR (2018) Effects of vegetable oil based cutting fluid in machining Kevlar composite material. Am J Mech Eng 6:54–60

Wang F, Qian B, Jia Z, et al (2017) Effects of cooling position on tool wear reduction of secondary cutting edge corner of one-shot drill bit in drilling CFRP. Int J Adv Manuf Technol

Zainal NA, Zulki NWM, Gulzar M, Masjuki HH (2018) A review on the chemistry, production, and technological potential of bio-based lubricants. Renew Sustain Energy Rev 82:80–102
Zulkifli NWM, Kalam MA, Masjuki HH et al (2013) Wear prevention characteristics of a palm oil-based TMP (trimethylolpropane) ester as an engine lubricant. Energy 54:167–173

Tribological Properties of Natural Fibre Reinforced Polymer Composites

Qumrul Ahsan, Zaleha Mustafa, and Siang Yee Chang

Abstract In this chapter, authors aim to highlight the prospect of natural fiber reinforced polymer composites (NFRPC) as tribo-materials for different engineering system. Incorporation of fibers originate from plants in polymer composite is not new as they provide environmental friendly lighter composite demanded by automative sectors where the conservation of energy is concerned. First two section gives a brief understanding on treatment of fiber surface for efficient compatibility with polymer matrix and their arrangement in composites and followed by composite fabrication for thermoset and thermoplastic composites using natural fibers. Later authors compile some published research works in order to interprete the tribo properties based on the different test parameters and composite systems impregnated by different types fibers various arrangements. Finally prospect of using hybrid polymeric composite incorporated with natural fiber and synthetic micro and Nano fillers as tribo materials is highlighted.

Keywords Tribology · Natural fibre · Polymer composites

1 Introduction

Tribology, where the two contact surfaces are in sliding with each other, mainly focuses on studying metal/metal or metal/ceramic systems for vehicles, machinery and other industrial equipment. Later, polymers light in weight and easy to fabricate are increasingly replacing the metals or ceramics in tribology. Therefore, more attention is now surfacing on the use of metal/polymer and polymer/polymer tribo-contacts. Comparing the surface characteristics of polymers, it is completely different from metals and ceramics in terms of friction and wear mechanism. Polymers are susceptible to provide almost smooth and less friction motions when encounters moving counterfaces of any materials, therefore give low coefficient of friction. On the other hand, it abrades very fast when it comes into the contact of materials

Q. Ahsan (✉) · Z. Mustafa · S. Y. Chang
Faculty of Manufacturing Engineering, Universiti Teknikal Malaysia Melaka (UTeM), Hang Tuah, Jaya, Durian Tunggal, 76100 Melaka, Malaysia
e-mail: qumrul@utem.edu.my

M. T. Hameed Sultan et al. (eds.), *Tribological Applications of Composite Materials*, Composites Science and Technology, https://doi.org/10.1007/978-981-15-9635-3_14

i.e. it is very weak resistant to wear. Hence, to enhance the tribo behaviour of polymeric materials by avoiding adhesion and increasing strength and stiffness (Friedrich 2018), fillers are frequently incorporated in the various polymeric systems (Aldousiri et al. 2013; Friedrich 1997; Reinicke et al. 1998) which include thermoset, thermoplastic and elastomer. Fillers are usually introduced as in the form of particles or fibers (short/long or continuous) from synthetic origin. Usually graphite and PTFE fillers act as internal lubricants in matrix which reduces the adhesion with counterpart material by forming friction reducing transfer films (Hager and Davies 1993). Whereas, artificial or synthetic fibers mainly glass and carbon fiber or special fiber aramid having high strength and stiffness are commonly added as reinforcing fillers that may retain the polymer matrix systems from tribological failure resulted from secondary crack in matrix, bended, broken and debonded fibers and generation of wear debris. One of the major concerns in tribo system is the generation of heat due to friction between two mating surfaces and because of non-conducting nature of polymer and fiber, they raise the temperature which in turn reduces the mechanical properties of composites. Therefore, polymer composites could attain resistance to thermal degradation with the introduction of carbon nano tube (CNT) and graphene particles which are termed as nano fillers with high thermal conductivity. Additionally, these nano fillers combined with other micro fillers result minimising the friction and improving the wear resistance of composite (Zhang et al. 2004).

In the recent time, environmental regulations limit the use of synthetic materials for engineering components especially in automation industries. Synthetic fillers have nondegradable constituents and adverse ecological effect for global warming. By taking advantages of the growing demand, utilisation of natural fibers lead to development of green tribo materials which are renewable and biodegradable materials (Elkhaoulani et al. 2013; Menezes et al. 2011) but have strength equivalent to synthetic fibers. These tribo materials are termed eco-friendly as the natural fibers are lighter in weight which ultimately reduce the energy consumption and pollution during use. However, natural fibers are difficult to wet by polymeric resins and compatability between fibers and resins are very low. Fibers also absorb moisture from environment due to their hydrophilic in nature. All these are affecting dimensional instability of the components with tribological failure mainly from weak interfacial adhesion between fiber and matrix (Omrani et al. 2016). However, these drawbacks are partly minimised by using chemically treated fibers and including additives. Considering the above issues, using natural fibers as an alternative to synthetic fibers could not provide better performance. Therefore, development of hybrid composites is introduced where both natural and synthetic fibers are blended in order to provide better adhesion and bonding either between fibers or between fiber and matrix (Karthikeyan et al. 2017). Eventually, the extent of performance of tribo engineering polymeric materials controls by the distribution and form of fibers or fillers as reinforcement in the polymer matrix. Components related to the properties and morphology of composites in a tribo-system are shown in Fig. 14.1.

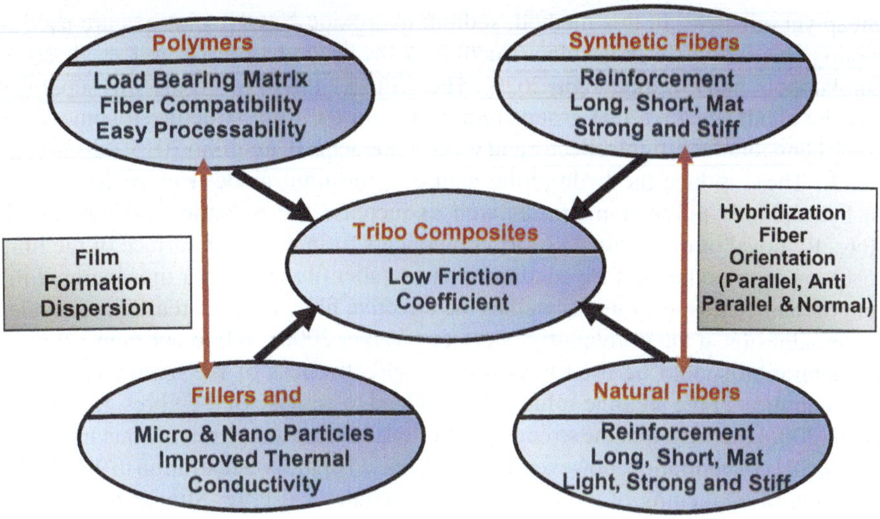

Fig. 14.1 Components for tribo composite system

2 Fiber Characteristics for Tribo Composite Preparation

2.1 *Surface Modification of Fiber*

Fabrication of the fibre reinforced polymer composite may pose challenges due to incompatibility between these materials. Natural fibers are polar and hydrophilic, sensitive to the moisture absorption while polymeric matrix quite often is hydrophobic, Raw natural fibres contain non-cellulosic component consist of pectin, lignin and hemicelluloses sensitive to hydroxyl and carboxylic acid groups that make them prone to water absorption (Lilholt and Lawther 2000; Zulkafli et al. 2019). Poor fibre/matrix interface bonding may hinder efficient stress transfer from matrix to the fibre which may produce inferior polymer composite with poor mechanical properties and low life span. However, surface modifications can be adopted to change the interfacial chemistry and physical nature of the fibre and reduced the incompatibility of the materials (Fadzullah and Mustafa 2016). While many approaches could be utilized to improve the surface characteristic of the natural fibers, amongst only three most widely used surface treatment process such as alkaline treatment, silane treatment and maleated coupling agent treatment will be discussed in next sections.

2.1.1 Alkaline Treatment

Chemical modification using alkaline solution is the most utilized method to improve the bonding of the natural fibres in the polymeric matrix. The method is simple, cheap yet effective. In this method, sodium hydroxide NaOH solutions are used to modify the cellulosic molecular structure of the natural fibre (Ramli et al. 2017; Gholampour and Ozbakkaloglu 2020). The cellulose micro molecule was separated and the alkali sensitive OH groups removed from the fibre structure together with some hemicellulose, lignin, pectin and wax components (Campilho 2015; Ahsan et al. 2019). These reduce the hydrophilic nature of the fibre made it more compatible in hydrophobic polymer matrix as well as increase the moisture resistant of the fibre. Removal of the lignin and wax components from the outer surface of the fibre resulting in smoother surface conditions and smaller fibre diameter thus increase the length/diameter aspect ratio as well as the effective fibre surface area. This provides better adhesion at the fibre/matrix interface (Walker 2006). While optimum alkaline concentration should be used to ensure the effectiveness of the treatment, higher concentration of the alkaline solutions could lead to damage of the fibres. Edeerozey et al. (2007) reported that the strength of the kenaf fibre was optimum when treated in 6% NaOH concentration, however further increased of the concentration to 9% NaOH resulting in weakening of the fibre, lower than untreated fibre. Similar observation was reported by Boopathi et al. (2012) in their study of alkaline treated Borassus fruit fiber. They reported that the fibre strength was optimum when treated at 5% NaOH and further increased of the NaOH concentration (10% and beyond) significantly weaken the fibre. Brígida et al. (2010) reported that usage of the alkaline treatment on the coconut fibre not only able to retain their native hydrophilic characteristic but improve their thermal stability. In study by Nam et al. (2011) reported that the interfacial shear strength of the coir fiber reinforced poly (butylene succinate) was optimum when treated with 5% NaOH for 72 h. Morphology analysis revealed that the 5% NaOH coir fibre reinforced composite failed due to breaking of the fibre instead of interfacial failure indicating better adhesion was achieved in the alkaline treated composite in comparison to the untreated composite. Other study reported that the inclusion of the surface treatment using alkaline solutions not only able to improve the mechanical but the tribological characteristic of the composite as well (Swain and Biswas 2017; Sampath and Kumar 2019). Valášek et al. (2018) studied the influence of the alkali solutions onto the mechanical and abrasive wear of the coir reinforced epoxy composites and reported that alkaline treatment effectively improved the tensile strength as well as wear resistance of the composite.

2.1.2 Silane Treatment

In this method, the fibre is immersed in the silane solution for a period of time allowing the coupling agent in the solutions to coat the micro pore on the surface of the natural fibre. Silanol then formed using the hydrolyzable alkoxy group that present. The bonding between natural fibre and matrix forms via a link known as soloxage bridge. One end of the forms sialon chain attached to surface of the natural fibre surface by creating a formation with their cellulose OH group and the other end links with functional group in the matrix thru a condensation process (Faruk et al.

2012). Subsequently, molecular continuity is created across the fibre/matrix interface (Campilho 2015; Gholampour and Ozbakkaloglu 2020). Reports have shown that mechanical properties of the composite using silane treatment could be significantly improved, often better than alkaline treatment. This is due to alkaline treatment removed impurity on the natural fibre and provided bonding via mechanical interlocking while silane provides chemical links attachment from the coupling agents. Yallew et al. (2014) reported that while all treated jute fibre showed improved wear resistance when imbedded in polylactide matrix composite, silane treated jute exhibited highest wear resistance resulting from the strong interfacial adhesion produced by coupling linkage. Mayandi et al. (2018) reported that the mechanical strength of natural fibre exhibits better mechanical properties and thermal stability when subjected to silane treatment in comparison to alkaline treatment. Liu et al. (2019) investigate the impact of silane treatment on mechanical, tribological properties of the corn stalk fiber (CSF) reinforced bio-polymer composites while selecting four different silane concentration (1, 5, 9 and 13 wt%). This work revealed that while silane-treated CSF could not effectively improve the friction performance, however it is able to significantly increase the wear rate of the polymer composites.

2.1.3 Maleated Coupling Agent Treatment

In this approach, maleic anhydrite (MA) is grafted with the polymer which is used as a matrix material; then it is allowed to react at the interface of the natural fibre and matrix. MA reacts with hydroxyl group that is located in the amorphous region of the cellulose structure to form a covalent bond while the aliphatic chain of MA gets diffused and links with polymer chains of matrix via carbon-carbon bonds (Fuqua and Ulven 2008; Anbupalani et al. 2020). The MA chemically attached to the surface of the fibre serves as a bridge between the fibre and matrix in order to improve the interfacial adhesion. Hong et al. (2008) attempted to modify the surface of jute fibres by maleic anhydride treatment and to improve the interaction between jute fibre and matrix phase and to enhance the mechanical properties of the jute fibres reinforced polypropylene. Morphology analysis revealed that treated jute fibre were broken without complete pull-out in comparison to the untreated composite. This indicated improved fibre-matrix adhesion at the interface as the result of the MA modification. Kakou et al. (2014) investigated the dispersion of the oil palm fibre in the high density polyethylene (HDPE) matrix in the presence of maleated coupling agent. They reported that the treatment has effectively boost the stiffness due to better adhesion between the fibres and matrix. Catto et al. (2014) evaluated the effect of various MA concentration on the tensile strength and stiffness of the maleated polyethylene grafted recycled HDPE and eucalyptus fibre. They reported that addition of 3% of MA in composite shows the highest mechanical properties indicated better compatibility and improved interfacial adhesion achieved in the presence of MA.

2.2 *Size and Texture of Fiber Pattern (Short, Long and Mat)*

The fibre can be arranged in the matrix in various designs such as continues or discontinues, woven arrange in form unidirectional or bidirectional, random as well randomly distributes (Fig. 14.2). The fibers are very significant part of a fibre-reinforced composite material, since the fibre characteristic such as length, orientation and their loading significantly alter the mechanical properties of the composite (Jusoh et al. 2015; Fadzullah et al. 2016) and may as well its ultimate tribological behaviour.

In order to study the influence of the fibre orientation onto the wear properties of the composite, the sliding direction could be applied in three conditions as shown in Fig. 14.3 such as:

- Parallel (P-O): fibre mat and fibre orientation are in the parallel direction to the sliding direction
- Normal Orientation (N-O): fibre mat and fiber are in normal orientation to the applied force and parallel to the sliding direction
- Anti-Parallel (AP-O): fibre mat and fiber are in the perpendicular direction to the sliding direction.

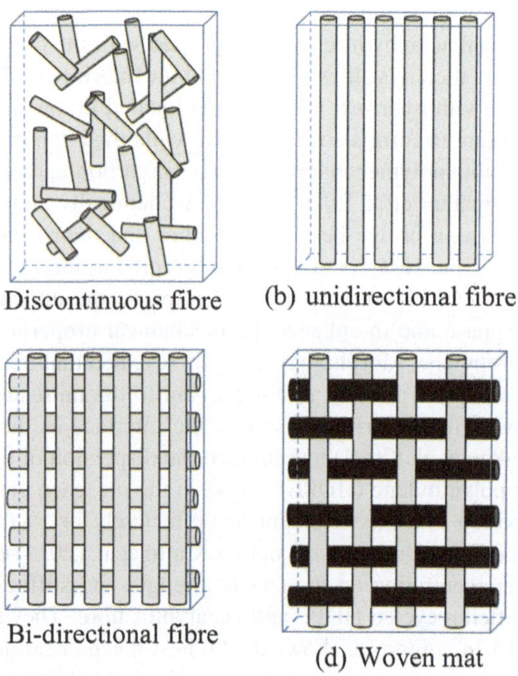

Fig. 14.2 Various types of fibre reinforcement pattern

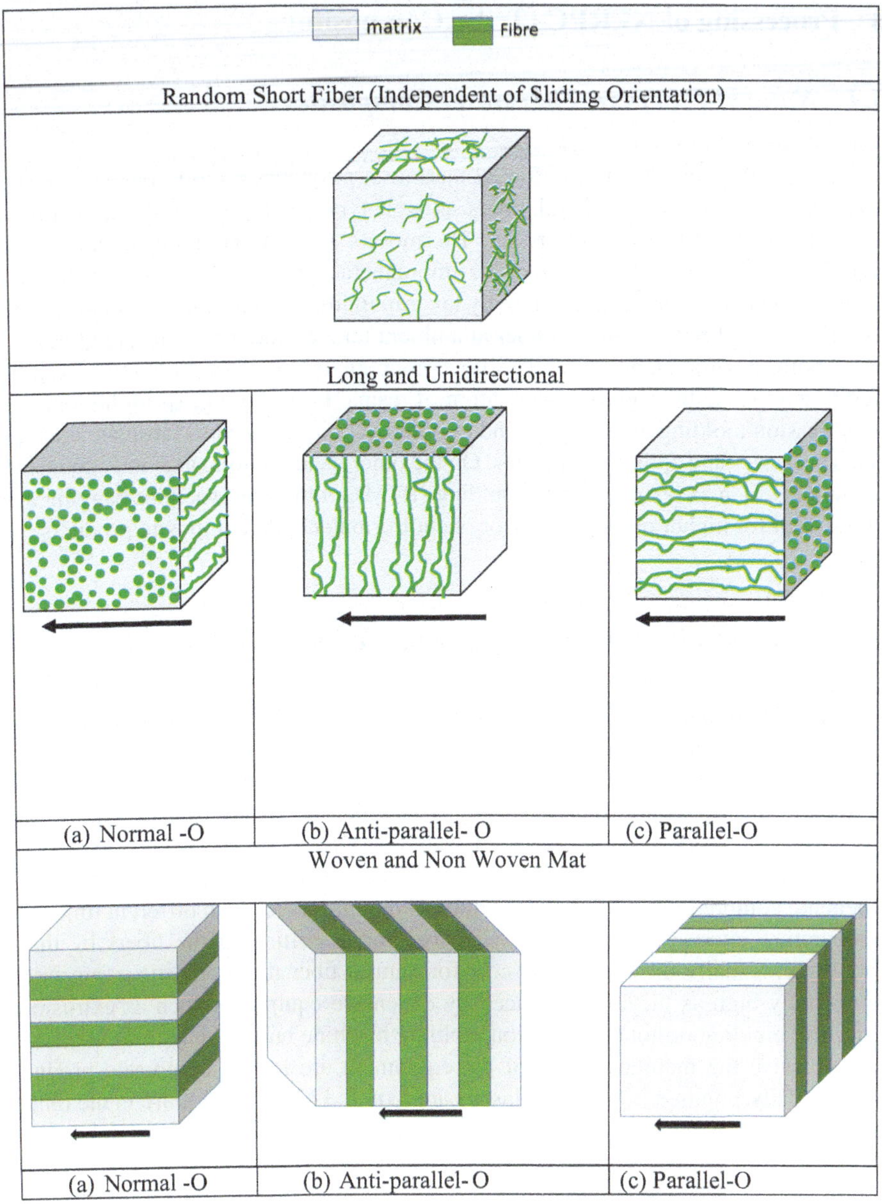

Fig. 14.3 Schematic illustration of composite orientations and sliding direction

3 Processing of NFRPC (Tribo Composite)

3.1 Natural Fibers in Thermoset Composites

Processing of natural fibers reinforced thermoset polymers mainly depends on the size type and texture of natural fibers as well type of thermoset. Usually fibers are in the form of short, long or mat. Thermosets also known as resins are in the liquid form where the short fibers are impregnetaed randomly or resin-saturated long continous fiber are placed layer by layer in unidirectional, bidirectional woven mat form and allow for curing either at ambient temperature or at elevated temperature with or without pressure. The fiber impregnated resin is completely solidified to composite through polymerisation of resins. Hand lay-up, spray lay-up, hot compression molding are low cost and popular techniques used to fabricate unsaturated polyester and epoxy composites. On the other hand complicated and expensive techniques such as resin transfer moulding (RTM), pultrusion, vacuum bagging are being used for higher quality epoxy or phenolformaldehyde composite products.

3.2 Natural Fibers in Thermoplastic Composites

In case of thermoplastic composites, dispersion and distribution of short fibers in plastic matrix or the wetting of long continous fiber by thermoplastic are the critical issues in composite fabrication as the plastic resins are in the solid form. Short fibers are mixed with plastic resins in cyclone vortex or internal mixer where both fibers and thermoplastic are premixed and converted to pellet shape. Later these pellets are used in injection or hot compression molding. Whilst for long or mat fibers hot compression molding or thermoforming processes with different time and temperature interval stages are utilisied for proper wetting of the fibers by thermoplastic. Usually the fabrication cost for natural fiber thermoplastic composites are usually high as they are produced by expensive equipment such as extrusion, injection molding or hot compression molding machine but as the natural fibers are less abrasive the maintance cost of the equipment are low in composite production. Usually, engineering thermoplastics are required high temperature in the range of 200–350 °C which is not suitable for natural fibers as most of them degrade at or above 180 °C. This processing factor, therefore, limits the use of engineering thermoplastics whereas PE, PP and PU which have low processing temperature are being used with lignocellulosic fibers. Some typical natural fiber compatiable thermoset and thermo plastic products using different processing techniques are shown in Table 14.1 (Karthikeyan et al. 2017).

Table 14.1 Processing methods of various natural fiber reinforced thermoset and thermoplatic composites in tribilogical applications (Karthikeyan et al. 2017)

Fiber	Matrix	Test conducted	Processing method
Banana and Kenaf	Polyester	Mechanical wear test	Hand lay-up
Rice straw dust/rise husk dust	Phenolic	Friction assessment and screening test	Hot pressing
Sugarcane/glass	Polyester	Adhesive friction and wear	Hand lay-up
Linen/jute	Polyester	Dry friction wear test	Casting process
Sisal	Phenolic	Constant speed tester	Hot compression
Betelnut	Polyester	Mechanical pull-out wear and friction	Hand lay-up
Sea shell nano powder	Poly-methyl methacrylate	Wear micro-hardness	Mold
Jute	PP		Injection molding machine
Kenaf	PU		Hot compression mold

4 Tribological Characterization of NFRPC

Natural fiber reinforced thermoset and thermoplastic composite materials are now attempting to fabricate as tribo materials specially to perform in dry and non-lubricated condition. This is due to the composite's low density contributed from using light weight natural fiber and in some cases, these composites have better specific stiffness and strength provided tribological properties comparable to many tribo materials that are currently used. Tribology of natural fiber reinforced polymer composites (NFRPC) usually provide information on tribo properties and wear mechanisms in composite system. First term is closely related to material removal and energy dissipation at the surface through transfer layer formation, contact temperature, and degradation of composites whereas the last term mainly focuses on damages of fibers and polymer matrix, debonding of fibers characteristics of wear debris.

4.1 Testing and Evaluation of NFRPC

All of the above features are characterised by considering the typical test parameters that are relative movement of test piece and counter surface, surface nature of counter face, sliding speed, sliding distance, applied load, test temperature and environment etc. used in tribo testing. So far research test data for NFRPC obtained from dry sliding wear test using pin on disk (POD), block on disk (BOD) or block on ring (BOR) test systems in order to determine the coefficient of friction μ, wear loss, w_L and the specific wear rate w_s by the following equations:

$$\mu = F_R/F_N \quad (1)$$

where F_R is the friction force and F_N is the normal load measured in N.

$$w_L = Dm = w_i - w_f, \text{ in g} \quad (2)$$

where w_i is the initial weight and w_f is the final weight taken before and after the test respectively

and

$$w_s = \Delta m/(\rho F_N L) = \Delta V/F_N L; \text{ expressed in } mm^3/Nm \quad (3)$$

where Δm is the weight loss, ρ is the density of the composite material, ΔV is the volume loss, and L is the sliding distance.

The specific wear rate can be termed as "material property" or "wear factor (k^*)" of the material (Friedrich 2018) in tribology and could be comparable between tribo materials as long as test conducted under equivalent condition. From test data, researchers also calculated the wear rate depth $w_{t.}$ (reduction of specimen per unit time) and plotted against the *pave* product (normal pressure times sliding velocity) and found that there exists a nearly straight line relationship with a slope of w_s or k^* ($= w_t/pv$) provided w_s is independent and not influenced by changes in either contact pressure (p) or sliding velocity in the test system and the relationship is plotted schematically in Fig. 14.4 (Friedrich 2018). Figure 14.4 clearly shows that w_s remains constant up to certain pv product beyond that w_s is no longer constant

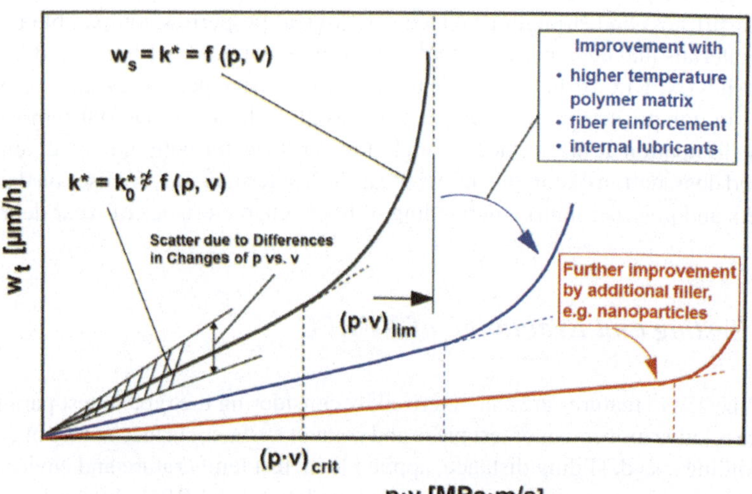

Fig. 14.4 Test parameters for the enhancement of wear behaviour of polymer composites (Friedrich 2018)

and indicates the unstable behaviour of the material. Therefore, *pv* limit is used to indicate the onset of catastrophic failure of composites due to thermal softening and fracture. It is also evident from Fig. 14.4 that using different reinforcements and fillers in given polymer reduce the slope (lowering w_s) and push the limiting *pv* values towards higher values (Friedrich et al. 2018). Therefore, it is expected that incorporation of natural fibers in polymer matrix would manifest lower slope in w_t vs *pv* plot construction as compared to that in matrix alone. The tribological properties such as friction and wear of natural fiber reinforced polymer composites (NFRPC) are controlled by the adhesion between the polymer matrix and metallic counterpart and worn out of natural fiber respectively in performance system.

4.2 Friction in NFRPC

It is associated with the release of energy due to chemical-mechanical damages take place at the mating surfaces as adhesion and/or for the loss of material due to ploughing (Briscoe and Friedrich 1993). The term friction force and its value are generally used to monitor the adhesion of surfaces of polymer composites and metallic counterpart. The initial rise of the friction force in tribo system is due to unstable adhesion film formation (unsteady state) on the contaminated surfaces. With further progress of sliding, deformed asperities cause increase of surface areas which increases adhesion and stabilise the friction force shortly after starting the test. Previously theory of friction was adhesion based where friction force associated with friction originated from contact area of surface asperities. But experimental values of friction coefficient in most cases do not tally with theory. Hence bulk properties such as ploughing is also considered in addition to adhesion to narrow down the discrepancy between the experimental values and theoretical values of friction coefficient.

Evidence of lower friction force or friction coefficient for the presence of natural fiber either in nondegradable or biodegrables polymers were reported by researchers (Yallew et al. 2014; Bajpai et al. 2013). The friction force and coefficient of friction in pure polypropylene matrix and polymer matrix wth jute fiber mat at three different sliding speeds and applied loads were studied for a total sliding distance of 3000 m (Yallew et al. 2014). In both cases steady state friction force values reached within a very short period of travel and it is indepedent of applied load but the friction force increases as the applied load increases and at any applied load (Fig. 14.5). The friction force and coefficient of friction of pure PP is always higher than that of PP-Jute Fiber composite. In case of pure polypropylene, plastic debris once sticks to the sliding interface, instead of getting thicker it deforms by consuming energy and thus increases the friction force as it can be seen in SEM of polypropylene surface. On the other hand, in NFRPC, usually steel surface is very hard and rough which usually ploughs the softer polymer surface, removes the polymeric materials and causes wear of sliding surface. That is why ridges are observed along the sides of ploughed grooves under the scanning electron microscope.

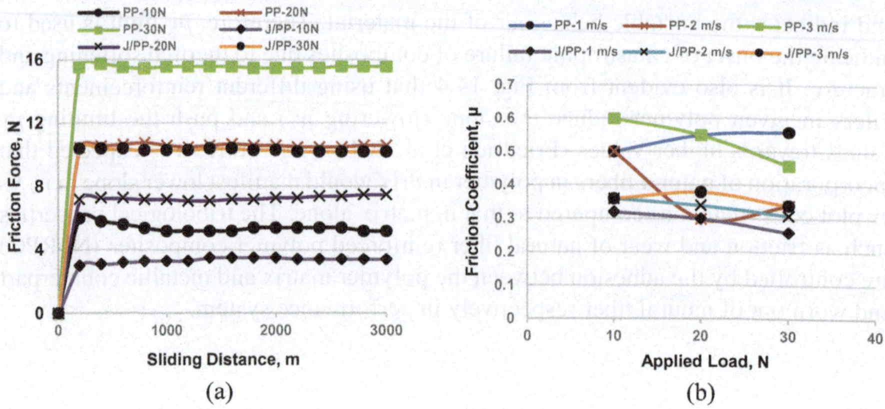

Fig. 14.5 Variation of **a** friction force against the sliding distance for PP and Jute/PP at sliding speed of 3 m/s and **b** friction coefficient against applied load for PP and Jute/PP (Yallew et al. 2014)

Similar effects were also observed by Bajpai et al. (2013) when they studied tribo behaviour poly lactic acid (PLA), PLA-glass and PLA-sisal composites (Fig. 14.6). Interestingly, sisal fiber in composites has observed a lower value of friction force and COF as compared to pure PLA and PLA-Glass composite at any contact load. CoF is not an intrinsic property of materials, it also depends on the test type, temperature, type of film formed and nature of fiber used. Unlike metals, polymer softens with the rise of temperature and for this, fiber easily debonded from the polymer matix and mix with polymer debris which alter the charecteristics of polymer adhesion later. Obviously natural fiber like sisal stiffens the polymer layer but not like as glass polymer layer. Hence polymer protective layer with the presence of natural fibers deforms platically more which reduces the friction co efficient (El-Tayeb 2008a, b; Yousif and EI-Tayeb 2008).

4.3 Wear in NFRPC

In tribo system, wear is defined as the loss of materials of two bodies when their contact surfaces are moving relative to each other [Chand and Fahim (2008)]. Although wear does not cause sudden failure of material, but certainly reduces the performance characteristics from changing and deforming the shape of the components or from surface damage. Hence volume changes in materials causes vibration and surface damages formed secondary cracks that may lead to collapse of the components. Although the common forms of wears are adhesive (sliding), abrasive, corrosive, erosive and fretting but first two forms prevail mainly in NFRPC. Schematic illustration of the major form of dry sliding wear mechanisms are summarized in Fig. 14.7.

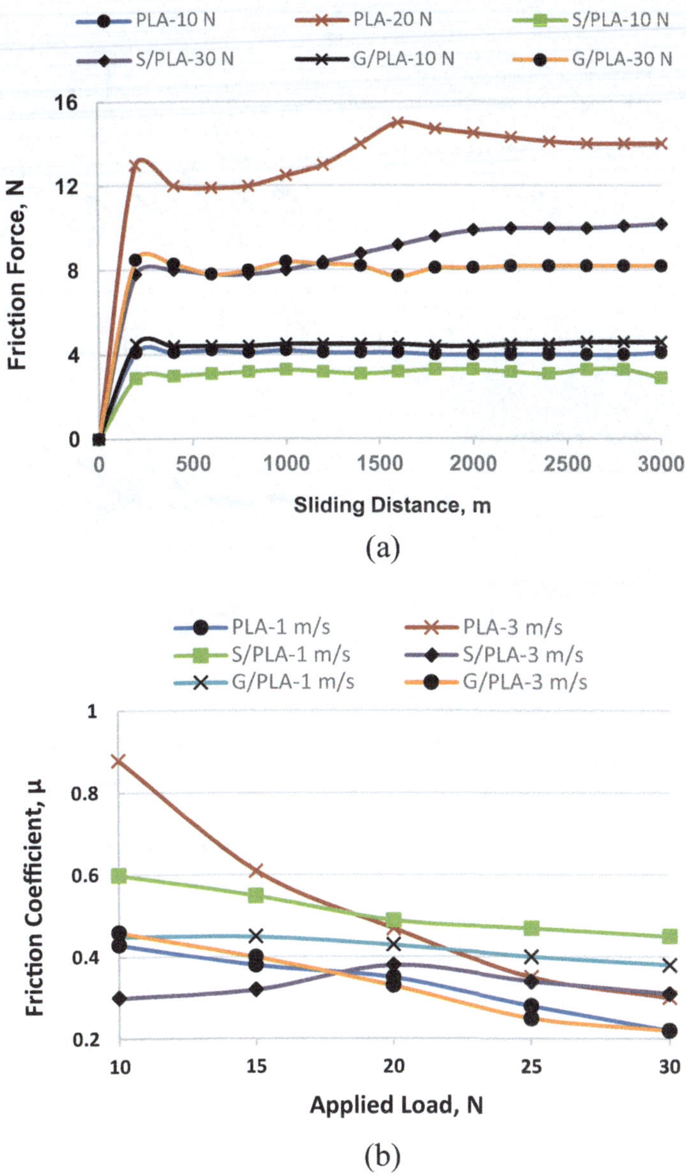

Fig. 14.6 Variation of **a** friction force against the sliding distance for PLA GF/PLA and Sisal/PLA at sliding speed of 2.8 m/s and **b** friction coefficient against applied load for PLA, GF/PLA and Sisal/PLA (Bajpai et al. 2013)

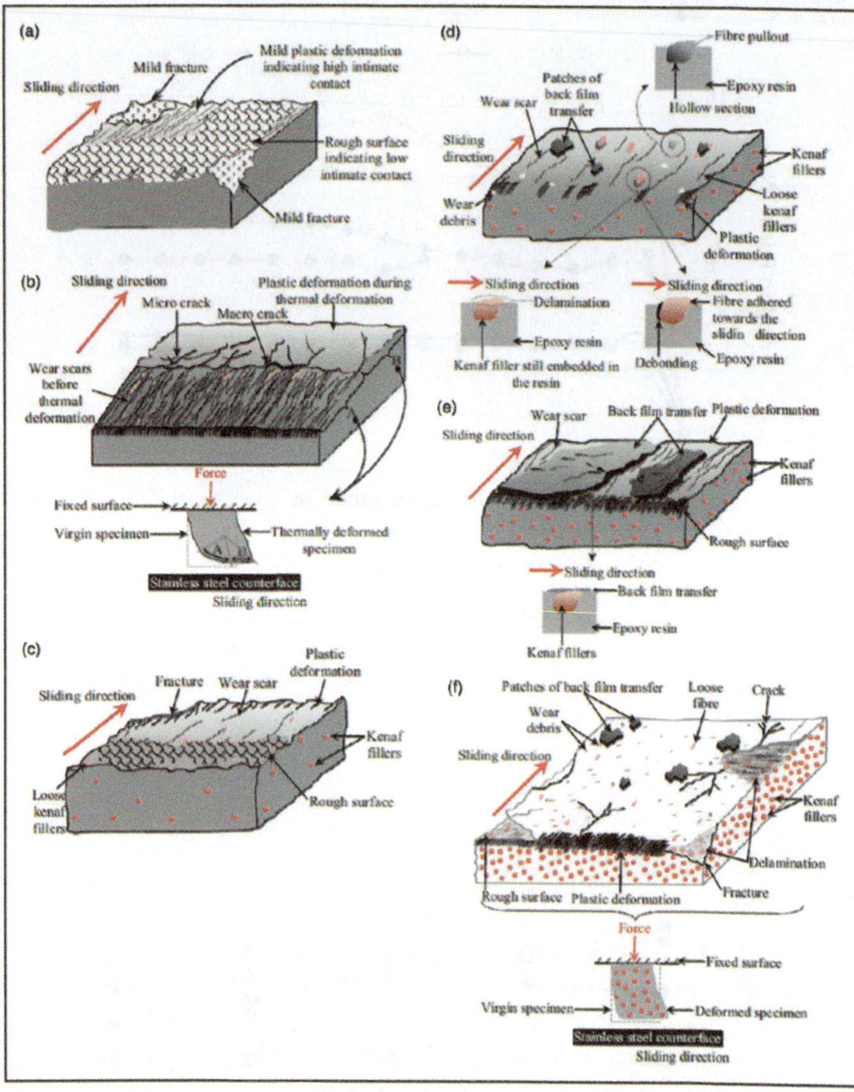

Fig. 14.7 Schematic illustration of different dry sliding wear mechanisms (Rajini et al. 2012)

The tribo properties such friction and wear are dependent on the total test system and not sole related to material properties and vary by changing the test parameters. Although no direct relationship could be made between tribo properties and mechanical properties (strength, stiffness, hardness etc.) of materials, but mechanical properties are utilised to explore the correlation between wear resistance and friction coefficient. It is widely established that synthetic reinforcements (fibers and fillers) are stronger and more wear resistant than polymer matrix therefore, the presence of fiber not only utilise to transfer load from matrix to fiber and increase the strength of the composites but also minimise the exposition of the polymer for counterface. The wear performance of fiber reinforced polymer composites depend not only on the size, shape, volume and orientation and type of fiber but also the extent of adhesion between fiber and matrix. Proper wetting of fibers by polymers is one of the biggest issue in processing of composites and beyond the optimum loading, usually fibers starves from matrix wetting resulted in low mechanical and tribo properties as the fibers can readily debonded from the matrix, Generally, long fibers usually impregnated in the form of bi- or uni-directional mat in polymer matrix improve the wear resistance of composites very significantly, whereas short fibers are easily wetted by polymer which may improve the friction and rapid mouldability. Usually natural fibers are hydrophilic (attracted by water) and polymers are hydrophobic (repels water) which may cause poor interfacial bond between fiber and matrix and may lead to poor wear performance of the composite. In addition, natural fibers easily absorb water which may cause swelling of the composites with debonding of fibers. It can also easily be attacked by the microorganism and become weak which may reduce the composite service life.

Therefore, chemical treatments on natural fibers are usually carried out to increase the hydrophobicity and to improve the interface bonding between fibers and polymers. One of the prime consideration in polymer and its composites is load-temperature interaction. Hence, temperature at wear surface is usually measured during wear test in order to monitor the thermal softening of polymer that can lead to increase in the real areas of contact resulted in rapid increase of both the coefficient of friction and the wear.

In case of natural fibre reinforced composites, various kinds of fibers (jute, sisal, coir, oil palm, flax, hemp, bamboo) are reinforced in thermosets and thermoplastics either as chopped short or continuous fibers. The chopped short fibers are randomly distributed in polymer matrix as reinforcement, whereas continuous fibers not as long as synthetic fibers are placed in matrix as uni- or bi-directional woven fabric or mat. The direction of sliding force to orientation (parallel, anti parallel and normal) of natural fibers also control the properties of composites. Table 14.2 summarizes the tribo test results obtained from published research work for various polymer composites impregnated with different types natural fibers. The table also provides information on the fiber treatments and their corresponding consequence on friction and wear rates that are plotted in Fig. 14.8.

It is apparent from Fig. 14.8 that pristine thermoplastic polymers with low specific wear rates manifest significant improvement in wear resistance when impregnated with reinforcing natural fibers. Jute has been reported as high strength and high modulus (Chand and Fahim 2008), therefore jute fiber reinforced polypropylene exhibits the lowest wear rate a lower w_s of the order of 10^{-7} m^3/Nm and a low friction coefficient of 0.3 while coir reinforced polyester composites exhibit a high w_s of the order of 10^{-4} m^3/Nm and a high friction coefficient of 0.6. Improvement on wear resistance and friction for jute composite compared to coir composite is attributed to higher stiffness and strength with better selflubricating (lesser abrasive) ability of jute fibers. Once again jute or oil palm fibers reinforced epoxy also exhibit

Table 14.2 Tribological data for the natural fiber reinforced thermoset and thermoplastic composites

Matrix	Fiber	Fiber vol.%	Type of testing	Counter face	Test parameters			Tribological properties			References
					Applied load, N	Sliding speed, m/s	Sliding distance, km	CoF, μ	Wear rate, $mm^3/Nm \times 10^{-5}$	Interface temp., °C	
Polyester		0	POD	Stainless Steel	30–100	2.8	0–5 km	0.4–0.6	0.4–2.6	25–54	Yousif and El-Tayeb (2008)
	UT-oil palm short random	48						0.46–0.2	1.0–1.6	25–45	
	T-oil palm short random							0.21–0.4	0.4–1.4	25–45	
Polyester		0	POD	Steel EN 32	10–100	1.75	0–6	0.5	1.0–2.0	40–60	Chand and Dwivedi (2007)
	UT-Sisal short random	27						0.83	3.6–2.4	55–100	
	T-Sisal short random							0.9	1.8–1.0	70–100	
Polyester		0	BOD	Stainless Steel	20	2.8	0–5	0.62–0.96	12.0–15.0	25–45	Yousif (2008)
	Oil palm short mat	25						0.5–0.85	18.0–12.0	22–25	
	Glass long mat	25						0.2–0.5	4.0–6.0	23–34	
Polyester	UT-coir long unidirectional	25	POD	Stainless Steel	10–30	2.8	4.2	0.5–0.65	37.0–8.0	40–70	Yousif et al. (2009)
	AT-coir-long unidirectional							0.4–0.6	18.0–76.0	30–60	
	B-coir long unidirectional							0.3–0.65	30.0–7.0	30–60	

(continued)

Table 14.2 (continued)

Matrix	Fiber	Fiber vol.%	Type of testing	Counter face	Test parameters			Tribological properties			References
					Applied load, N	Sliding speed, m/s	Sliding distance, km	CoF, μ	Wear rate, $mm^3/Nm \times 10^{-5}$	Interface temp., °C	
Epoxy	UT-jute woven mat	10–40	TBAW	Dry Sand (100–400 μm/Rubber Wheel	10–40	Not Reported	50–80 m	0.26–0.55	3.0–5.75	NR	Swain and Biswas (2017)
	AT-jute woven mat							0.4–0.6	3.0–4.5	NR	
	BT-jute woven mat							0.3–0.75	2.4–3.75	NR	
Polypropylene		0	POD	Steel EN 32	10–30	3.0	0–3	0.3–0.6	0.022–0.32	NR	Yallew et al. (2014)
	Jute mat	40						0.3–0.4	0.0035–0.0098	NR	
Polyurethane		0	BOD	Stainless Steel	30–60	2.8	0–2.7	0.2–0.57	0.01–0.15	NR	
	P-Kenaf Mat (bi directional)	38						0.21–0.62	0.005–0.055	40–130	Narish et al. (2011)
	AP-Kenaf Mat (bi directional)							0.27–0.52	0.003–0.075	50–140	
	N-Kenaf Mat (bi directional)							0.32–0.63	0.005–0.034	40–120	

UT—untreated, AT—alkali treated, B—bleached, BT—benzoyl treated, P—parallel, AP—antiparallel, N—neutral, POD—pin on disc, BOD—block on disc, TBAW—three body abrasive wear, NR—not reported

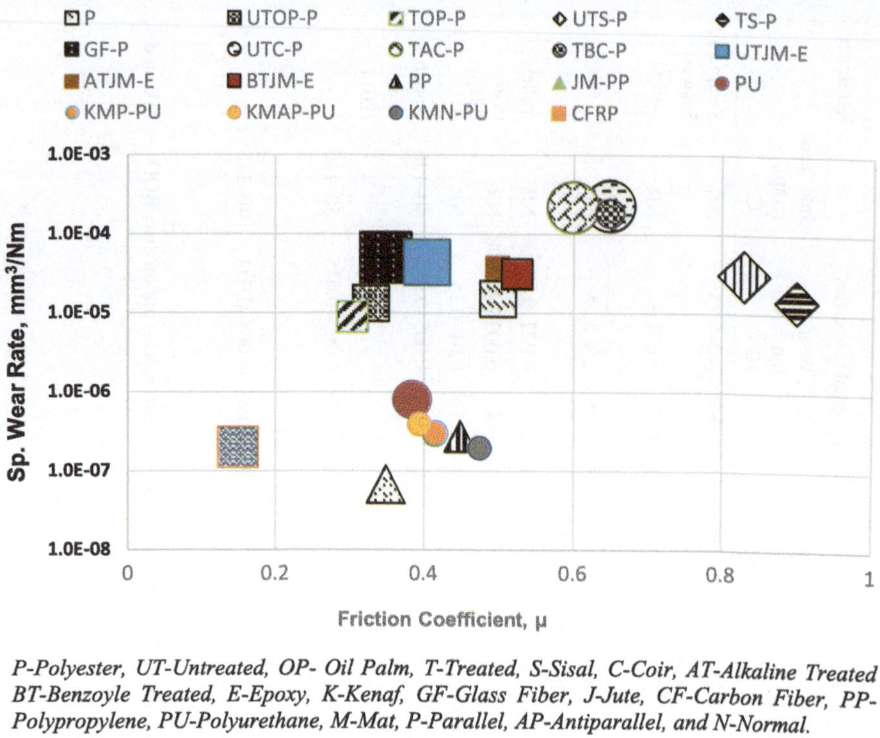

Fig. 14.8 Friction coefficient and specific wear rate of various natural fiber reinforced polymer composites in sliding wear mode. (Data are collected from research works included in Table 14.2)

equivalent tribo properties compared to glass fiber reinforcement in different wear modes. This may attribute to the fact that the high fiber contents in thermoset resins act as load bearing agents that control tribological properties by fiber reinforcement.

Figure 14.9 shows the specific wear rate results of glass fiber, untreated and alkaline treated oil palm and coir fibers under similar tested conditions (Yousif and El-Tayeb 2008; Yousif et al. 2009) and their test details are provided in Table 14.2. Specific wear rate of the composites predominantly varies by the sliding distance except for glass fiber composites which shows steady wear rate and almost independent of sliding distance. Results also manifest that for unidirectional coir epoxy composite, longer the sliding distance, higher the specific wear rate whereas specific wear rate of the randomly distributed oil palm fibers in epoxy composites increases with increase in sliding distance. On comparison, for alkali treated both fibers reveal better wear resistance than the composites with untreated fibers. Result from fiber treatment with NaOH resulted in removal of gummy contents (pectin, lignin etc.), which may rise the roughness of fiber surface and develop better mechanical locking with high-interfacial adhesion between fibers and matrix. The SEM studies of worn surfaces for two treatment conditions are different where the surface cracking, fiber

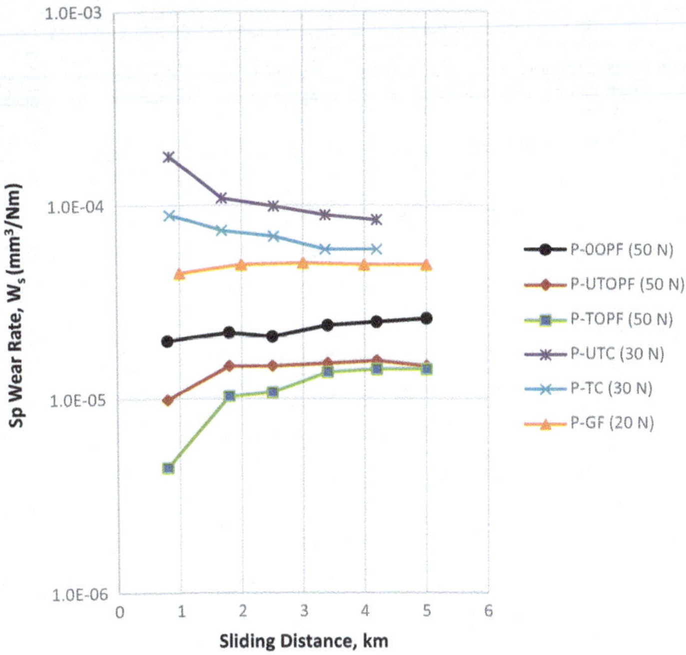

Fig. 14.9 Effect of chemical treatment on specific wear rate of different sliding distances for oil palm and coir fiber reinforced polyester composites at a sliding speed of 2.8 m/s in similar test condition. (Plots are reconstructed from research works of Yousif et al. 2009; Yousif and El-Tayeb 2008; Yousif 2008)

debonding with abrasion (ploughing) are predominant in untreated fiber reinforced composite compared to that in treated fiber reinforced composites.

4.4 Fiber Orientation on Wear

The role of fibers as reinforcements in composites affect the composite wear especially when the sliding force direction of sliding plane encounters the fibers at different orientations i.e. parallel, anti-parallel and normal respect to the sliding plane. Usually short fibers imbedded in composite are in all directions therefore wear in composite is independent of sliding direction but for long continuous fiber depending on its orientations the effect of wear is substantial as the combination of different wear mechanisms are controlled by the fiber orientation. The effect of the type of fiber orientations on specific wear rate and friction coefficient under different test conditions are summarized in Table 14.3 and Fig. 14.10.

Among all the composites in Fig. 14.10, carbon fiber reinforced in a thermoset resin exhibits the lowest wear rate and friction coefficient in sliding wear mode

Table 14.3 Tribological data for the natural fiber reinforced thermoset and thermoplastic composites

Matrix	Fiber	Fiber vol.%	Type of testing	Counter face	Test Parameters			Tribo Properties			References
					Applied load, N	Sliding speed, m/s	Sliding distance, km	CoF, μ	Wear rate, $mm^3/Nm \times 10^{-5}$	Interface Temp., °C	
Epoxy		0	BOD	Stainless Steel	30–100	2.8	0–5	0.65–0.72	14.0–16.0	42–68	Chin and Yousif (2009)
	Kenaf (unidirectional) P	48						0.35–0.58	6.0–3.5	41–61	
	Kenaf (unidirectional) AP							0.36–0.42	3.8–2.0	35–49	
	Kenaf (unidirectional) N							0.52–0.68	1.0–2.0	43–65	
Epoxy		0	POD	Stainless steel	30	1.7–3.96	0–4	0.9–1.0	11.0–7.0	30–95	Nirmal et al. (2012)
	Bamboo (random)	45						0.85–0.9	5.0–7.0	30–92	
	Bamboo (unidirectional) P							0.55–0.75	3.5–5.5	30–78	
	Bamboo (unidirectional) AP							0.5–0.7	3.0–5.0	30–75	
Epoxy		0	BOD	Sic particles	1–7	2.56	6–25.6 m	NR	26.0–11.0	NR	Chand and Dwivedi (2008)
	Sisal (unidirectional) P	50							18.0–10.0	NR	

(continued)

Table 14.3 (continued)

Matrix	Fiber	Fiber vol.%	Type of testing	Counter face	Test Parameters			Tribo Properties			References
					Applied load, N	Sliding speed, m/s	Sliding distance, km	CoF, μ	Wear rate, $mm^3/Nm \times 10^{-5}$	Interface Temp., °C	
	Sisal (unidirectional) AP								13.0–8.0	NR	
	Sisal (unidirectional) N								12.0–7.0	NR	
Epoxy	Jute short fiber mat P	35	BOR	Stainless Steel	30	3	10	0.6	0.007–0.004	NR	Alshammaria (2018)
	Jute short fiber mat AP							0.6–0.9	0.0068–0.003	NR	
	Jute short fiber mat N							0.26–0.55	0.003–0.007	NR	

P—parallel, AP—antiparallel, N—neutral, POD—pin on disc, BOD—block on disc, BOR—block on ring, NR—not reported

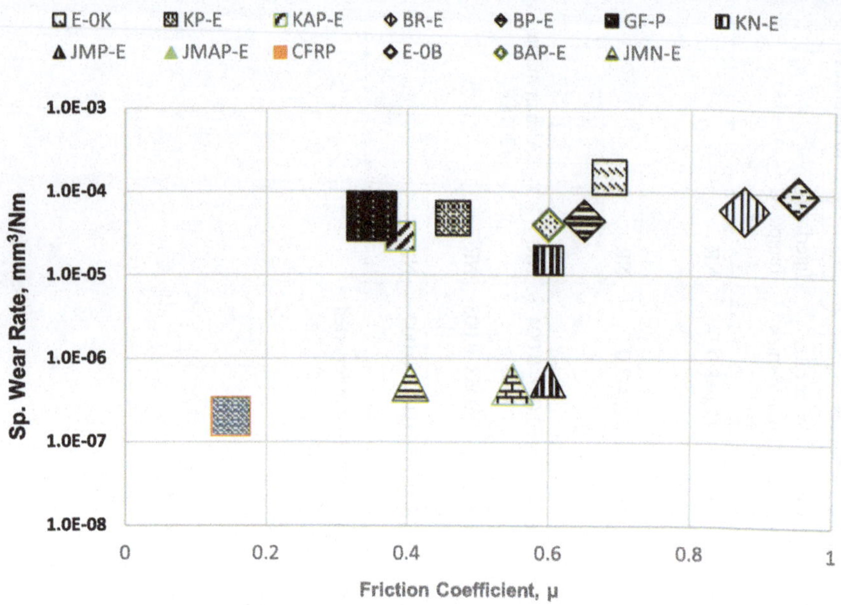

E-Epoxy, K-Kenaf, B-Bamboo, GF-Glass Fiber, J-Jute, CF-Carbon Fiber, R-Random, M-Mat, P-Parallel, AP-Antiparallel, and N-Normal.

Fig. 14.10 Effect of fiber orientation on friction coefficient and specific wear rate of various natural fiber reinforced polymer composites in sliding wear mode (Data are collected from research works included in Table 14.3)

where bamboo fiber reinforced composites have the highest wear rate and friction coefficient. Although, for any group of composites fiber orientations have substantial contribution on wear rate, friction coefficient of composite exhibit low to high value in the order of N < AP < P orientation. El-Tayeb (2008a, b) conducted a series of studies on the tribological properties of natural fibre. In order to evaluate the effect of the fibre length onto their wear properties, sugarcane fibre with length ranging from 1, 5 and 10 mm were arranged in random orientation (R–O) and subjected to 20–80 N of applied normal loads, at constant sliding velocity of 2.5 m/s, and constant sliding distance of 2.25 km. The main finding indicated that wear rate were decrease with increase of fibre length. In his second study (2008), he focused on influence of the fibre orientation by using two conditions where the sliding forced direction was applied in anti-parallel orientation (AP–O) and parallel orientation (P–O) with respect to the mat orientation. He reported that P–O shown a significantly higher weight loss than AP–O. Morphological analysis revealed that no obvious fracture in the AP-O due to high interfacial adhesion of the fibres and matrix as compared to the P-O direction. This is because that in P-O orientated fibre, there was a higher chance for the failure due to fibre pull-out as the sliding force is applied in parallel to the fibre orientation. Narish et al. (2011) studied the wear properties of treated kenaf fibres reinforced polyurethane (PU) composites at different applied loads (30–60 N), at constant sliding velocity of 2.8 m/s at different fibre orientations of anti-parallel orientation (AP-O), parallel orientation (P-O) and normal orientation (N-O). They reported that wear performance compared to neat PU, giving an improvement in specific wear rate (Ws) at 78% in comparison to neat composite. In regards to PU, adhesion film forms during wear and manifests low friction due to smoother sliding resulted from thermal softening of PU. Considering the case for composites, fibers debond from matrix but it adheres with the soft PU film and makes the film stiffer that may improve the wear performance but with high friction. Alshammaria et al. (2018) aims to correlate the influence of fibre orientation on tribological performance of the alkaline treated jute mat in the epoxy matrix. They were using three different orientations with respect to sliding force such as N-O, AP-O and P-O. The main finding of their work revealed that fibre orientation has very significant contribution to the wear and friction properties of the composite. The optimum condition of the wear resistance was obtained when the composite was tested in antiparallel condition.

Figure 14.11 shows the effect of fiber orientation on specific wear rate of kenaf and bamboo fiber composites at different sliding distances in similar test conditions (Chin and Yousif 2009; Nirmal et al. 2012). Obviously presence of kenaf or bamboo fibers in composite shows the improved wear performances as compared to that of neat epoxy under same tribo conditions and the specific wear rate of the composites decrease when sliding distance increases. For normally oriented (N) kenaf fibers wear rate is lower when compared with P or AP conditions. This is attributed to more film

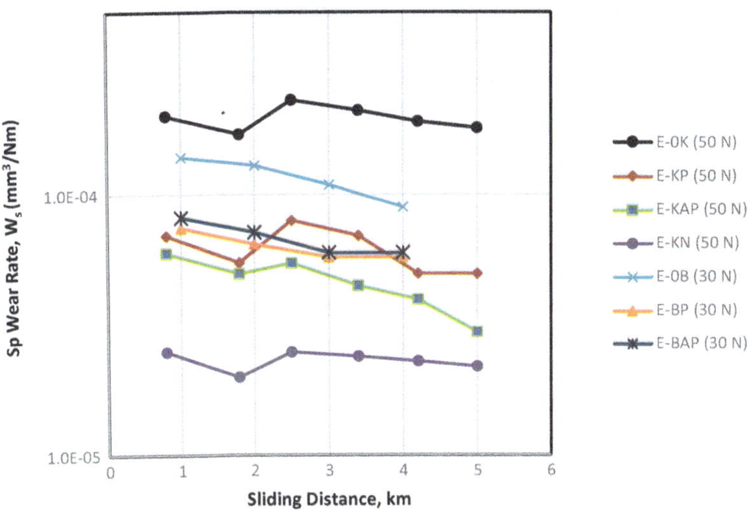

E-Epoxy, K-Kenaf, B-Bamboo, P-Parallel, AP-Antiparallel and N-Normal

Fig. 14.11 Effect of fiber orientation on specific wear rate for different sliding distances of kenaf and bamboo fiber reinforced epoxy composites at a sliding speed of 2.8 m/s in similar test conditions (Plots are reconstructed from research works of Chin and Yousif 2009 and Nirmal et al. 2012) formation by normal (N) oriented fibers as compare to normal AP/P conditions where debonded fiber volumes are higher and worn out of the matrix rapidly instead of forming film. Usually in normal orientation fibers are perpendicular to worn surface so the contact area is only the diameter of each fiber whereas in parallel or anti parallel orientations fibers in length wise are exposed. Therefore, for a given volume fraction of fiber, the effective fiber contact area is less for N orientation as compare to A or AP orientation.

Chin and Yousif (2009) proposed the wear mechanism for the three tested orientations shown in Fig. 14.12 in terms of damages of fibers that are different for P, AP and N orientations. In case of P orientation (cf. Figure 14.12a), long length of fibers encounter sliding force which could bend and break the fibers along the sliding direction if the sliding force exceeds the interfacial adhesion strength. Even at certain condition when the sliding force is lower than the adhesion strength, splitting of fibers may occur and finally fibers could tear and break. For AP orientation, in addition to debonding and bending, detachment of fibers may occur when the sliding force is higher the interfacial adhesion as shown in Fig. 14.12b. On the other hand, there is less possibility of fiber debonding when the fibers are in N orientation (Fig. 14.12c) where the fibers diameters are only exposed and the fibers are embedded inside the

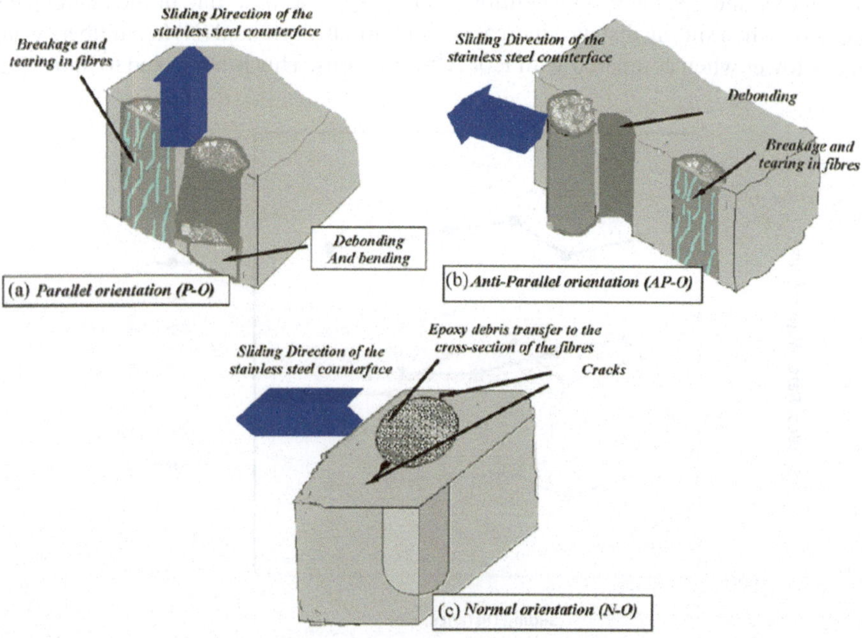

Fig. 14.12 Proposed wear mechanism of the composites in **a** parallel, P, **b** anti-parallel, AP and **c** normal, N orientations (Chin and Yousif 2009)

Table 14.4 Polyethylene-based composites (Satov 2008)

Function	Material	Dosage range (%)
Matrix component	Polyethylene resin	Difference from total of other components of 100%
Reinforce component	Natural fiber	30–60
Coupling agent	Maleated polyolefin	2–5
Lubricant(s)	Stearates/esters/EBS/other	3–8
Antioxidants	Phenolics/phosphites	0–1
Acid scavengers	Stearates/hydrotalcites	0–1
UV protection	HALS/benzophenones/benzotriazoles	0–1
Mineral filler	Talc	0–10
Biocide	Zinc borate	0–2
Density reduction	Microspheres\chemical or physical blowing agent	0–5
UV protection/aesthetics	Pigments	As required
Flame retardants/smoke suppressants	Various	As required

matrix along their length. But the sliding force may generate secondary crack perpendicular to matrix around the fibers due to shearing force. Worn surface morphology characterised by SEM further clarified the proposed mechanism.

4.5 *Role of Additives in NFRPC*

Additives are part of the important recipe in fabricating tribo-composites owning to its role in manufacturing process and end product performance. Generally, additives are used based upon typical formulas and industrial norms, and are modified whenever necessary. Satov (2008) revealed that composite industries tend to use additives to make up for process deficiencies due to the lack of coordination between end product requirements and materials-process technology, resulting in an increase of operational cost in a long run. It is, hence necessary to optimize the usage of additives through proper integration of materials, process technology and product performance. Tables 14.4 and 14.5 summarize the functions, materials and dosages for polyethylene- and polypropylene-based composites.

4.5.1 Solid Lubricants

In order to develop a good tribo-composite component, incorporating the composites with specific fillers and/or additives is essential to achieve special requirements

Table 14.5 Polypropylene-based composites (Satov 2008)

Function	Material	Dosage range (%)
Matrix component	Polyethylene resin	Difference from total of other components of 100%
Reinforce component	Natural fiber	30–60
Lubricant(s)	Stearates/esters/EBS/other	3–8
Antioxidants	Phenolics/phosphites	0–1
Acid scavengers	Stearates/hydrotalcites	0–1
UV protection	HALS/benzophenones/benzotriazoles	0–1
Mineral filler	Talc	0–10
Biocide	Zinc borate	0–2
Density reduction	Microspheres\chemical or physical blowing agent	0–5
UV protection/aesthetics	Pigments	As required
Flame retardants/smoke suppressants	Various	As required

for tribological applications, such as exhibiting low friction characteristics while maintaining good mechanical strength. Solid lubricants such as graphite, molybdenum disulfide (MoS_2), poly-tetrafluoroethylene (PTFE) (Chang and Friedrich 2010; Konovalova and Suchanek 2012; Panda et al. 2016; Shalwan and Yousif 2013, 2014; Sharma et al. 2017; Subramanian et al. 2016), hexa-boron nitride (hBN), tungsten disulphide (WS_2) (Panda et al. 2016) and boric acid (H_3BO_3) (Mutlu et al. 2007; Reeves et al. 2013) have been widely employed in polymer composites to enhance wear performance by reducing wear rate and friction coefficient of the composites.

These solid lubricants are generally useful in developing a uniform transfer layer on the surface of hard metallic counterpart, which hamper the direct contact between the two sliding bodies and thus protect the composites from severe abrasive wear (Chang and Friedrich 2010; Shalwan and Yousif 2013). The molecular structure of these solid lubricants which is composed of layers of atoms bonded together by weak van der Waals force rendered the formation of transfer film upon dry sliding (Reeves et al. 2013). Figure 14.13 illustrates schematic representations of the layered crystal structure of graphite, MoS_2, hBN and H_3BO_3 molecules. Addition of these solid lubricants in the composites leads to the development and usage of a new class of materials called self-lubricating polymer matrix composites, which are highly sought after in mechanical and tribological components such as gears, cams, wheels and impellers. These self-lubricating composites fulfilled the increasing industrial demand to operate in machineries subjected to relative movement where no external and/or fluid lubricant shall be used (Suresha et al. 2010).

In general, the presence of solid lubricants demonstrated significant reduction of coefficient of friction and wear rate in various polymer matrix composites (Bijwe and Indumathi 2004; Hashmi et al. 2007; Shalwan and Yousif 2014; Zhang et al.

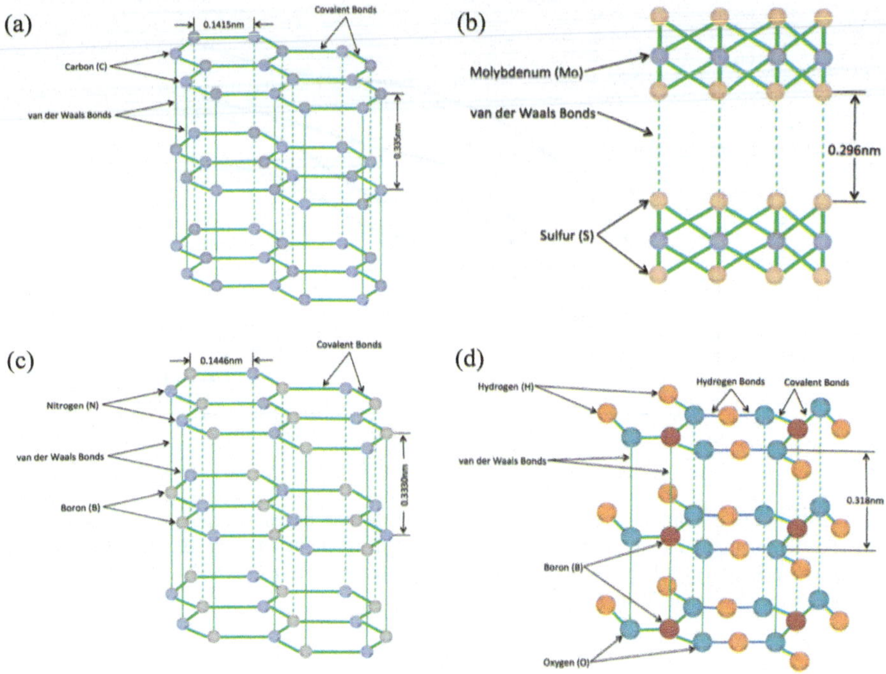

Fig. 14.13 Illustration of layered crystal structure of **a** graphite C, **b** molybdenum disulfide MoS_2, **c** hexa-boron nitride hBN and **d** boric acid H_3BO_3 molecules (Reeves et al. 2013)

2008). Several researchers reported on the reduction of coefficient of friction and wear rate as a function of solid lubricant concentration (Hashmi et al. 2007; Suresha et al. 2010; Zhang et al. 2008). Coefficient of friction as low as 0.1 was reported in poly-(phthalazinone ether sulfone ketone) (PPESK) composites filled with 30 wt% graphite (Zhang et al. 2008) under 500 N sliding load. Nevertheless, it is also noted that the presence of high solid lubricant content results in the deterioration of mechanical properties of the respective composites (Shalwan and Yousif 2014; Zhang et al. 2008). However, Suresha et al. (2010) inferred that treating the graphite filler with silane coupling agent led to improved interfacial adhesion and proper dispersion of graphite filler in carbon-epoxy composite, and subsequently enhanced the tensile strength, Young's modulus and hardness of the composite.

Whilst these solid lubricants exhibited desirable frictional characteristics, thermal conductivity of these solid lubricants must also be considered as frictional heat generated could harshly affect the friction and wear mechanisms. Solid lubricants having high thermal conductivity are preferable in tribo-composites applications, allowing heat dissipation from the tribo-surface and hence minimizing damage to the component surface. Hashmi et al. (2007) reported that contact interface temperature reduced with increasing amount of graphite filler in their work on graphite/cotton fibre reinforced polyester composites under sliding wear condition. Similar trend was also

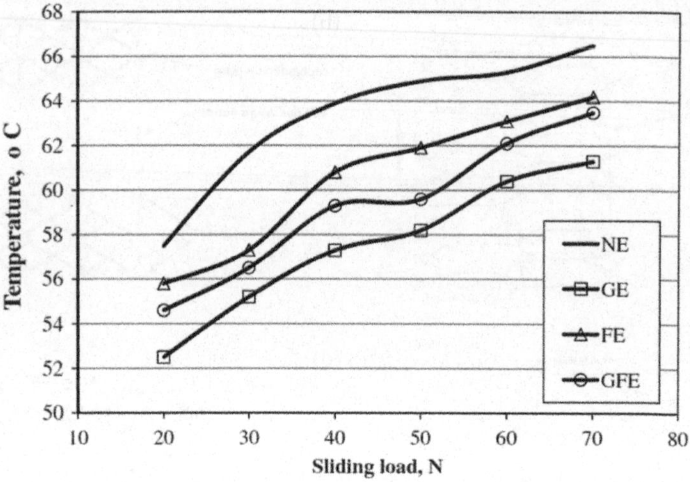

Fig. 14.14 Interface temperature of different epoxy composites based on graphite and/or date palm fibre at different applied loads after 5.04 sliding distance using block-on-ring (BOR) technique. *NE: neat epoxy; GE: 3% graphite/epoxy; FE: Date palm fibre/epoxy; GFE: 3 wt% graphite/date palm fibre/epoxy* (Shalwan and Yousif 2014)

observed by Shalwan and Yousif (2014) wherein the lower interface temperatures were recorded in wear test when 3% graphite was added into both neat epoxy and date palm fibre reinforced epoxy composites, as illustrated in Fig. 14.14.

However, among all the solid lubricants, graphite and hexa-boron nitride (Panda et al. 2016) are the promising candidates and their related properties are detailed in Table 14.6. In fact, Bijwe and Indumathi (2004) found that the PTFE is not the right choice for high temperature lubrication in tribo-composites, unless PTFE is used in a combination with graphite and MoS_2. More recently, Panda et al. (2017) highlighted that synergistic effect of two solid lubricants (15% graphite and 5% hBN) incorporated in the glass fibre reinforced polyaryletherketone (PAEK) exhibits the lowest friction coefficient (0.04) and specific wear rate (5.68×10^{-16} m^3/Nm) compared to

Table 14.6 Characteristic properties of high performance solid lubricants (Panda et al. 2016)

Material	h-BN	Graphite	PTFE
Density (g/cm^3)	~2.1	~2.1	2.15
Bulk modulus (GPa)	36.5	34	1.8
Thermo-oxidative stability (°C)	1000	570	260
Thermal conductivity (W/m K)	600^a; 30^b	200–2000^a; 2–800^b	0.25
Thermal expansion (10^{-6}/°C)	−2.7^a; 38^b	−1.5^a; 25^b	−148
Specific heat capacity (J/kg K)	710–830	840–1610	1000

Effect of fiber orientation on tribo properties
^{a}Parallel to planes/layers; ^{b}Perpendicular to planes

that of single solid lubricant. It is expected that such remarkable synergism at 15% graphite and 5% hBN was due to improved thermal conductivity and fibre-matrix adhesion (Panda et al. 2017).

4.5.2 Nanoparticles

In line with the advancement in nanomaterials, incorporation of various nano-sized inorganic particles such as ZrO_2, Al_2O_3, SiO_2, SiC, TiO_2 and CuO in polymer matrices has attracted great attention due to significant enhancement in tribological performance (L. Guo et al. 2017). Konovalova and Suchanek (2012) emphasized that significant size reduction of these particles down to nano-scale level leads to a completely distinct wear behaviour and better properties under dry sliding conditions with advantages of:

i. Generally lower abrasiveness due to reduced angularity;
ii. Enhanced strength, modulus and toughness due to defect-free structure;
iii. Higher specific surface areas and thus, improved filler-matrix adhesion; and
iv. High effectiveness at very low contents.

Several researchers (Cho and Bahadur 2005; Guo et al. 2009; Shao et al. 2004) pointed out that the enhanced bonding between the nanoparticles and polymer matrix due to the high surface area-to-volume ratio of the nanoparticles, led to strengthening of the transfer film. The transfer film formed during the dry sliding wear was fairly thinner and smoother, hence providing a better coverage on the steel counterpart surface, resulting in lower friction coefficient and wear rate of the composite materials. Furthermore, it was highlighted that the smaller the particles, the better was the wear resistance of the composites; and the optimum filler content of these nanoparticles ranges between 1 and 4 vol.% to avoid the tendency of particle agglomeration (Zhang and Friedrich 2005).

On the other note, Chang and Friedrich (2010) investigated the effect of TiO_2 nanoparticles on the dry sliding wear of short carbon fibre reinforced polymer composites with addition of conventional tribo-fillers, i.e. graphite and PTFE. It was found that addition of 5 vol.% of nano-TiO_2 could significantly reduce the friction coefficients and contact temperature, especially under high pv (the product of p (pressure) and v (velocity)) conditions. The results clearly indicated that the presence of nano-TiO_2 which acts as 'spacers' at the contact region effectively reduce the adhesion between the transfer film and polymeric specimen, rendering a lower coefficient of friction. Figure 14.15 compares the wear mechanisms for sliding wear of short fiber reinforced polymer (SFRP) composites between with and without inclusion of nanoparticles.

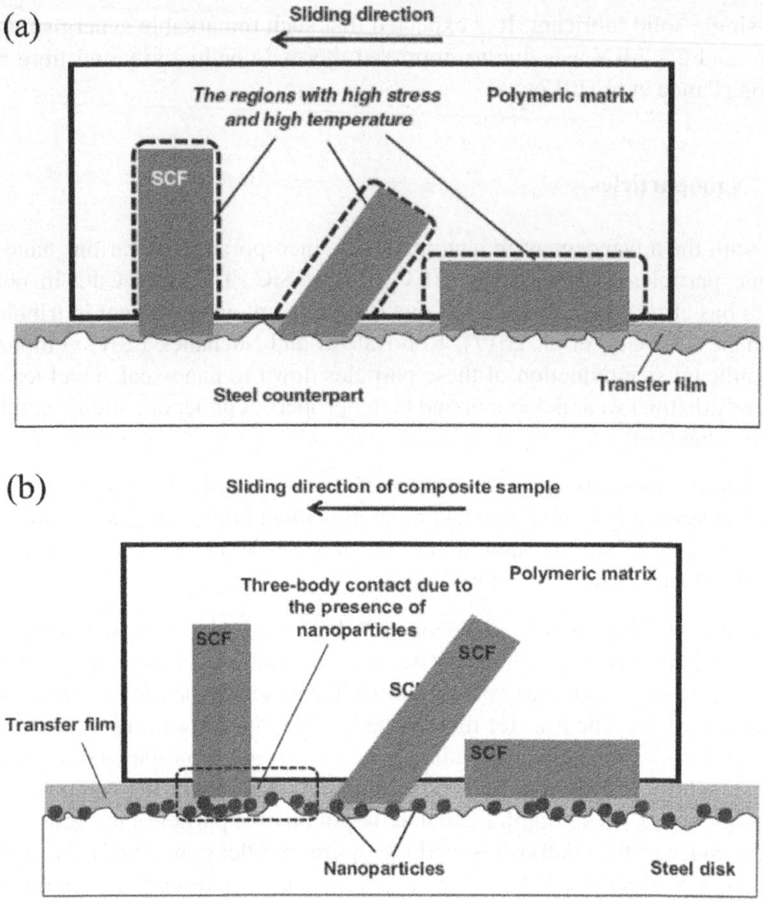

Fig. 14.15 Schematic illustration of the failure mechanism for the sliding wear of SFRP composites **a** with and **b** without nanoparticles (Chang and Friedrich 2010)

5 NFRPC as Tribo Materials

NFRPCs have emerged as a potential alternative to synthetic fiber reinforced composite as they minimize harmful pollutants and reduce the environmental impact. In some occasions natural fibers are expected to perform close to that of synthetic fibers especially brake pads for automotive. With great care fabrication must ensure the homogeneous distribution of fibers with a high thermal resistance to withstand severe temperature during braking. Several research teams are developing automotive components dedicated to tribo system by replacing a substantial amount of synthetic fibers by natural fibers (ECOPAD). Various natural fibers reinforced polymer hybrid composites with limited tribological applications have been illustrated in Table 14.7.

Table 14.7 Various natural fiber reinforced hybrid composites in tribological applications (Karthikeyan et al. 2017)

Fiber	Matrix	Fiber treatment	Tribological application
Banana and Kenaf	Polyester	NaOH, sodium lauryl sulphate	Clutch
Rice straw dust/rise husk dust	Phenolic	Untreated	Brake pad
Sugarcane/glass	Polyester	Untreated	Bearing
Linen/jute	Polyester	Untreated	Bearing
Sisal	Phenolic	Silane coupling	Brake pad
Sea shell nano powder	Poly-methyl methacrylate	Untreated	Dental

6 Conclusion and Future Perspective

The concept of completely replacing synthetic fillers by natural fiber does not provide solutions for using materials as engineering components while considering its properties and dimensional stabilities. Although natural fibers polymer composites are lighter as compare to synthetic composites but degrades drastically when come to contacts during service even though the natural fibers are modified chemically. Incorporating synthetic fillers in NFRPC not only enhance load-carrying capacity, but also considering for the improvement of wettability and compatibility between the fiber and the matrix. The role of hybridisation by blending natural and synthetic fillers in polymer composites has not been elaborated substantially in tribo testing of composites. Therefore, the design of hybrid composites using natural fibers and synthetic fillers mainly focuses on the reduction of water or moisture absorption with adequate adhesion between fiber and matrix for the development of improved wear and friction properties. Cotton fiber and graphite filler in modified polymers were designed initially to develop hybrid composite with significant reduction in the specific wear rate as well as reduction in contact temperature at the interface due to presence of conductive graphite filler. Subsequently, nanomaterials such as carbon nanotube and graphene are found to be efficient nanofillers for polymer composites due to their good mechanical strength with high thermal conductivity which resist the thermal gradation of composites. These synthetic nanofillers not only transport the heat efficiently but also reduce the abrasion as they have high aspect ratio with reduced angularity edges. Hence the wear performance of nano filler composites may be significantly improved better from that of micro-particle filled systems.

Introduction of nano fibrillated cellulose (NFC) synthesised from any plant fibres rich in cellulose (e.g. oil palm empty fruit bunches, bamboo pulps, jute fibers, kenaf and others) as nano fillers in tribo composite materials could be the key challenge for future research projects. NFC may display thermo-mechanical properties comparable to available synthetic and non-biodegradable graphene/CNT nano fillers and ensure to fabricate tribo-composites which till date has not been reported in research. Eventually, the overall research activities will emphasize the benefits of fabricating natural and environmentally friendly tribo materials in order to convert renewable resources to high value end products which possess lesser or no harm to environment.

References

Ahsan Q, Carron TSS, Mustafa Z (2019) On the use of nano fibrillated kenaf cellulose fiber as reinforcement in polylactic acid biocomposites. J Mech Eng Sci 13(2):4970–4988

Aldousiri B, Shalwan A, Chin CW (2013) A review on tribological behavior of polymeric composites and future reinforcements. Adv Mater Sci Eng 8. Article ID 645923

Alshammaria FZ, Saleha KH, Yousif BF, Alajmib A, Shalwanc A, Alotaibib JG (2018) The influence of fibre orientation on tribological performance of jute fibre reinforced epoxy composites considering different mat orientations. Tribol Ind 40(3):335–348

Anbupalani MS, Venkatachalam CD, Rathanasamy R (2020) Influence of coupling agent on altering the reinforcing efficiency of natural fibre-incorporated polymers—a review. J Reinf Plast Compos. https://doi.org/10.1177/0731684420918937

Bajpai PK, Singh I, Madaan J (2013) Tribological behaviour of natural fiber reinforced PLA composites. Wear 297:829–840

Bijwe J, Indumathi J (2004) Influence of fibers and solid lubricants on low amplitude oscillating wear of polyetherimide composites. Wear 257(5–6):562–572

Boopathi L, Sampath PS, Mylsamy K (2012) Investigation of physical, chemical and mechanical properties of raw and alkali treated Borassus fruit fiber. Compos B Eng 43(8):3044–3052

Brígida AIS, Calado VMA, Gonçalves LRB, Coelho MAZ (2010) Effect of chemical treatments on properties of green coconut fiber. Carbohydr Polym 79(4):832–838

Briscoe BJ, Friedrich K (eds) (1993) Advances in composite tribology, composite materials series, vol 8. Elsevier, Amsterdam, p 3

Campilho RDSG (2015) Natural fiber composites. CRC Press, Boca Raton

Catto AL, Stefani BV, Ribeiro VF, Santana RMC (2014) Influence of coupling agent in compatibility of post-consumer HDPE in thermoplastic composites reinforced with eucalyptus fiber. Mater Res 17:203–209

Chand N, Fahim M (2008) Chapter 2 Introduction to tribology of polymer composites, Tribology of natural fiber polymer composites, Chand N, Faheem, M (eds) CRC Press and Woodhead Publishers

Chand N, Dwivedi UK (2008) Sliding wear and friction characteristics of sisal fibre reinforced polyester composites: effect of silane coupling agent and applied load. Polym Compos pp 80–84

Chand N, Dwivedi UK (2007) Influence of fiber orientation on high stress wear behavior of sisal fiber-reinforced epoxy composites. Polym Compos pp 437–441

Chang L, Friedrich K (2010) Enhancement effect of nanoparticles on the sliding wear of short fiber-reinforced polymer composites: a critical discussion of wear mechanisms. Tribol Int 43(12):2355–2364

Chin CW, Yousif BF (2009) Potential of kenaf fibres as reinforcement for tribological applications. Wear 267:1550–1557

Cho MH, Bahadur S (2005) Study of the tribological synergistic effects in nano CuO-filled and fiber-reinforced polyphenylene sulfide composites. Wear 258(5–6):835–845

Edeerozey AM, Akil HM, Azhar AB, Ariffin MZ (2007) Chemical modification of kenaf fibers. Mater Lett 61(10):2023–2025

Elkhaoulani A, Arrakhiz F, Benmoussa K, Bouhfid R, Qaiss A (2013) Mechanical and thermal properties of polymer composite based on natural fibers: Moroccan hemp fibers/polypropylene. Mater Des 49:203–208

El-Tayeb NSM (2008a) A study on the potential of sugarcane fibers/polyester composite for tribological applications. Wear 265(1–2):223–235

El-Tayeb NSM (2008b) Tribo-characterization of natural fibre-reinforced polymer composite material. Proc Inst Mech Eng, Part J: J Eng Tribol 222(7):935–946

Fadzullah SSM, Mustafa Z (2016) Fabrication and processing of pineapple leaf fiber reinforced composites. In: Green approaches to biocomposite materials science and engineering. IGI Global, pp 125–147

Fadzullah SSM, Mustafa Z, Ramli SNR (2016) The effect of fiber length on the mechanical properties of pineapple leaf (PALF) fiber reinforced PLA biocomposites. Proc Mech Eng Res Day 2016:123–124

Faruk O, Bledzki AK, Fink H-P, Sain M (2012) Biocomposites reinforced with natural fibers: 2000–2010. Prog Poly Sci 37(11):1552–1596

Friedrich K (2018) Polymer composites for tribological applications. Adv Ind Eng Polym Res 1:3–39

Friedrich K (1997) (1997). Wear performance of high temperature polymers and their composites. In: Luise RR (ed) Application of high temperature polymers. CRC Press, Boca Raton, USA, pp 221–246

Fuqua MA, Ulven CA (2008) Characterization of polypropylene/corn fiber composites with maleic anhydride grafted polypropylene. J Biobased Mater Bioenergy 2(3):258–263

Gholampour A, Ozbakkaloglu T (2020) A review of natural fiber composites: properties, modification and processing techniques, characterization, applications. J Mater Sci, pp 1–64

Guo L, Qi H, Zhang G, Wang T, Wang Q (2017) Distinct tribological mechanisms of various oxide nanoparticles added in PEEK composite reinforced with carbon fibers. Compos A Appl Sci Manuf 97:19–30

Guo QB, Rong MZ, Jia GL, Lau KT, Zhang MQ (2009) Sliding wear performance of nano-SiO_2/short carbon fiber/epoxy hybrid composites. Wear 266(7–8):658–665

Hager AM, Davies M (1993) (1993). Short-fiber reinforced, high temperature resistant polymers for a wide field of tribological applications. In: Friedrich K (ed) Advances in composite tribology. Elsevier Scientific Publishers, Amsterdam, The Netherland, pp 107–157

Hashmi SAR, Dwivedi UK, Chand N (2007) Graphite modified cotton fibre reinforced polyester composites under sliding wear conditions. Wear 262(11–12):1426–1432

Hong CK, Kim N, Kang SL, Nah C, Lee YS, Cho BH, Ahn JH (2008) Mechanical properties of maleic anhydride treated jute fibre/polypropylene composites. Plast Rubber Compos 37(7):325–330

Jusoh MSM, Yahya MYM, Mustafa Z, Ahmad HAI (2015) Effect of layering pattern on mechanical and water absorption properties of glass/flax reinforced epoxy. Jurnal Teknologi 79:5–2

Friedrich K, Theiler G, Klein P (2018) Polymer composites for tribological applications in a range between liquid helium and room temperature. In: Sinha SK (ed) Handbook of polymer tribology. World Scientific, Singapore, p 307e343

Kakou CA, Arrakhiz FZ, Trokourey A, Bouhfid R, Qaiss A, Rodrigue D (2014) Influence of coupling agent content on the properties of high density polyethylene composites reinforced with oil palm fibers. Mater Des 63:641–649

Karthikeyan S, Rajini N, Jawaid M, Jappes JTW, Thariq MTH, Siengchin S, Sukumaran J (2017). A review on tribological properties of natural fiber based sustainable hybrid composite. Proc IMechE Part J: J Eng Tribol 0(0) 1–19, IMechE 2017. https://doi.org/10.1177/1350650117705261

Konovalova O, Suchanek J (2012) Significance of polymer nanocomposites in triboengineering systems. In: NANOCON 2012—Conference proceedings, 4th international conference

Lilholt H, Lawther JM (2000) Natural organic fibers. In: Kelly A, Zweben C (eds) Comprehensive composite materials, vol 1. Elsevier Science, Amsterdam

Liu Y, Xie J, Wu N, Wang L, Ma Y, Tong J (2019) Influence of silane treatment on the mechanical, tribological and morphological properties of corn stalk fiber reinforced polymer composites. Tribol Int 131:398–405

Mayandi K, Rajini N, Pitchipoo P, Jappes JTW, Rajulu AV (2018) Properties of untreated and chemically treated Cissus quadrangularis natural fibers and their composites with polyester as the matrix. Polym Compos 39(3):876–886

Menezes PL, Rohatgi PK, Lovell MR (2011) Tribology of natural fiber reinforced polymer composites. In: ASME/STLE 2011 international joint tribology conference, American Society of Mechanical Engineers, 2011, pp 341–343

Mutlu I, Oner C, Findik F (2007) Boric acid effect in phenolic composites on tribological properties in brake linings. Mater Des 28(2):480–487

Nam TH, Ogihara S, Tung NH, Kobayashi S (2011) Effect of alkali treatment on interfacial and mechanical properties of coir fiber reinforced poly (butylene succinate) biodegradable composites. Compos Part B: Eng 42(6):1648–1656

Narish S, Yousif BF, Rilling D (2011) Adhesive wear of thermoplastic composite based on kenaf fibres. Proc Inst Mech Eng, Part J: J Eng Tribol 225(2):101–109

Nirmal U, Hashim J, Low KO (2012) Adhesive wear and frictional performance of bamboo fibres reinforced epoxy composite. Tribol Int 47:122–133

Omrani E, Menezes PL, Rohatgi PK (2016) State of the art on tribological behavior of polymer matrix composites reinforced with natural fibers in the green materials world. Eng Sci Technol, Int J 19:717–736

Panda JN, Bijwe J, Pandey RK (2016) Role of treatment to graphite particles to increase the thermal conductivity in controlling tribo-performance of polymer composites. Wear 360:87–96

Panda JN, Bijwe J, Pandey RK (2017) Attaining high tribo-performance of PAEK composites by selecting right combination of solid lubricants in right proportions. Compos Sci Technol 144:139–150

Rajini N, Jappes JTW, Rajakarunakaran S et al (2012) Effect of processing variables on mechanical properties of montmorillonite clay/unsaturated polyester nanocomposite using Taguchi based grey relational analysis. J Polym Eng 32:555–566

Ramli SNR, Fadzullah SHSM, Mustafa Z (2017) The effect of alkaline treatment and fiber length on pineapple leaf fiber reinforced poly lactic acid biocomposites. Jurnal Teknologi 79(5–2)

Reeves CJ, Menezes PL, Lovell MR, Jen TC (2013) Tribology of solid lubricants. In Tribology for scientists and engineers: from basics to advanced concepts, pp 447–494. https://doi.org/10.1007/978-1-4614-1945-7_13

Reinicke R, Haupert F, Friedrich K (1998) On the tribological behavior of selected, injection molded thermoplastic composites. Compos Part A Appl Sci Manuf 29:763–771

Sampath P, Kumar VS (2019) Mechanical and tribological behaviour of treated and untreated moringa oleifera pods fiber reinforced epoxy polymer composite for packaging applications

Satov DV (2008) Additives for wood-polymer composites. In Wood-polymer composites, pp 23–40. https://doi.org/10.1533/9781845694579.23

Shalwan A, Yousif BF (2013) In state of art: mechanical and tribological behaviour of polymeric composites based on natural fibres. Mater Des 48:14–24

Shalwan A, Yousif BF (2014) Influence of date palm fibre and graphite filler on mechanical and wear characteristics of epoxy composites. Mater Des 59:264–273

Shao X, Liu W, Xue Q (2004) The tribological behavior of micrometer and nanometer TiO_2 particle-filled p (phthalazine ether sulfone ketone) composites. J Appl Polym Sci 92(2):906–914

Sharma M, Sharma H, Shannigrahi S (2017) Tribology of advanced composites/biocomposites materials. In Biomedical composites. pp 413–429. https://doi.org/10.1016/b978-0-08-100752-5.00018-4

Subramanian K, Nagarajan R, De Baets P, Subramaniam S, Thangiah W, Sukumaran J (2016) Eco-friendly mono-layered PTFE blended polymer composites for dry sliding tribo-systems. Tribol Int 102:569–579

Suresha B, Ramesh BN, Subbaya KM, Chandramohan G (2010) Mechanical and three-body abrasive wear behavior of carbon-epoxy composite with and without graphite filler. J Compos Mater 44(21):2509–2519

Swain PTR, Biswas S (2017) Abrasive wear behaviour of surface modified jute fiber reinforced epoxy composites. Mater Res 20(3):661–674
Valášek P, D'Amato R, Müller M, Ruggiero A (2018) Mechanical properties and abrasive wear of white/brown coir epoxy composites. Compos B Eng 146:88–97
Walker JCF (2006) Primary wood processing: principles and practice. Springer, Berlin
Yallew TB, Kumar P, Singh I (2014) Sliding wear properties of jute fabric reinforced polypropylene composites. Procedia Eng 97:402–411
Yousif BF (2008) Replacing of glass fibres with seed oil palm fibres for tribopolymeric composites. Tribology 2(2):99–103
Yousif BF, El-Tayeb NSM (2008) Adhesive wear performance of T-OPRP and UT-OPRP composites. Tribol Lett 32:199–208
Yousif BF, Leong OB, Ong LK, Jye WK (2009) The effect of treatment on tribo-performance of CFRP composites. Recent Patents Mater Sci 2:67–74
Zhang X, Liao G, Jin Q, Feng X, Jian X (2008) On dry sliding friction and wear behavior of PPESK filled with PTFE and graphite. Tribol Int 41(3):195–201
Zhang Z, Friedrich K (2005) Tribological characteristics of micro- and nanoparticle filled polymer composites. In: Polymer composites, pp 169–185. https://doi.org/10.1007/0-387-26213-x_10
Zhang Z, Breidt C, Chang L, Haupert F, Friedrich K (2004) Enhancement of the wear resistance of epoxy: short carbon fiber, graphite, PTFE and nano-TiO_2. Part A Appl Sci Manuf 35:1385–1392
Zulkafli N, Malingam SD, Fadzullah SHSM, Mustafa Z, Zakaria KA, Subramonian S (2019) Effect of water absorption on the mechanical properties of cross-ply hybrid pseudo-stem banana/glass fibre reinforced polypropylene composite. Mater Res Express 6(9):095326

Friction and Wear Properties of Natural Fiber Reinforced Composites

T. P. Mohan and K. Kanny

Abstract The objective of this chapter is to review the recent progress on the tribological properties of natural fiber reinforced composites (NFRC). Specific emphasis is given to plant based fibers as they are abundantly available and share major portion of natural fibers than other natural fibers extracted from animal and mineral sources. The various factors affecting the friction and wear properties under dry and wet medium conditions of NFRC materials are discussed. An outline of the tribological test in NFRC materials are discussed with emphasis on test methods, NFRC materials and types of fibers. A general trend on the tribological properties with influencing parameters is represented in graphical format for readers to understand the interplay of various effects. The wear mechanism of NFRC and nanoparticle treated (such as nanocellulose and nanoclays) materials with respect to transfer film forming capabilities and measurement techniques were discussed. Commonly used theoretical analysis such as artificial neural networks (ANN) models for predicting frictional properties of NFRC were discussed. Overall, this chapter provides the reader a conscience and succinct information about friction and wear properties of NFRC materials that are studied in last two decades.

Keywords Natural fibers · Plant based fibers · Natural fiber reinforced composites · Frcition · Wear

T. P. Mohan (✉) · K. Kanny
Composites Research Group (CRG), Department of Mechanical Engineering, Durban University of Technology, Durban, South Africa
e-mail: MohanP@dut.ac.za

K. Kanny
e-mail: KannyK@dut.ac.za

M. T. Hameed Sultan et al. (eds.), *Tribological Applications of Composite Materials*, Composites Science and Technology, https://doi.org/10.1007/978-981-15-9635-3_15

1 Introduction

During the past five-decade advancement of materials have resulted in rapid growth of technology and application. Several type of materials were developed that suit various application ranging from electronic to aerospace industries (Faruk et al. 2012; May-Pat et al. 2013). Among these application, search for light weight material occupied the heart of research, growth and innovation. Composite materials were developed that were very low weight but superior properties than that of steel and ceramics were obtained. A typical density of glass fiber reinforced epoxy composite have about two times higher density than water (1 g/cc) and has almost equal strength to that of plain carbon steel. Composite material is a type of material that has two distinct physical phases, one being 'matrix' and other being 'reinforcement'. The composite material is just a mere physical and mechanical combination of two distinct phase material develop for specific objectives and goals. Matrix phase is a continuous phase while reinforcement phase is a discontinuous phase. Matrix are commonly, polymers, ceramics and metallic materials. Reinforcement also made of polymer, ceramics or metallic, however, the property of composite depend upon the type of architecture of the reinforcement. Reinforcement can be in the form of fibers, particles, laminates or sandwich types, with dimensional size ranging from macro to nano-size (Gbadeyan et al. 2017; Gu et al. 2017). Among these various types of composite materials, polymer reinforced fiber composites have attracted significant interest due to their light weight, low cost and high mechanical properties. In one hand rapid advancement in light weight composite technology were happened in the past five-decade, on other hand there caused a significant damage to environmental pollution and scarcity of natural resources. Most of the reinforcement and polymers are extracted from synthetic sources such as oil and heavy industrial processes of natural minerals. Currently human kind never seen such an environmental damage due this industrial development and came to tipping point where several policy makers and scholars to focus urgent attention on alternative green materials. During the past two-decade, researchers have started focusing attention on green materials to reduce the impact of environmental damages (Boopathi et al. 2012; Mohammed et al. 2015). Several type of green materials were started manufacturing and already finding application to certain extent in commercial, consumer and household applications. As composite materials are extensively used in mechanical application, this paper focus on the tribological application of natural fiber reinforced composites (NFRC). Virtually all the parts and materials undergo wear and tear during application. The study of wear of NFRC materials are relatively scare and this report present a precise progress in friction and wear properties of NFRC materials.

2 Natural Fibers

The term natural fibers meant the fibers that are extracted from natural sources. Extraction from natural source is a speculative terms, where different authors view in different ways. However, they have an overriding common consideration about 'extraction', which includes, organic, regenerative, naturally available, cultivable, biodegradation, environmentally friendly, cause no or meagre harm to environment, green, etc. Natural sources are plants, animals, minerals, etc. However, focus of the natural fibers are skewed more towards plant sources than that of animals and mineral sources. This is due to the higher percentage of natural fibers produced from plant source (~80 to 85% of world natural fiber production) (Tong et al. 1995; Yi and Yan 2007; Cai et al. 2015; Chegdani et al. 2016). Although natural fibers can be extracted from animal and mineral sources, it involves certain level of destruction to the eco system when compared with natural fiber extracted from plant sources. Figure 1 shows the classification of natural fibers and various types of natural fibers.

It can be seen that different type of fibers can be obtained from different natural resources. This is the significant advantage of natural fibers. For instance, fibers obtained from silk and sisal fibers possess entirely different microstructure and chemical constituents, which in turn adds benefit to tailor make the properties for various applications.

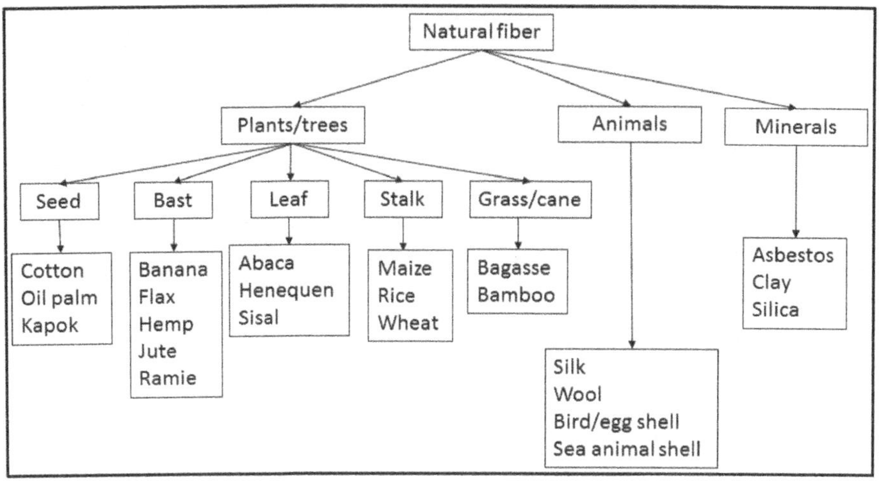

Fig. 1 Sources of natural fibers

2.1 Mechanical and Structural Properties of Natural Fibers

Table 1 shows the tensile properties of selected natural fibers and synthetic fibers (Tong et al. 1995; Yi and Yan 2007; Cai et al. 2015; Chegdani et al. 2016). It can be seen that the tensile properties are comparable to that of synthetic fibers. In particular, specific modulus (ratio of modulus to density) of natural fibers are several times higher than that of glass fiber, which indicate a promising potential in the light weight high strength application. In one study, the hardness of the natural fiber (hemp and flax) reinforced composites reported higher than that of glass fiber reinforced polymer composite.

Few drawbacks of NFRC are high water mass absorption, reduced flammability, inconsistent chemical composition of fibers, the stage of growth and maturity of the plant during fiber extraction. These factors heavily affect the thermal and mechanical properties of NFRC materials. The hydrophilic characteristics of the natural fibers attracts moisture and deteriorates hardness and surface properties. This in turn affects the tribological properties (Tong et al. 1995; Chand and Dwivedi 2006; Yi and Yan 2007; Cai et al. 2015; Chegdani et al. 2016, 2015).

Structural of natural fibers based on plant sources are represented by a typical cell wall type layered structure as shown in Fig. 2. Figure 2 shows the microstructure of a plant based fibril structure. Almost all plant fibers consist of hollow lumen structure surrounded by cell wall structure. Cell wall structure composed of primary

Table 1 Comparison of the tensile properties of various natural fibers with synthetic fibers

Fiber	Density (kg/m^3)	Tensile strength (MPa)	Tensile modulus (GPa)	% Elongation
Jute	1460	300–800	10–30	1.5–2
Sisal	1450	200–400	8–20	2–15
Pineapple	1440	400–1700	30–90	0.8–1
Kenaf	1400	200–300	4–5	0.1–2
Flax	1500	300–1500	20–80	1–3
Hemp	1480	500–900	60–80	1.4–1.8
Banana	1350	500–800	5–15	1–4
Coir	1150	100–300	4–6	15–40
Bamboo	910	400–600	30–40	1–1.8
Palm	100–200	300–400	2–3	10–15
Bagasse	100–200	200–300	10–15	10–15
Cotton	1600	200–600	5–15	3–10
Ramie	1500	200–1000	40–130	2–5
E-glass	2550	3400	70–75	3–4
S-glass	2550	4500–4600	80–90	4–5
Carbon	1820	2500–2700	200	1.2–1.4
Aramid	1400	4000	220–250	1.4–1.8

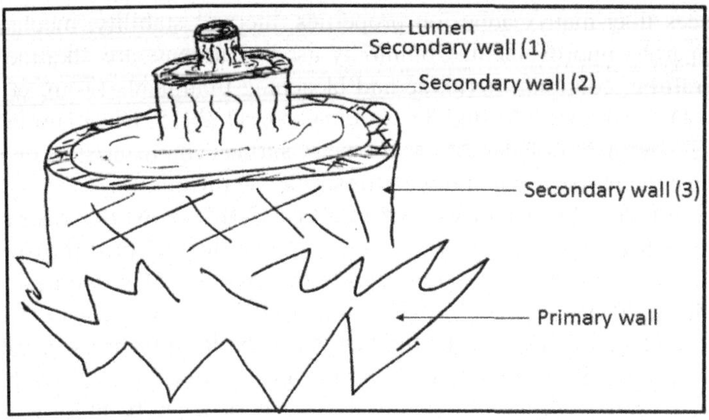

Fig. 2 Typical microstructure of a plant fiber and its segments

and secondary cell wall, where the secondary cell walls significantly influence the mechanical and structural properties of the material. The cell wall structure made up of cellulose, hemi cellulose, lignin pectin and other minor chemical constituents. Among these chemical constituents, cell wall is a strong and hard phase due to their crystalline structure and affects the mechanical properties. The size of the secondary and thickness of secondary wall along with cellulosic content of the plant fiber determines the mechanical properties of the fibers. A typical approximate chemical constituents of plant fibers is shown in Table 2.

Since all the fibers possess hydroxyl group (OH) on surface, the fibers were commonly surface treated to remove this hydroxyl group and coated with chemical

Table 2 Average chemical composition of selected natural fibers

Fiber	Cellulose (%)	Hemicellulose (%)	Lignin (%)	Pectin (%)	Moisture (%)
Jute	65	12	0.5	12	10
Sisal	65	12	10	1	10
Pineapple	65	8	8	2	Negligible
Kenaf	45	22	18	9	Negligible
Flax	65	16	2	2	10
Hemp	68	16	16	1	7
Banana	55	12	18	3	3
Coir	28	14	42	6	1
Bamboo	65	25	12	1	10
Palm (oil)	55	25	20	Negligible	Negligible
Sugarcane bagasse	42	25	22	Negligible	Negligible
Cotton	82	5.5	28	6	10
Ramie	68	13	1	2	10

that enhances fiber-matrix adhesion properties, thermal stability, mechanical and barrier properties improvement. Commonly used treatments are alkaline (mercerization), grafting, coupling, peroxide and bleaching treatments (Tong et al. 1995; Chand and Dwivedi 2006; Yi and Yan 2007; Cai et al. 2015; Chegdani et al. 2015, 2016). Each treatment induces different type of surface characteristics on the fibers and influence the properties mentioned for specific application.

A review article published in 2015 (Nirmal et al. 2015) shows that research article publication focusing specifically on tribological properties are steadily growing since year 2000. The number of research reports on tribological properties of NFRC increased from 30 in early 2000 to 250 in year 2014. This suggests the growing importance and significance of the application of NFRC materials in tribological application. The common prediction of the outcome was that the study on this subject continuous to grow as different type of NFRC materials can be tailor made to suit the performance and application.

3 Tribology of NFRC

The study of tribological properties of NFRC is a complex subject, as it involves multi-level understanding of many subjects, such as type of NFRC material, test method, friction, wear, lubrication and environmental effects (El-Tayeb 2008; Chin and Yousif 2009; Nordin et al. 2013; Zhang et al. 2014).

3.1 Material Type

NFRC material can be classified based upon type of matrix and reinforcements. Matrix materials are commonly thermoset, thermoplastics or elastomers. The processing method of NFRC composites are influenced by the type of matrix polymer. In general, casting and molding methods are suitable for thermosets, while elevated temperature induced processing are suitable for thermoplastics and vulcanization for elastomers based NFRC materials. Reinforcement are dependent upon the architecture of the fiber type. The reinforcement are of different types, namely, particles, short fibers, long fibers and laminar types.

Particle type reinforcement include spherical particles, filler, flakes or nanoparticles (nanoclay, carbon nanotube, nanocellulose, etc.). short fibers are typically 3–5 cm long, randomly dispersed on the matrix polymer. Thermoplastic based composites were commonly prepared by short fiber reinforced composites, due to temperature effect. Long fibers reinforced thermoplastic composite are in general difficult to produce due to the high temperature operation (>150 °C). Long fibers are typically longer than 5 cm, however, they are not continuously reinforced in the matrix polymer. Commonly they are produced for thermoset and chopped strand mat type of composites. Laminar type of fibers reinforcement is continuous reinforcement where the fiber reinforcement carried out in layered manner. In this work, natural fibers extracted from plant source are specifically focused. The study applies to all types of plant fibers that are extracted from stem, leaf or root section of the fibers.

3.2 Test Method

Tribological properties of material is about the study of friction, wear and lubrication of a material. Two type of friction and wear tests of NFRC material were focused, namely, dry and wet conditions testing. In dry testing condition, there is not any liquid or barrier medium between wear material and counter face. In wet condition, medium such as water, lubricant or barrier medium were present in between the rubbing surfaces. Friction and wear properties also depended upon the type of lubricant used in the system. In general carbon based lubricant induce reduced frictional properties than that of other lubricant material.

The friction and wear properties were commonly carried out using different type of tribometers. Different type of tribometers are available which predict the friction and wear properties. Commonly used tribometers, are pin on drum, pin on disc, block on ring, pin/block on wheel types. All these instruments are designed to study the basic tribological mechanisms and are focused in this study. The study and understanding the mechanisms leads to advancement in three main areas of tribology, viz.—friction, wear and lubrication.

3.3 Fiber Orientation and Nomenclature

Since tribology is study about rubbing surfaces and its mechanisms, the orientation of fibers with respect to sliding direction (or rubbing direction) plays a significant role in the tribology properties of composites. Figure 3 shows the different orientation of fibers orientation to that of sliding direction and its nomenclature. The nomenclature used in this manuscript for four different commonly oriented fibers is as follows:

P-O: Parallel orientation of fibers with respect to (w.r.t) sliding direction
AP-O: Anti-parallel orientation of fibers w.r.t sliding direction
N-O: Normal orientation of fibers w.r.t sliding direction
R-O: Randomly oriented fibers w.r.t sliding direction.

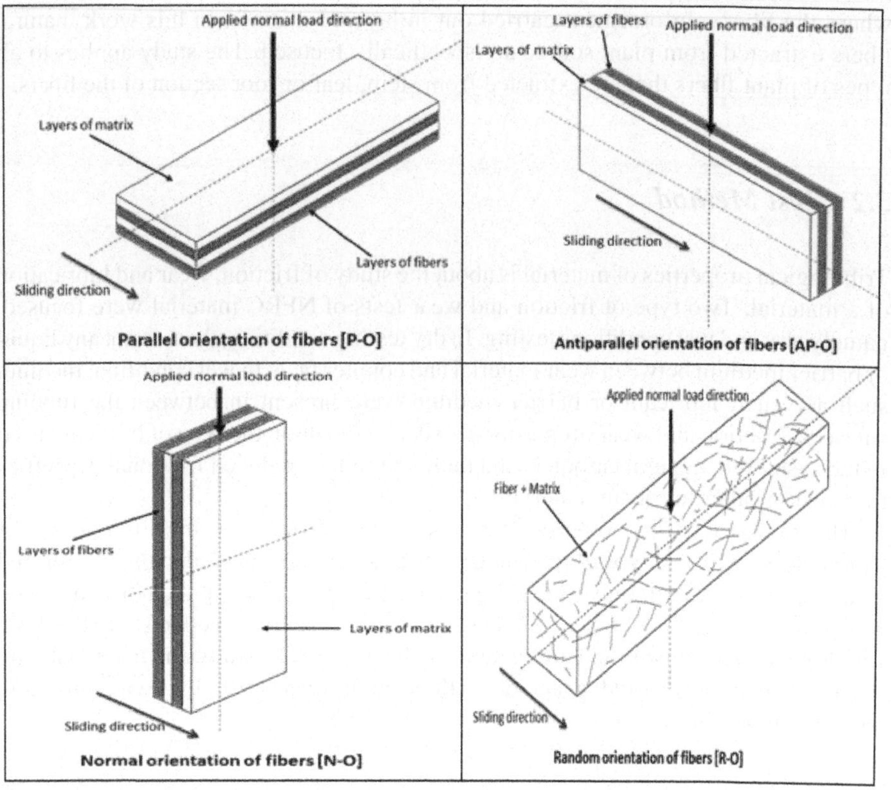

Fig. 3 Different types of fiber orientation and their nomenclature with respect to sliding direction

4 Friction

4.1 Static Friction of NFRC

Static friction is an important property that deals with the friction of material under static condition. Commonly determined static properties are angle of repose (Φ) and static coefficient of friction (μ_s). These properties determine the stability of braking and load bearing characteristics of vehicle or structure under stationary condition. Static frictional properties of natural fiber (banana fiber) reinforced epoxy composites were examined (Sinha et al. 2009; Dayma et al. 2011; Mohan and Kanny 2019) and the results are shown in Table 3. The result showed improved Φ and μ_s for the banana fiber reinforced epoxy composites than that of neat epoxy polymer. The result also shows that the μ_s and Φ values of composites are dependent on fiber content. As fiber content increases these values (μ_s and Φ) shows increasing trend. The fiber addition induces friction to the composites due to the reinforcement effect.

Table 3 Static friction properties of a typical NFRC material

Material	Fiber content, wt%	Angle of repose, degree (°)	Coefficient of static friction, μ_s
Neat epoxy	0	24	0.38
Banana fiber reinforced epoxy composites	20	27	0.43
	40	33	0.47
	60	37	0.51

4.2 *Dynamic Friction of NFRC*

Dynamic frictional properties of NFRC were examined under dry and wet conditions. In general it is reported that the dynamic friction coefficient (μ_d) increases due to natural fiber reinforcement. However, the level of improvement depends upon fiber type, fiber orientation, concentration and test parameters. Frictional properties were examined for several types of NFRC materials, and the general trend obtained in the result under dry condition are discussed herewith.

Commonly literature report shows that the frictional properties of NFRC were examined in sliding distance up to 3000 m, varying normal loading condition 10–50 N, varying sliding velocities up to 3 m/s. The relationship between sliding distance and frictional force is shown in Fig. 4. Frictional force increases as loading and sliding distance increases (Nirmal et al. 2010, 2012; Bajpai et al. 2013; Correa et al. 2017). The μ_d also depends upon sliding distance and loading conditions. Figure 5 shows the effect of μ_d at various loadings as a function of sliding distance. In general, μ_d increases as sliding distance and loading increases. The μ_d also depend upon sliding velocity and loading conditions, as shown in Fig. 6. As the sliding velocity increases, the μ_d tend to increase. Whereas, when the loading increases at a particular sliding velocity and initial stage, the μ_d decreases.

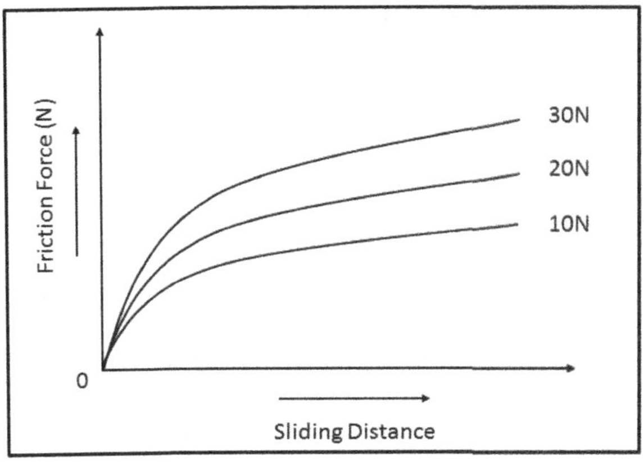

Fig. 4 Effect of frictional force on normal load and sliding distance of NFRC

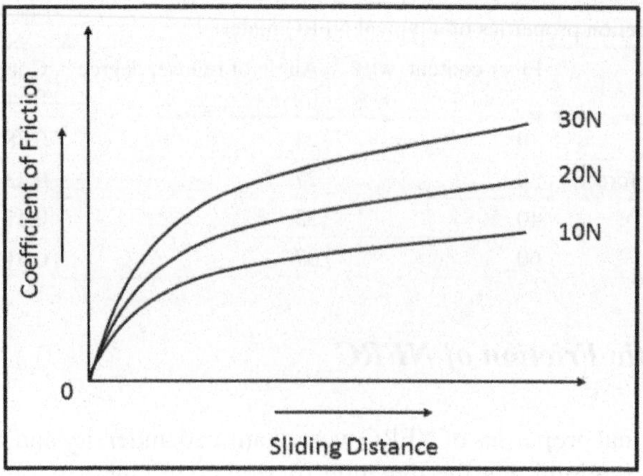

Fig. 5 Effect of dynamic friction coefficient (μ_d) on sliding distance and normal load of NFRC

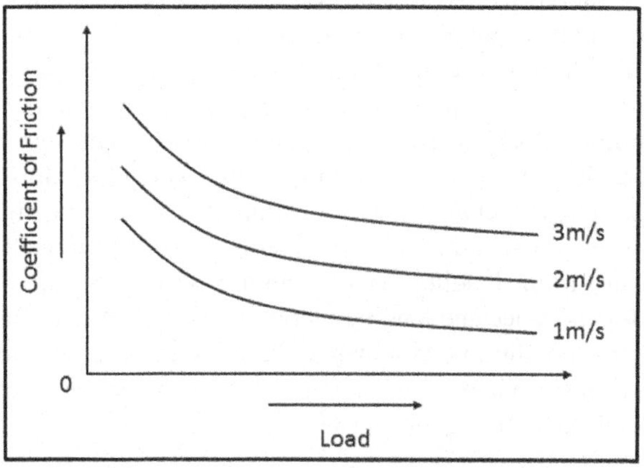

Fig. 6 Effect of μ_d on sliding velocity and normal load of NFRC

Friction also depended upon the type of fiber reinforcement in the composite. Different types of fiber reinforcement were examined, namely, randomly oriented (RO) and dispersed fibers, parallel oriented (PO) fibers with respect to sliding direction, anti-parallel orientation (AP-O) in the polymer matrix. It was observed the μ_d depended upon the type of fiber orientation. Figure 7 shows the effect of μ_d on the type of fiber orientation. In general, the fiber reinforcement decreases the μ_d when compared to neat polymer matrix. However, the variation of μ_d among the fiber orientation is different. AP-O oriented fiber composites resulted in least μ_d when compared with other fiber oriented series. Figure 8 shows the μ_d tested under wet condition. In wet condition, the composites show higher bearing of applied load than dry condition with increased frictional properties.

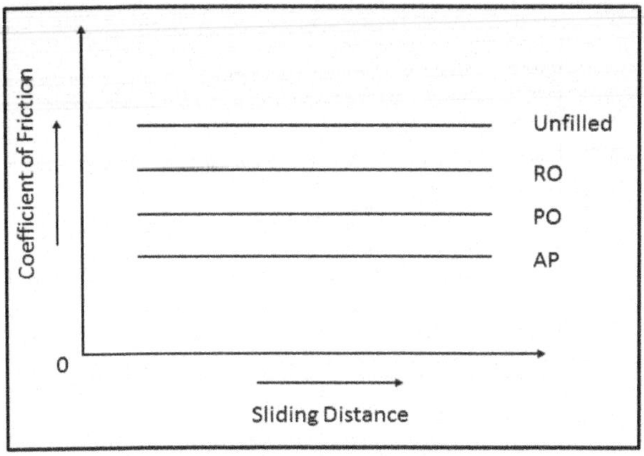

Fig. 7 Effect of μ_d on different orientation of fibers with respect to (w.r.t) sliding distance

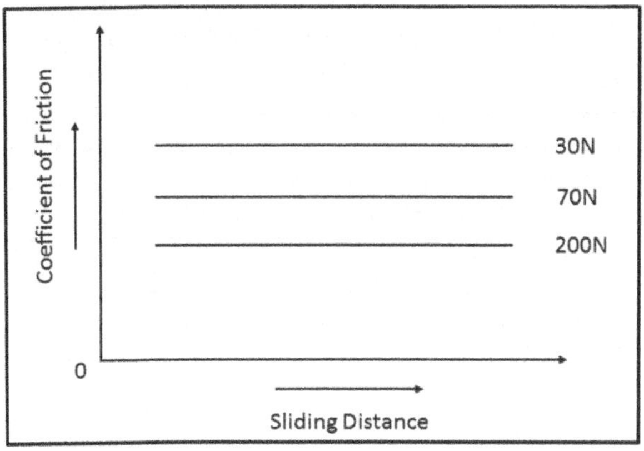

Fig. 8 Effect of μ_d on normal load under wet condition

5 Wear

5.1 Dry Sliding Condition

The wear behavior of NFRC materials was extensively analyzed in recent past as a function of various material constituent, experimental parameters and test methods (Tong et al. 2005; El-Tayeb 2008; Russo et al. 2015; Omrani et al. 2016).

Figure 9 shows the effect of effect wear resistance as a function of applied load and fiber length of sisal fiber. It is observed that wear resistance depends upon applied load and fiber length. Wear resistance tends to increase as the fiber length increases, up to critical length of fiber. Beyond the critical length of fiber, the wear resistance tend to decrease due to entanglement. Up to certain level of applied normal load

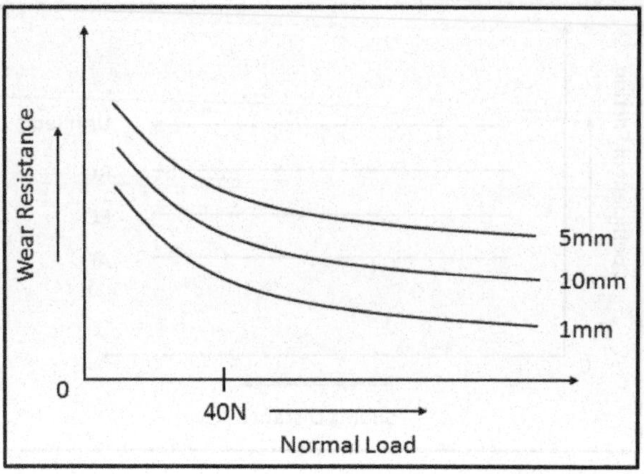

Fig. 9 Effect of wear resistance on fiber length under varying normal loads

(approx. 40 N) wear resistance tend to decrease, but there after increases. Possibly, the load transfer effect might be affected up to lower load level.

Figure 10 shows the effect of wear rate as function of fiber length and applied normal load. As applied load decreases, wear rate increases and amount of wear rate depends upon fiber length. Figure 11 shows the effect of wear rate of NFRC as a function of sliding velocity. The result shows that as sliding velocity increases, wear rate also increases. The wear rate also depends upon the amount of normal applied. Figure 12 shows the effect of wear rate on the type of fiber orientation. The normally oriented NFRC materials shows lowest wear rate than that of other fiber types. Figure 13 shows the trend of wear rate of NFRC material as an effect of fiber

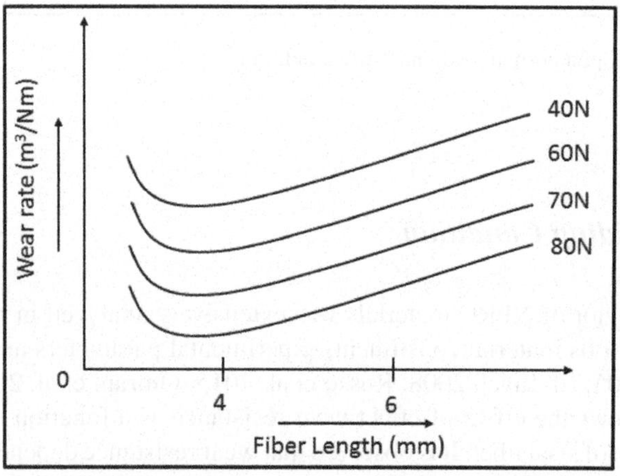

Fig. 10 Effect of wear rate on applied normal load and fiber length

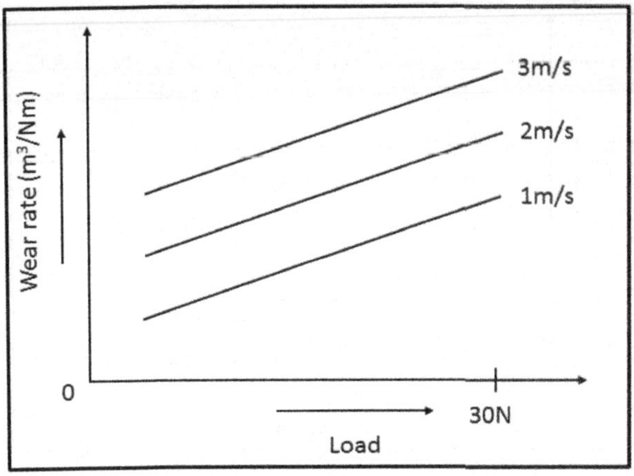

Fig. 11 Effect of wear rate on sliding velocity and applied normal load

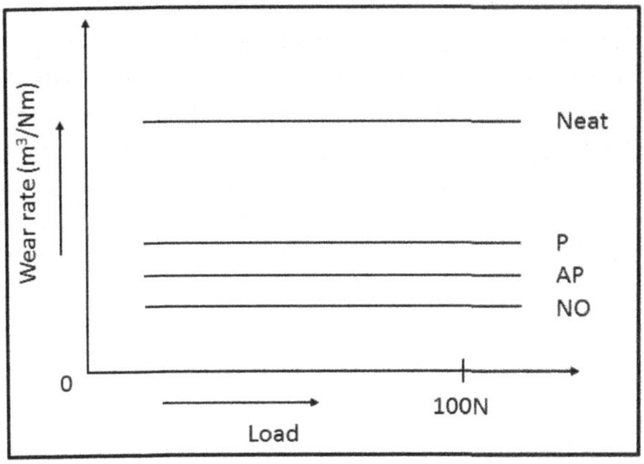

Fig. 12 Effect of wear rate on different types of fiber orientation under varying normal load

type and sliding velocity. The result shows sliding velocity is independent on wear rate, however, it affects unfilled near polymer matrix.

The wear mechanism of NFRC materials are extensively studied (Yousif et al. 2009, 2010; Correa et al. 2015; Mohan and Kanny 2019). It was observed that the material transfer phenomenon was the primary mechanism of wear. This material transfer is accompanied with the help of adhesion, microcracking and ploughing, which also depends upon loading and other experimental conditions. In general, adhesion and microcracking are accompanied at lower loading and sliding velocity conditions. Whereas, ploughing and cutting are accompanied at higher loading and sliding velocity conditions. Another important observation was the ability to form a stable transfer layer due to the natural fiber reinforcement. The natural fibers are able to form a stable transfer layers that are generated during wear. The wear debris

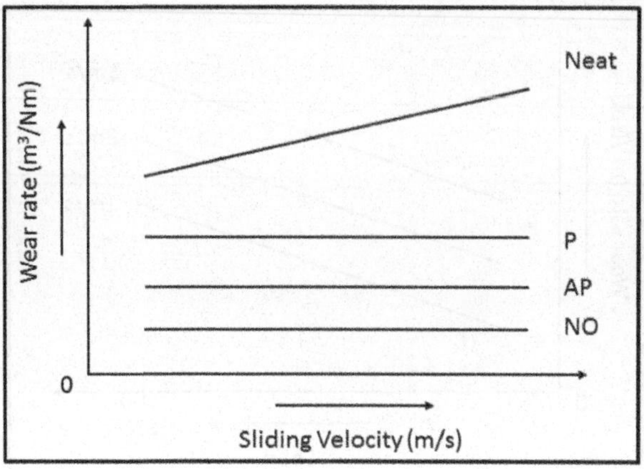

Fig. 13 Effect of wear rate on different types of fiber orientation under varying sliding velocity

of fiber, polymer and counter face constituents of transfer layer. Figure 14 shows the formation of transfer layer over the counter face surface during wear before and after the test of 40 wt% natural fiber (nanoparticle infused banana fiber reinforced composite), and the chemical concentration of counter face and wear track is shown in Table 4. This stable layer resists the material from further damage and wear and decreases the wear loss and friction. The EDX result of the selected area of Fig. 14b at counter face and wear track shows the transfer layer was formed by material constituents of banana fibers, nanoparticles and polymer matrix. Overall result shows that the characteristics of transfer layer depends upon type of natural fibers, orientation and architecture of fibers. Long fiber tend to have more stable transfer layer formation than that of short fiber reinforced composite materials.

5.2 *Wet Sliding Condition*

Figure 15 shows the effect of sliding velocity on wear rate of a natural fiber reinforced with varying normal load. In wet medium, the composite bears more load than that of dry medium test condition (Nirmal et al. 2010). The wet surface acts as a lubricant and reduces friction thereby increasing more load bearing. The result shows that the as the applied load increases, wear rate increases. In general, sliding velocity doesn't have any effect on the wear rate, however, at higher applied normal load level.

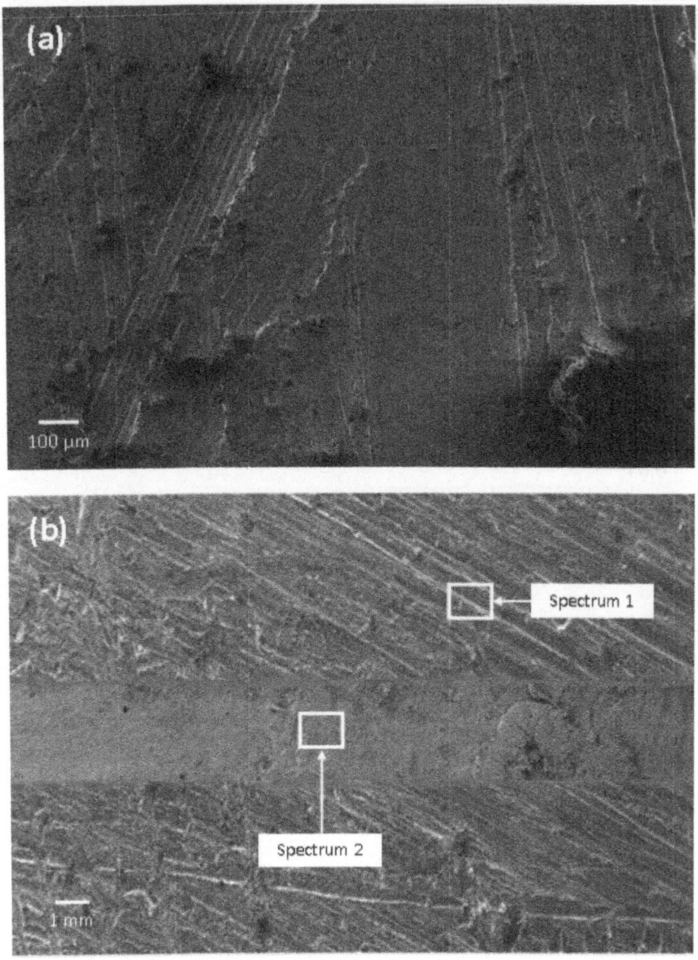

Fig. 14 Scanning electron microscopy (SEM) image of **a** steel counter face and **b** wear track of modified banana fiber reinforced epoxy composite. [Spectrums 1 and 2 are selected areas for energy dispersive x-ray (EDX) analysis]

6 Effect of External Factors

External factors such as counter face characteristics, temperature, lubricant and barrier medium affects the friction and wear of NFRC materials (Nasir 2013; Tahir et al. 2016; Sanjay et al. 2018). Figure 16 shows the effect of different type of metallic counter faces on the friction properties of natural fiber reinforced composites. Results shows the effect of μ_d as a function of fiber content with respect to two types of metallic counter face surfaces (Al and Cu), under same set of experimental and material conditions. The result shows that the μ_d depends upon metallic surfaces

Table 4 EDX analysis of steel counter face and modified banana fiber reinforced epoxy composite

Element	Steel counter face (Spectrum 1)	Wear track of 20 wt% nanoparticle treated banana fiber reinforced epoxy composite (spectrum 2)
	wt%	wt%
C	7.25	27.23
O	35.36	70.47
F	2.31	–
Al	–	0.51
Si	0.14	1.43
Cl	–	0.02
S	0.03	–
K	–	0.30
Cr	0.04	–
Mn	0.51	–
Ca	–	–
Fe	54.36	0.01
Rb	–	–
Mo	–	0.03
Total	100.00	100.00

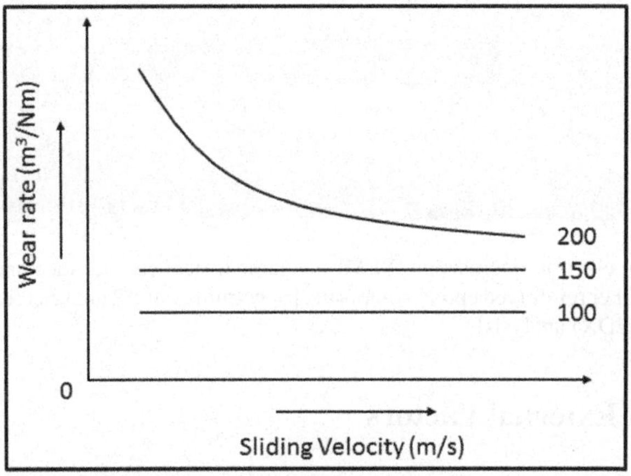

Fig. 15 Effect of wear rate on applied normal load and sliding velocity under wet condition

and fiber concentration. However, at optimized fiber content (~30 vol/wt%), the μ_d decreased considerably, irrespective of counter face metallic surfaces being used. Figure 17 shows the effect of counter face surface roughness on the wear rate of the NFRC material. The coarse surface counter face tends to show higher wear rate when

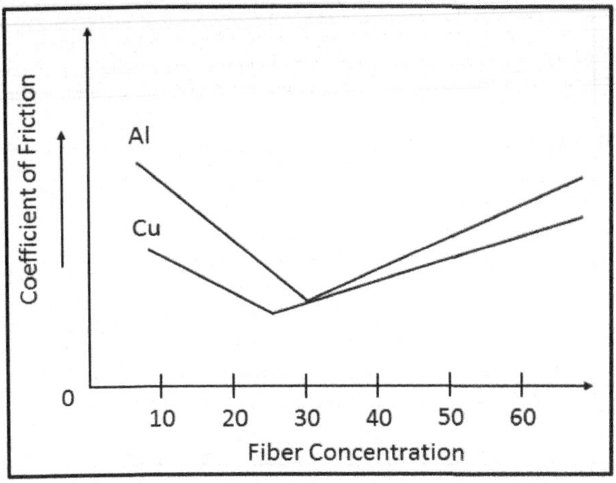

Fig. 16 Effect of dynamic friction coefficient (μ_d) on different metallic surfaces

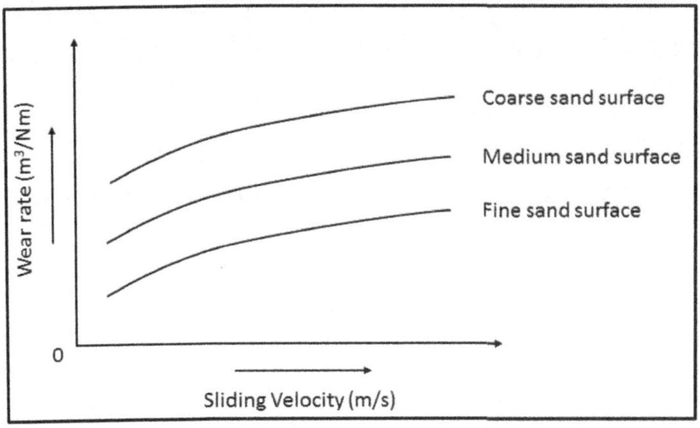

Fig. 17 Effect of dynamic friction coefficient (μ_d) on different metallic surfaces

compared with fine counter face roughness, due to high asperities. Wear properties also depends upon the hardness of the test materials and operating temperature. As the temperature increases, the hardness of the material decreases and hence the wear resistance. The variation of hardness versus temperature of a typical NFRC material is shown in Fig. 18.

It was observed that lubricant also plays a significant role in the tribological properties of NFRC materials. The NFRC materials had resulted in in improved tribological properties (decreased friction, wear and volume reduction) in the presence of lubricant medium when compared with neat polymer matrix. Wear of NFRC materials are also affected with barrier medium. NFRC material are exposed to moisture and water medium shows mass uptake and affects the tribological properties. The

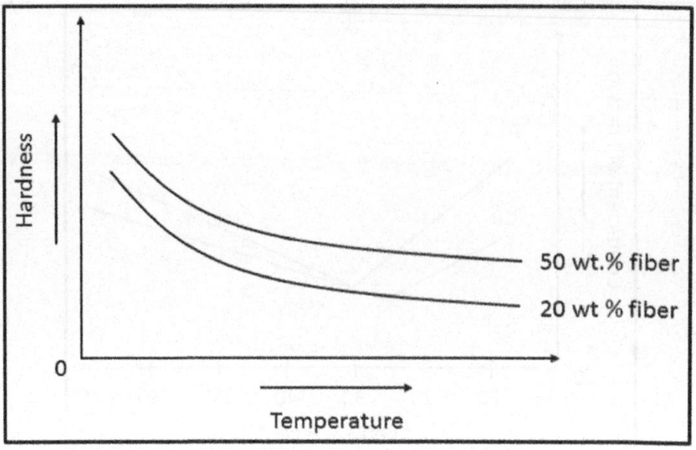

Fig. 18 Effect of hardness on temperature of NFRC

mass uptake increases as fiber content increases in the matrix. The high mass uptake resulted with decreased hardness in the NFRC materials, which in turn reduces the wear resistance.

7 Wear of Nanocellulose Reinforced Polymer Composites

During the past two decades, nanoparticles were extracted from plant sources (Dufresne 2013; Phanthong et al. 2018; Mohan and Kanny 2019). Commonly extracted nanoparticles are nano cellulosic fibrils, nanocrystalline cellulose and lignin nano particles. These nanomaterials are a smallest entities of the plant fibers. The properties of these material are stronger than that of metallic steel material. Dramatic properties were observed in thermal, physical, chemical and mechanical properties when nanocellulose are filled in polymer matrix. Few studies on wear properties of nanocellulosic (NC) particle filled polymer composites are available in the literature. The NC filled polymer composites shows improved frictional and wear properties when compared with polymer matrix. In some instances, the wear properties are even better than that of conventionally filled synthetic nanoparticle filled polymer composites, such as nanoclays, nanosilica, CNTs, etc. The level of property improvement in this NC filled polymer composites depends on the purity of the cellulosic phase, aspect ratio of the nano fibril and dispersion in the polymer matrix.

It is observed that the nanoparticles (such as nanoclay, nanocellulose) when reinforced or filled into the polymer, they are capable of forming a stable transfer layer, which significantly improves the friction and wear properties. A schematic illustration and mechanism of transfer film forming capabilities of nanoparticle infused natural fiber reinforced composite is shown in Fig. 19.

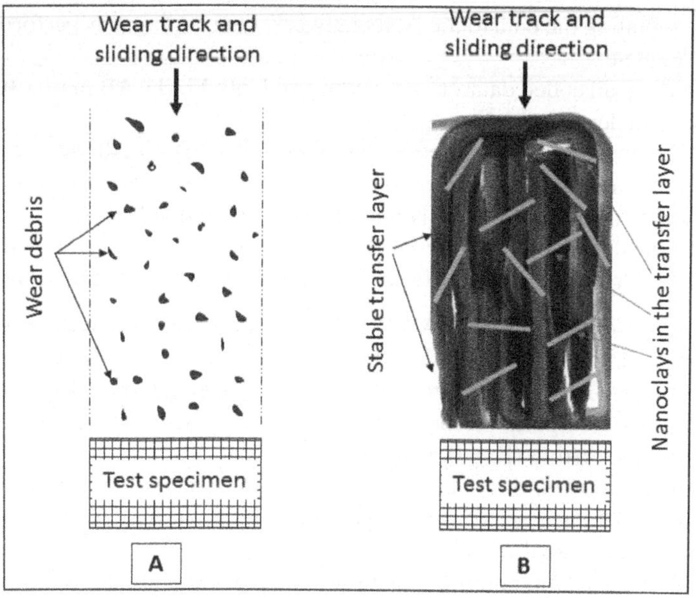

Fig. 19 Schematic illustration of formation of **a** wear debris in a untreated banana fiber reinforced composites and **b** stable transfer layer in the nanoparticle (such as nanoclay or nanocellulose) infused banana fiber reinforced composites

The nanoparticles (such as nanocellulose) due to their hard characteristics are able to hold the wear debris and form a stable transfer layer. These layers are formed at certain sliding distances, and after the formation of the layer, the further material loss and damage are prevented during dynamic friction and wear conditions.

8 Artificial Neural Networks (ANN) Models

Tribological properties of NFRC materials are examined theoretically using ANN models (Rapetto et al. 2009; Nirmal 2010). The ANN model has an accuracy of up to about 99% in prediction experimental friction coefficient values. ANN model is developed and inspired based upon the biological nerve system having the ability to learn and be trained. Significance of this model is that the lengthy experiments can be avoided and applied in ANN model to get experimental values. The development of ANN model involves following steps, namely

- Collection of experimental data related to the work.
- Prediction of friction coefficient of a typical NFRC material.
- The ANN model is subjected to three different input parameters.
- A database consisting of ~500 sets of experimental friction coefficient values of different set of NFRC materials is used to develop the ANN model.

- Prior to inputting the data to the ANN network, data coding was performed to the input parameters.
- Upon coding, all coded datasets were converted into MATLAB matrix files which were used to develop ANN models.
- The ANN model was configured based on trial and error methods.
- Predicted values that were closer to the experimental values were chosen.
- The network is a four layer ANN model with two hidden layers.
- Logical transfer functions were incorporated in both hidden layers while a pure linear transfer function was incorporated in the fourth layer; output layer.

The developed multi layered ANN model showed high accuracy in predicting the frictional coefficient values of NFRC composites, (e.g. treated banana fiber reinforced plastic (T-BFRP) composite in three different orientations). Remarkably, the developed ANN configuration for the current work is equally competitive in terms of its performance rating (B = 0.994077).

A schematic illustration of the operation of ANN model is shown in Fig. 20, which consists of three set of operations, namely, input, hidden and output. The work sequence of these three sets are as follows:

Input—External data (number of neurons dependent on user)

Hidden—Processing takes place (number of neurons based on optimum performance)

Output—The end result (number of neurons dependant dependent on user)

An example of a ANN model is 4-[10-20]-3, Whereby the '4' represents the number of neurons in the input layer, the [] show hidden layers, The '10' represents the number of neurons in the first hidden layer, the '20' represents the number of neurons in the second hidden layer and the '3' indicates the neurons in the output

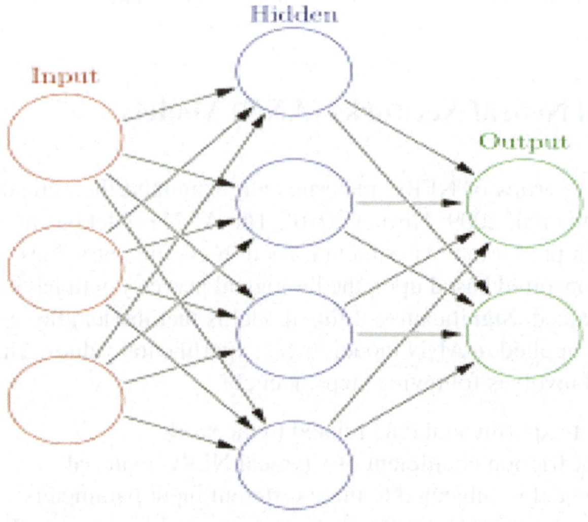

Fig. 20 Schematic illustration of ANN model sequence

layer. The connections among the neurons is based on the transfer function and weight embedded within the neurons. The formula for this relationship is:

$$X_f^{(n+1)} = F\left(\sum_i W_{ji}^{(n)} X_i^{(n)}\right)$$

$X_f^{(n+1)}$ Vectors output of unit j in the nth layer of composite
F Transfer function
$W_{ji}^{(n)}$ Weights from unit i in the nth layer to unit j in the (n + 1)th layer
$X_i^{(n)}$ Vectors input of unit I in the nth layer.

Sum Squared Error (SSE)

$$SSE = \frac{1}{2}\sum_{N=0}^{N}(A_i - O_i)^2$$

A_i is the *i*th ANN output data
O_i is the *i*th experimental data
n is the *n*th number of data set.

Commonly, there are three different types of fiber orientations, viz, Anti-parallel (AP), Parallel (P) and Normal (N). Subjected to different loads between 5 and 30 N and sliding distances between 0 and 6.72 km under dry contact condition.

The ANN prediction on friction coefficient resulted in high accuracy and closely compared to experimental values for the three different orientations. Where there is a sudden increase/decrease in experimental friction coefficient ANN is able to predict the exact values of these friction coefficients points but instead follows a general trend of the experimental friction coefficient values throughout the predicting process. ANN shows poor predicting performance during 'running in' process (transition period from static to kinetic friction) but good predicting characteristics during 'steady state' transition region (e.g. kinetic coefficient of friction). To show the level of accuracy for the developed ANN model, the coefficient of determination B is equal to:

$$B = 1 - \frac{\sum_i^n = 1\left[A_i^i - O_i^i\right]^2}{\sum_i^n = \left[O_i^i - O\right)^2}$$

n is the number of test data
$A_i^{(i)}$ is the *i*th predicted friction coefficient
$O_i^{(i)}$ is the *i*th measured value
O is the mean value of $O_i^{(i)}$.

As a result of the various literature work on ANN of NFRC materials, it was found that the ANN model showed promising results in predicting the friction coefficient of the T-BFRP composite in three different orientations (AP, P and N orientations) of fibers and subjected to different applied normal loads and sliding distances.

9 Future Perspectives

In recent years, there are much talk about 4th industrial revolution (4IR). In our view, developing natural materials in tune with 4IR will hold a promising future for economy and societal benefits.

Apart from cost saving, the impact on damages to environmental and natural resources will be greatly reduced. NFRC can be processed using some of the futuristic technologies such as 3D printing, automation, nanotechnology, etc. to cater products ranging from automotive and aerospace industries.

Apart from this, there are several issues that need to addressed on NFRC material itself. Issues such as obtaining uniform and consistent concentration and properties will also hold a key issue on development of this material. A reliable extraction method of nanocellulosic (particles, crystals and fibrils) also play a significant role on the future development of these materials.

10 Conclusion

This chapter presents the review on tribological properties of NFRC materials. The static and dynamic friction and wear properties are discussed. A conscience and general trend on the various effects of experimental condition on friction and wear properties are examined. The experimental condition for tribologial studies of NFRC materials with emphasis on material type, test methods and fiber orientation is outlined. Commonly used theoretical models such as ANN in prediction frictional properties of NFRC materials is also discussed. The wear mechanism in NFRC and nanoparticle filled NFRC materials are discussed with illustration, which shows the ability of transfer layer formation.

References

Bajpai PK, Singh I, Madaan J (2013) Tribological behavior of natural fiber reinforced PLA composites. Wear 297:829–840

Boopathi L, Sampath PS, Mylsamy K (2012) Influence of fiber length in the wear behaviour of borassus fruit fiber reinforced epoxy composites. Int J Eng Sci Technol 4:4119–4129

Cai P, Li Z, Wang T, Wang Q (2015) Effect of aspect ratios of aramid fiber on mechanical and tribological behaviors of friction materials. Tribol Int 92:109–116

Chand N, Dwivedi UK (2006) Effect of coupling agent on abrasive wear behaviour of chopped jute fibre-reinforced polypropylene composites. Wear 261:1057–1063

Chegdani F, Mezghani S, ElMansori M, Kaddem AM (2015) Fiber type effect on tribological behaviour when cutting natural fiber reinforced plastics. Wear 332–333:772–779
Chegdani F, Mezghani S, El Mansori M (2016) On the multi scale tribological signatures of the tool helix angle in profile milling of woven flax fiber composites. Tribol Int 100:132–140
Chin CW, Yousif BF (2009) Potential of kenaf fibres as reinforcement for tribological applications. Wear 267:1550–1557
Correa CE, Betancourt S, Vázquez A, Gañan P (2015) Wear resistance and friction behavior of thermoset matrix reinforced with Musaceae fiber bundles. Tribol Int 87:57–64
Correa CE, Betancourtb S, Vázquezc A, Gañand P (2017) Wear performance of vinyl ester reinforced with Musaceae fiber bundles sliding against different metallic surfaces. Tribol Int 109:447–459
Dayma N, Satapathy BK, Patnaik A (2011) Structural correlations to sliding wear performance of PA-6/PP-g-MA/nanoclay ternary nanocomposites. Wear 271:827–836
Dufresne A (2013) Nanocellulose: a new ageless bionanomaterials. Mater Today 16:220–227
El-Tayeb NSM (2008) A study on the potential of sugarcane fibers/polyester composite for tribological applications. Wear 265:223–235
Faruk O, Bledzki AK, Fink H-P, Sain M (2012) Biocomposites reinforced with natural fibers: 2000–2010. Prog Polym Sci 37(11):1552–1596
Gbadeyan OJ, Kanny K, Pandurangan MT (2017) Tribological, mechanical, and microstructural of multiwalled carbon nanotubes/short carbon fiber epoxy composites. J Tribol 140(2):022002–022002-6. https://doi.org/10.1115/1.4037357
Gu D, Jue J, Dai D, Lin K, Chen W (2017) Effects of dry sliding conditions on wear properties of al-matrix composites produced by selective laser melting additive manufacturing. J Tribol 140(2):021605–021605-12. https://doi.org/10.1115/1.4037729
May-Pat A, Valadez-González A, Herrera-Franco PJ (2013) Effect of fiber surface treatments on the essential work of fracture of HDPE-continuous henequen fiber-reinforced composites. Polym Test 32(6):1114–1122
Mohammed L, Ansari MNM, Pua G, Jawaid M, Saiful Islam M (2015) A review on natural fiber reinforced polymer composite and its applications. Int J Polym Sci 2015. Article ID 243947, 15 pages http://dx.doi.org/10.1155/2015/243947
Mohan TP, Kanny K (2019) Tribological properties of nanoclay infused banana fiber (NC-BF) reinforced epoxy composites. ASME J Tribol 141(5):052003 (9 pages)
Nasir RM (2013) Lubricated abrasion study of bio-fibre (paddy straw) and cockle-shell using pin-on-disk method. Procedia Eng 68:116–122
Nirmal U (2010) Prediction of friction coefficient of treated betelnut fibre reinforced polyester (T-BFRP) composite using artificial neural networks. Tribol Int 43:1417–1429
Nirmal U, Yousif BF, Rilling D, Breverna PV (2010) Effect of betelnut fibres treatment and contact conditions on adhesive wear and frictional performance of polyester composites. Wear 268:1354–1370
Nirmal U, Hashim J, Low KO (2012) Adhesive wear and frictional performance of bamboo fibres reinforced epoxy composite. Tribol Int 47:122–133
Nirmal U, Hashim J, Megat Ahmad MMH (2015) A review on tribological performance of natural fibre polymeric composites. Tribol Int 83:77–104
Nordin NA, Yussof FM, Kasolang S, Salleh Z, Ahmad MA (2013) Wear Rate of Natural Fibre: Long Kenaf Composite. Procedia Eng 68:145–151
Omrani E, Menezes PL, Rohatgi PK (2016) State of the art on tribological behavior of polymer matrix composites reinforced with natural fibers in the green materials world. Eng Sci Technol, Int J 19:717–736
Phanthong P, Reubroycharoen P, Hao X, Xu G, Abudula A, Guan G (2018) Nanocellulose: extraction and application. Carbon Resour Converst 1:32–43
Rapetto MP, Almqvist A, Larsson R, Lugt PM (2009) On the influence of surface roughness on real area of contact in normal, dry, friction free, rough contact by using a neural network. Wear 266(5–6):592–595

Russo P, Simeoli G, Acierno D, Lopresto V (2015) Mechanical properties of virgin and recycled polyolefin-based composite laminates reinforced with jute fabric. Polym Compos 36(11):2022–2029

Sanjay MR, Madhu P, Jawaid M, Senthamaraikannan P, Senthil S, Pradeep S (2018) Characterization and properties of natural fiber polymer composites: a comprehensive review. J Clean Prod 172:566–581

Sinha SK, Song T, Wan X, Tong Y (2009) Scratch and normal hardness characteristics of polyamide 6/nano-clay composite. Wear 266:814–821

Tahir NAM, Abdollah MFB, Hasan R, Amiruddin H (2016) The effect of sliding distance at different temperatures on the tribological properties of a palm kernel activated carbon–epoxy composite. Tribol Int 94:352–359

Tong J, Fien L, Li J, Chen B (1995) Abrasive wear behaviour of bamboo. Tribol Int 28(5):32–327. 2005

Tong J, Ma Y, Chen D, Sun J, Ren L (2005) Effects of vascular fiber content on abrasive wear of bamboo. Wear 259:37–46

Yi G, Yan F (2007) Mechanical and tribological properties of phenolic resin-based friction composites filled with several inorganic fillers. Wear 262:121–129

Yousif BF, Alvin D, Yusaf TF (2009) Adhesive wear and frictional behaviour of multilayered polyester composite based on betelnut fiber mats under wet contact conditions. Surf Rev Lett 16:407–414

Yousif BF, Lau S, McWilliam S (2010) Polyester composite based on betelnut fibre for tribological applications. Tribol Int 43:503–511

Zhang X, Li K-Z, Li H-J, Fu Y-W, Fei J (2014) Tribological and mechanical properties of glass fiber reinforced paper-based composite friction material. Tribol Int 69:156–167

The manufacturer's authorised representative in the EU is Springer Nature Customer Service Centre GmbH, Europaplatz 3, 69115 Heidelberg, Germany. If you have any concerns regarding our products, please contact ProductSafety@springernature.com

Printed and bound by CPI Group (UK) Ltd, Croydon, CR0 4YY
30/09/2026
02225288-0011